P9-EDR-577

Homeland Security Technology Challenges

From Sensing and Encrypting to Mining and Modeling

For a complete listing of the *Artech House Intelligence and Information Operations Series*,
turn to the back of this book.

Homeland Security Technology Challenges

From Sensing and Encrypting to Mining and Modeling

Giorgio Franceschetti
Marina Grossi

Editors

ARTECH
HOUSE

BOSTON | LONDON
artechhouse.com

Library of Congress Cataloging-in-Publication Data
A catalog record for this book is available from the U.S. Library of Congress.

British Library Cataloguing in Publication Data
A catalogue record for this book is available from the British Library.

Cover design by Igor Valdman

ISBN-13: 978-1-59693-289-0

685 Canton Street
Norwood, MA 02062

10 9 8 7 6 5 4 3 2 1

Homeland Security, la Patrie Défense,
Honi soit qui mal y pense.

Sensing, Encrypting, Trying to Guess,
Mining the Data, oh what a Mess!

May the book guide you all through this dance.

Contents

Preface

This book has emerged in part from the contributions to the international workshop, *Homeland Defence and Security*, Sorrento, Italy, October 18–21, 2006, organized by us editors. The origin is reflected in five (out of nine) chapters of the book, which consist of revised and edited versions of the lectures delivered at the workshop. Here we present the philosophy underlying the book project, pointing out the key technological features that are relevant to the Homeland Security (HS) scenario.

As a very general definition, HS consists of the application of a set of suitable measures and technologies to protect citizens and public property from hostile attackers that carry out nonauthorized intrusions in the homeland, employing, for example, chemical and biological agents and information viruses. In the broader sense, HS may be extended also to natural disasters, with the scope of mitigating their effects.

There are many vulnerabilities and threats to an organized society. They include, for instance, outages due to equipment breakdown and power failures, damages from environmental factors, physical tampering, and information gathering. Information and communication technology (ICT) can dramatically change the manner in which countries protect their citizens. An incomplete list of related ICT technologies that support HS, with full attention to their performance limits and requirements, is the following: networked sensing, time-varying scenarios modeling and their real-time simulation, wireless and Internet communications protocols, energy efficient sensors design, detection of unauthorized activities, new trends in localization techniques, and environment habitat monitoring. It is evident that the technical implementation of HS procedures poses an impressive multifacet technological challenge, where interdisciplinary competences are a must, and scientific and technical vision is an imperative requirement.

It should be noted that other fundamental challenges certainly appear in the HS scenario, for example, organizations and levels of government (federal, state, local, tribal) involvement; coordination and information sharing/exchange with the operators sector (private and commercial areas); and handling the very

1

The Homeland Security Scenario

Paolo Neri

1.1 The Scenario

Homeland security usually addresses activities, solutions, systems, and issues related to natural and man-made events that can cause threat to life and/or bring national economy losses. This definition primarily includes civil and paramilitary contexts, while typical military applications are termed with homeland defense. These two main domains are grouped under the homeland protection umbrella.

The focus of this book is on homeland security, although many concept developed for the security can be migrated to the defense and vice versa. The development of this specific discipline has been driven by the following two main facts.

Fact (1): The conclusion of the Cold War, with the associated opposition of red and blue forces, has determined the refocusing of the defense activities and strategies toward the development of means that can comply with rapid-force deployment in worldwide-defined areas to cope with a variety of missions spanning from peacekeeping to "cover-up" operations. This transformation process has been dramatically accelerated by the raising of the terrorism threat; usually military strategy calls this *asymmetric war*, meaning a new environment and scenarios where a limited number of people can develop a wide variety of potential threats through the implementation of new tactics and technologies; in many situations the traditional defense assets have been revealed to be ineffective both as a deterrent and as countermeasures.

A key characteristic of this new type of threat is that it is directed to people, places, and infrastructures related to the civilian world: for this reason, today we refer to it as an *asymmetric threat*.

Fact (2): Another significant drive for force transformation can be identified with the goal to get a better planning capacity and technology roadmap in order to manage the complexity, cope with the general defense budget reduction, and improve the dual use of the systems. This last point is particularly true for homeland protection application. Keeping this in mind, it is easier to understand the reason why today a huge assimilation of a number of technologies and means derived from the business to business (B2B) and civilian world practices is taking place within the defense sector.

On the other side, the guidelines for the defense rearrangement and the development of new concepts and doctrines for acting, as well as their reflections on organization, have been assimilated by civil protection and paramilitary forces to improve their performance in conducting homeland security tasks.

It should be clear that today we have a clear path for the development of dual-use technology significantly larger than one or two decades ago (see Figure 1.1). Homeland security–related application can benefit from the maturity of well-suited new doctrines as emerged from force transformation, as well as consolidated architectural framework for system architecture (e.g., DoD architectural framework, activity-based modeling) to describe and develop system of system architecture. Homeland defense application can benefit from reduced development cycles through the adoption of well-tested COTS technologies and open architectures.

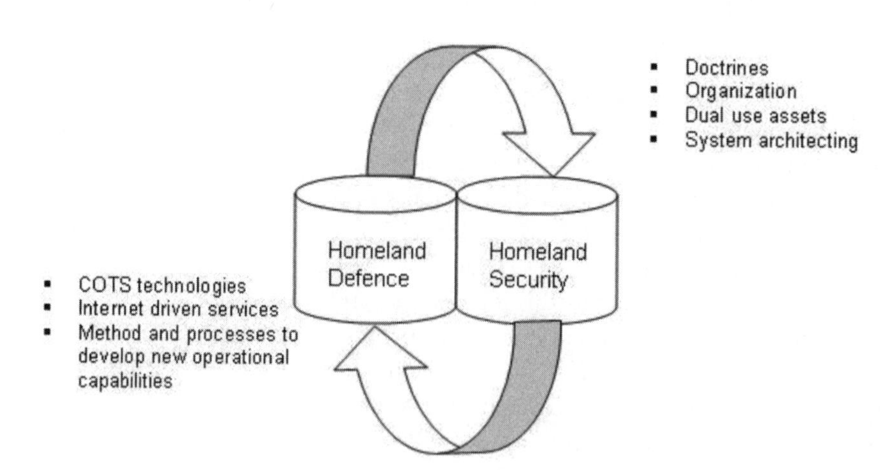

Figure 1.1 Dual use of current technologies and approaches for homeland protection.

In view of the above, a categorization of the different applications falling within the homeland security domain is referred to in Figure 1.2.

As stated above, *homeland protection* comprises the two blocks of defense and security. Homeland defense (HD) can be further split to address large systems, where the integration of the three armies takes place at the national level, while a second group refers to joint forces, where the system development is challenged by flexibility and interoperability needs. The homeland security constituent blocks have the features to resolve a wide variety of different needs: they deal with typical civil protection solutions that can be improved by the adoption of emerging technologies (which also greatly helps reorganization of the operational procedures) to reach the goal in a more effective way; or address emerging requirements as results from the asymmetric threats. Further splitting of land control is needed to comply with different needs related to the protection of different infrastructure and the environment at the country level.

To go into details, a first classification considers the human-driven application, including critical infrastructure protection, border control, crisis management related to large events, and to various threats with special focus on the urban environment, and all the other side effects on man-made infrastructures that can be threatened by illegal activities.

A second class is represented by natural disaster management, which deals with typical civil protection scenarios; for convenience, all naturally related phenomena have been grouped within the crisis management box in the above picture. Most of the typical civil protection scenarios can be structured as events, where early warning, management of the occurrence of the disaster, and postevent mitigation interventions must be carried out. The management of earthquakes, floods, wildfires, avalanches, drought, and heat waves falls within this box; processes and technology aids that have been implemented can be widespread.

Border control deals with the problems of impeding the entrance in the national territory of unauthorized people and materials. It typically deals with illegal immigration and smuggling, but can also concern more serious items such as the incoming of weapons of mass destruction. It can be split into blue (sea side) and green (land side) border control, depending on the environment where the monitoring takes place, bringing the usage of different technologies for the two areas.

Land security deals with the monitoring of the interior of a country and addresses a wide variety of applications. It includes territorial management with special focus on infrastructure and critical sites for industrial production and national facilities (communication, energy, and so forth), as well as population-related items like aqueducts or Olympic Games monitoring. Within land security a significant part is also dedicated to transportation security [1–3].

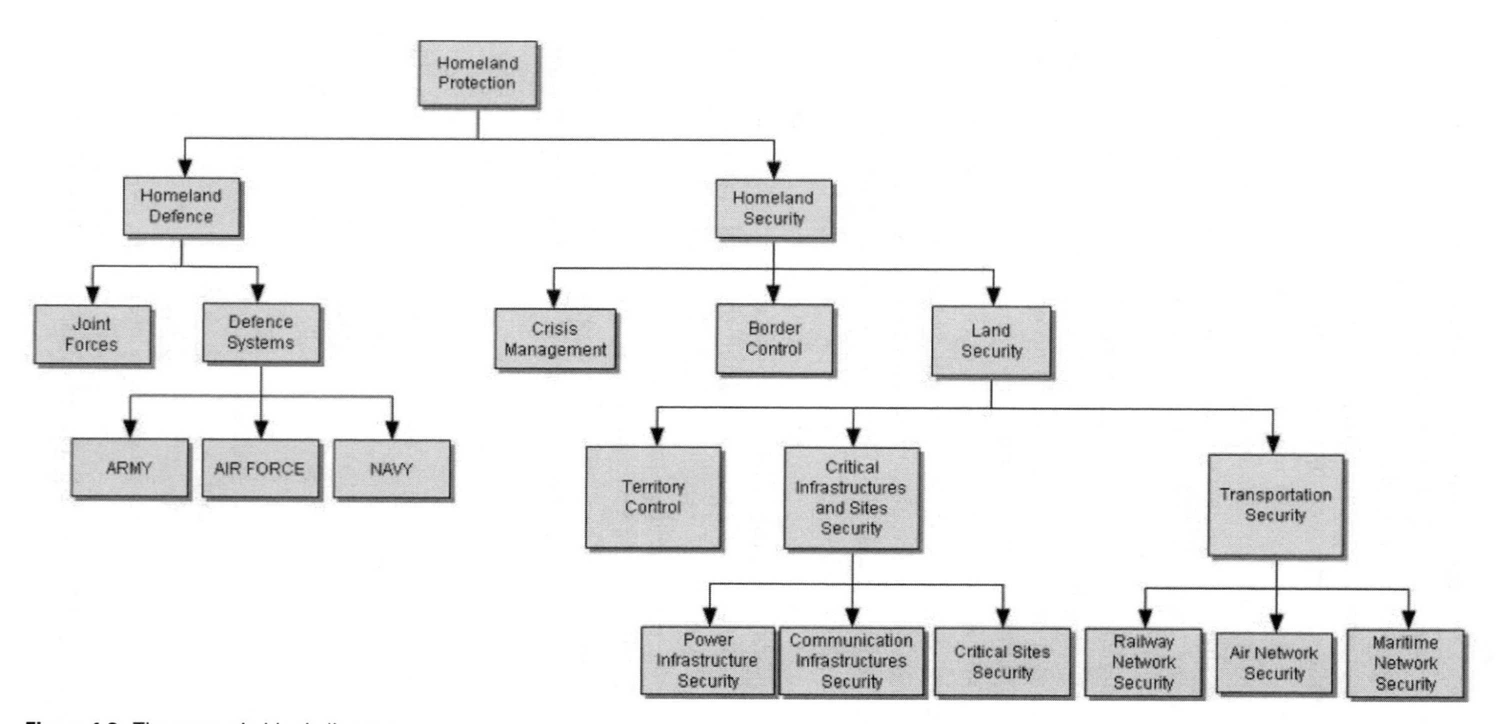

Figure 1.2 The scenario block diagram.

1.2 Composing Expertise and Technologies

Approaching the variety of application areas presented in the previous sections requires the combination and use of a significant number of heterogeneous technologies. Looking at Figure 1.3, it can be noted that the required spectrum is quite wide.

This scheme has been structured to classify technologies and expertise into four main parts: sensor, command and control, reaction systems, and a horizontal skill set that represents a common dispersed skill set for homeland security solutions.

The *sensor box* comprises a rich variety of technologies to increase early warning capabilities in a variety of application disciplines. Depending on the environment and the phenomena to detect, the homeland security sensor families span all the electromagnetic spectrum from microwave, infrared, and visible, up to X- and gamma-rays. From another perspective, sensors can be distinguished into wide area sensors (e.g., radar, cameras, smart dust) and gate sensors (scanners, detectors).

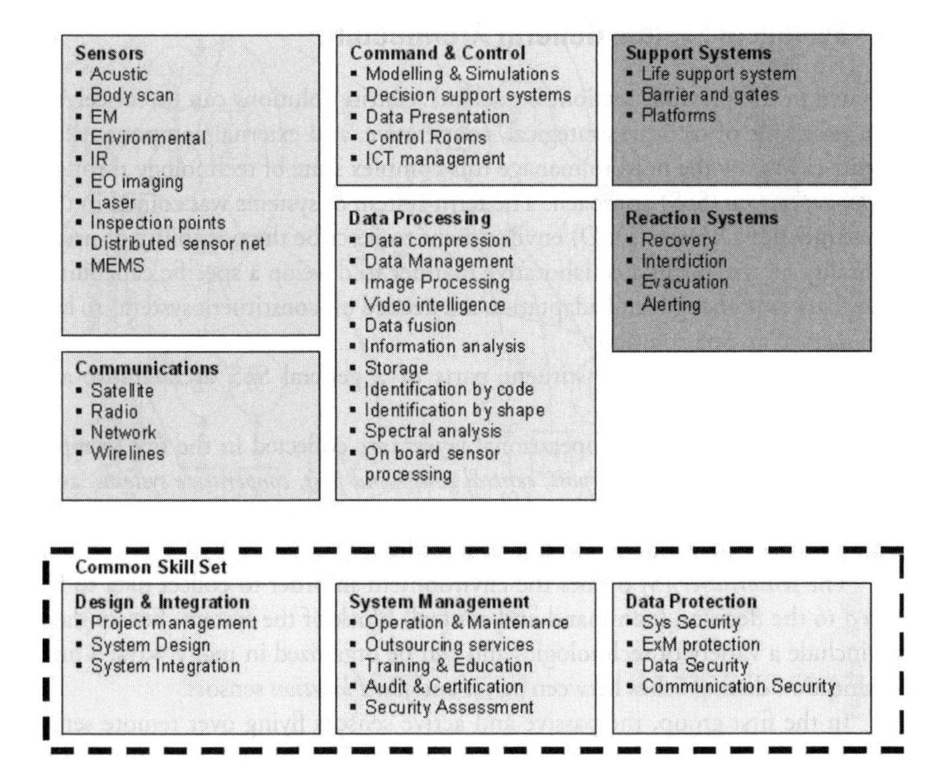

Figure 1.3 Expertise and technologies.

increasing their capability in terms of aerial coverage (swath and revisit time) and quality of data, thanks to the increasing performance of the basic technology. These kinds of sensors are vital in all the phases related to the control of an emergency, providing useful data for risk assessment, monitoring phenomena, and providing useful data for the postassessment phase [7, 8].

The second group of sensors are mainly related to fixed installations, like *radar, optical multispectral scanners (including also nuclear bacteriological chemicals*—NBC *technology)* that are designed to detect a well-defined class of phenomena. They can also be mounted on board vehicles.

This sensor compendium also includes the *interrogative technologies*, which can provide instant detection of mobile units equipped with emitters (e.g., **L**ong **R**ange **I**dentification and **T**racking and **A**utomatic **I**dentification **S**ystems).

Most of the sensor technologies are equipped with value-added services, implemented via *preprocessing operations* [PP], to improve the information dissemination to the other system components. Typical functionalities of this [PP] are data georeferencing, data filtering, data compression, and raw data processing, which provide different levels of products to be further exploited in the value-added chain.

A number of *remote command posts* [C2] are geographically distributed to guarantee proper coverage where the formation of a first local operational picture takes place. Thanks also to the availability of an incremental processing capability (data fusion, software agents, and simulation), it is possible to develop value-added understanding of the phenomena. This remote command post can autonomously manage the action of the *effectors* [E], here intended as countermeasures against the threat [9].

Remote command posts are linked together in a peer-to-peer architecture to share information and to guarantee system resilience.

Coupling of sensors [S] with local remote command post [C2] and effectors [E] is an enabling capability necessary to guarantee a fast reaction responsiveness of the homeland security systems: the threat dynamics cannot be managed with a traditional hierarchical architecture, which can fail in terms of executing a command within a useful time frame. In this context, a significant aid is provided by the application of the *software agents* (complex software entity that is capable of acting with a certain degree of autonomy in order to accomplish tasks on behalf of its user), which can mitigate the load of the decision process on humans, providing the decision maker (who is in charge of acting on a countermeasure) with synthesized, distilled information to speed up the decision process.

A limited number of *central command posts* [C2] are usually designed where formation of the common operating picture [COP] is realized, and where major tactical/strategical command decisions are taken. This kind of central command post is needed to group all the information together, especially for wide area applications. At variance of past systems, these new [C2] centers are organized to manage a bidirectional information flux. They gather all the information from the periphery of the systems (to the sensor level if needed), organize it in a suitable manner for the decision maker, and send back to the periphery that pertinent information that can enable a better reaction and awareness of the situation. These centers are robustly linked to both of the other system components and *external sources*, whose data are accessed via service-oriented architecture methods. The external system can be the provider of the so-called *organic and nonorganic* source of information: the central command post has the main task of integrating multilevel and multi-domain information for achieving superior understanding of the scenario. An example of a nonorganic source could be large data mining operations on public and private data to detect *weak signals* of threat activity.

An interesting point is the assessment of changes that an SoS design approach may enforce on current architectures. For the *sensor layer* it can be considered that these systems initially remain the current ones. Then their network interface must be modified in order to operate as services. Further modifications are necessary to increase the flexibility of reconfiguration upon request of the user.

Cooperative systems and *intelligence* are blocks that can be activated on demand depending on task that the system has to accomplish [10].

In addition to the previously listed systems, linked together in the SoS, other capabilities are diffused in all the SoS. This is the case of command and control [C2] capabilities that are more widely distributed in all the composing systems. Network interface and man-machine functions are added. New applications are activated, through intelligence agents, to support the decision-making process with specific suggestions, or to relieve the operator from a purely automatic job. Other agents help in the formation of distributed dynamic teams. In the following years, autonomic computing functions are expected to be inserted in the operational environment. New mechanisms to improve the trustworthiness of the information are also a key area under development.

As a general statement, the *network* is the key growing area that enables the paradigm of information superiority (the so-called network-centric operation slogan "the right information to the right people at the right time"). Networking is expected to be instrumented at the fabric level with a variety of interconnection links (radio, fixed cable, wireless, WiMax), which allow set up of a significant suite of value-added service with well-proofed and already tested tools, derived from the B2B world, like service-oriented architecture (SOA).

The *effectors*, even if they are the same used today, also undergo modification of the interface towards the network. Functionalities for the synchronized and autonomous control by operators are added subsequently.

This suite of technologies and systems collectively generates a set of well-organized information and a networking-processing capability.

1.4 The Applications

Depending on the application, the building blocks in Section 1.3 can have different weights within the overall system balance. In Table 1.1, a qualitative shaping is provided.

1.5 System of System Engineering

It appears clear that the solutions, to be developed to handle homeland security themes, address complex systems in terms of dimension, heterogeneity of technologies to be adopted, and elaborate management often delivered on a nationwide level. Industry is organizing to deal with this transformation process, accelerated by the growing connection in the global world economy and by the 9/11 events in New York.

However, the complexity of the engineering solutions to cope with homeland security themes, the need to reuse the existing national assets and to elaborate a scalable solution, as well as the multilevel and multistakeholder context, are all elements that have concurred to develop the science of complex system of system engineering (SoSE) [11, 12].

The SoSE is a complement, not a replacement, to system engineering. SoSE is another facet of engineering that is an enabler to understanding complex SoS challenges. Its purpose is to address the new world view of capability-based engineering. SoSE addresses a complex system in terms of relationships,

Table 1.1
Building Block Relevance Versus Application Domain

	Crisis Mgt.	Border Control	Territory Control	Critical Infrastructures	Transportation Security
Sensors	H	H	H	H	H
Remote C2	H	H	H	M	M
Central C2	H	M	H	M	H
Networks	H	H	H	M	H
Effectors	M	M	H	M	M
Cooperative Sensors	H	M	H	M	M
Social Network	H	M	H	M	M
Media Mgt.	H	H	M	L	L

Relevance: High, Medium, Low

politics, operations, logistics, stakeholders, patterns, policies, training and doctrines, contexts, environments, conceptual frames, geography, and boundaries. It takes context into account while addressing the desired capability. It can be considered the necessary tool to conceive, develop, deploy, and evolve homeland security.

Clearly, the homeland security domain is an ideal place to apply holistic application of the SoSE methods. Homeland security is an environment where technical, human, cognitive, and social domains are interacting with each other. Moreover, the common need for any governmental and industrial solution to solve different homeland security key requirements is the necessary ability and flexibility to adapt to a variable context, where solution agility is a key driver to match the needs.

In this view both end user and industrial providers are taking a step forward, making a quantum leap in approaching and solving problems. It is no matter to refine preexisting methods successfully adopted for the development of stove-piped system solutions; homeland security is a typical complex environment where a mixing of different skills must be considered, managed, and harmonized.

Exploitation of the above concept is summarized in Figure 1.5. It is noted that every homeland security application requires the involvement of cognitive, informative, social, and physical components. The relationship among these different domains has been analyzed several years ago by Alberts and Garstka [4], who developed a comprehensive framework (see Figure 1.5) highlighting how domains influence each other.

The figure shows the effectiveness of a network-based operation solution as a combination of material source (i.e., derived from sensors) and of quality information; how this information is passed through a communication network and gets by the operators; and models how human behavior manages this information at individual and shared levels. Consequently, the degree of decision is primarily conditioned by the previous steps and only at the end can we evaluate and assess the effectiveness of the systems in the physical domain.

It appears clear that today's challenge for industry is to develop the "art of system architecture" to get an SoS design able to maximize performance/price ratio, to improve system ability to dynamically provide current and emerging capabilities, and to minimize the undesirable and chaotic effects. Applications of appropriate value analysis and system dynamics evaluation metrics of system performance and effectiveness are needed to gauge the system design [13, 14].

The conclusion is that the development of homeland security solutions requires the management of a wide complex environment. Authorities in charge of managing emergencies and stakeholders in the decision-making process are challenged to provide to industry a clear view about the processes and the duty of several organizations contributing to emergency management. On the other

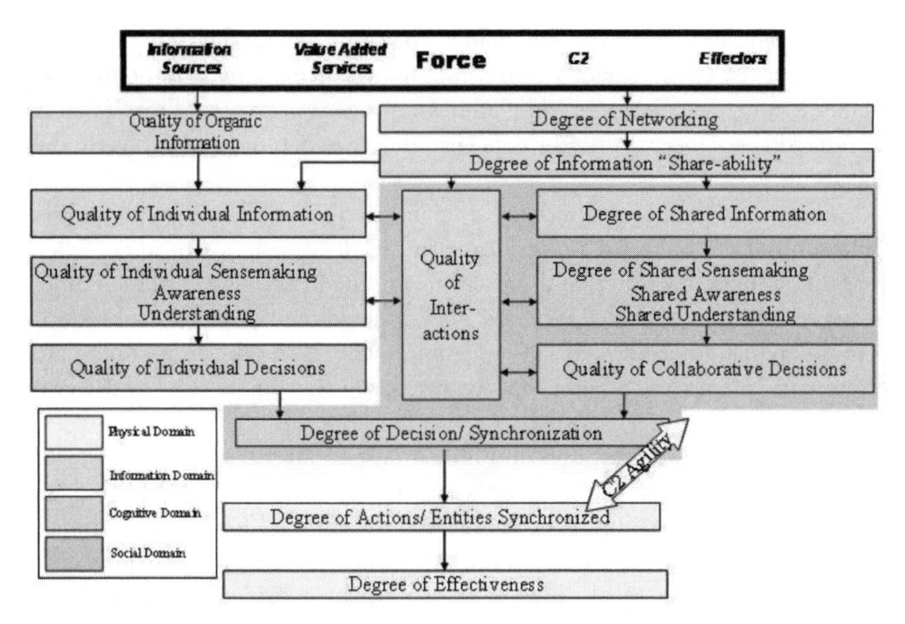

Figure 1.5 Network-centric operation conceptual framework. (*From:* [15]. Courtesy of Alberts and Garstka.)

side, industry is challenged to structure a sound solution in terms of effectiveness and costs.

Both sides are aware that they have to deal with complex and chaotic systems difficult to develop, deliver, and maintain. However, significant steps ahead have already been performed in terms of methodologies and technologies. The next chapters provide valuable insights on this matter.

References

[1] http://www.dodccrp.org/.

[2] http://www.mitre.org/.

[3] http://www.rand.org/.

[4] Alberts, D.S., Gartska, J.J., and Stein, F.P., *Network Centric Warfare: Developing and Leveraging Information Superiority,* Washington, D.C.: C4ISR Cooperative Research Program, August 1999.

[5] ASN(RDA) Chief Engineer, "CIAP Architecture Description," *INCOSE Las Vegas '02*, July 30, 2002. C4ISR Architecture Working Group. C4ISR Architecture Framework Version 2.0, December 18, 1997.

[6] Charles, P.H., and Dickerson, C.E., "Architectures Overview Brief," *RADM Sharp*, June 5, 2003.

[7] Charles, P.H., and Dickerson, C.E. "Use of Naval Architectures and C4I System Integration." *Littorals First Annual Conference*, May 21, 2003.

[8] Charles, P.H., and Dickerson, C.E. "Using Architectures in Naval Acquisition." Presentation to CAPT P.M. Grant, June 23, 2003.

[9] Charles, P.H., and Dickerson, C.E. "Using Naval Architectures to Demonstrate C4I Contributions to Warfighting Capabilities." Presentation to RADM Kenneth D. Slaght, May 19, 2003.

[10] Davis, P.K. "Analytical Architecture for Capabilities-Based Planning, Mission System Analysis, and Transformation." Office of the Secretary of Defense by RAND National Defense Research Institute.

[11] Department of Defense Systems Management College. *Systems Engineering Fundamentals*. Fort Belvoir, VA: Defense Acquisition University Press, January 2001.

[12] Dickerson, C.E., and Soules, S.M. "Using Architecture Analysis for Mission Capability Acquisition." *2002 Command and Control Research and Technology*.

[13] Crisp, H.E., and Chen, P. "Coalition Collaborative Engineering Environment." *INCOSE Insight* 5:3 (October 2002), 13–15.

[14] Dam, S. "C4ISR Architecture Framework: Myths and Realities." *Systems and Proposals Engineering Company*, Internal Report.

[15] Alberts, D.S., Gartska, J.J., and Stein, F.P. *Network Centre Warfare: Developing and Leveraging Information Superiority*, 2nd ed., Washington, D.C.: DoD C4ISR Cooperative Research Program, 2002.

2

Embedded Wireless Sensor Networks

Mani Srivastava

2.1 Introduction

High-fidelity and real-time observations of the physical world are critical for many issues facing science, government, military, first responders, business enterprises, agriculture, urban spaces, and individuals. Traditional means of sensing physical world information have relied on small numbers of powerful, high-quality, and often remote sensors, which fail to provide the sensing detail and density needed for many applications. With progress in microelectronics providing embedded processors, networked radios, and MEMS transducers with ever-improving performance at reduced cost, size, and power, the recent years have seen the emergence of the paradigm of *embedded wireless sensor networks*.

Unlike traditional embedded sensing, wireless sensor networks are based on networking a potentially large number of spatially distributed and wireless sensor nodes. The network of collaborating sensor nodes collectively acts as an instrument performing the required observations of the physical world at unprecedented detail and spatiotemporal sampling densities. Needless to say, it is this ability of sensor networks that has made them tremendously attractive for homeland defense as well as many other applications.

The sciences have been one of the early adopters of sensor networks as an instrument to investigate physical, chemical, biological, and social processes. To quote Sir Humphrey Davy, the exponent of scientific method, "Nothing tends so much to the advancement of knowledge as the application of a new instrument," and indeed at various times in human history the emergence of new instruments to enquire about the physical work has revolutionized science—such as the telescope for far away phenomena, the microscope for the tiny phenomena. Sensor networks represent a new type of instrument targeted at studying complex phenomena, and has been termed by David Culler [1, 2] as a manifestation of the macroscope [3, 4].

In homeland defense and military applications sensor networks are being used in applications such as reconnaissance, surveillance, target acquisitions, situational awareness, perimeter defense, critical infrastructure monitoring, and disaster response. A variety of wireless sensor node platforms with diverse sensing modalities are being used, including unattended ground sensors such as tripwires and trackers, and actuated sensors with a range of mobility from simple pan-tilt to large-scale autonomous navigation. A good example of a platform for defense applications is the U.S. Marines' small unit sensor system (SUSS) [5], which is a tactical, human-portable, unattended ground sensor system acting as the eyes and ears of small units by letting them observe tactical objects and danger areas beyond the line of sight and providing real-time "around the corner" sensing for increased situational awareness.

While distributed and large scale like the Internet, wireless sensor networks differ in being tightly coupled to the physical world, composed of nodes that are significantly more resource constrained than Internet hosts, and handling work load emphasizing real-time sensing and actuation as opposed to best-effort interaction. Designing such physically coupled, robust, scalable, networked systems is very challenging. This chapter provides an introduction to the sensor networking technology and its challenges in the context of homeland defense applications.

This chapter reviews the successes and lessons learned during the ten years that have passed since the early research efforts in sensor networks by Kaiser and Pottie at UCLA in the mid-1990s [6] that led to this vibrant area, and goes on to identify barrier challenges that remain. In particular, experience with this technology in recent years has led to the realization that the early view of sensor networks as "smart dust" [7]—a large and ad hoc cloud of simple devices—needs to be considerably expanded to a view of these systems as multiscale, multimodal, multiuser, rapidly deployable actuated observing systems. Moreover, as the embedded sensing technology moves from scientific, engineering, defense, and industrial contexts to the wider personal, social, and urban contexts, a new class of embedded applications is emerging that draws on sensed information about people, buildings, urban spaces, integrates with the global Internet and cellular infrastructure, and raises significant issues of privacy and data sharing.

2.2 Sensor Network Design Drivers: Resource Constraints and Autonomy

Based on a decade's worth of experience through research and commercial activity, the wireless sensor networking technology has now reached a level of maturity where there are many first generation reusable hardware and software components and tools available at all layers of the system, and nascent standards are emerging. For example, TinyOS [8] and 802.15.4 radios [9] have emerged as a commonly used system software environment and radio technology, respectively, for wireless sensor networks. Moreover, there is significant experience with sensor network deployments at many different spatial and temporal scales and in diverse application contexts. At the same time, sensor networks have proven to be quite multidimensional, involving complex multidisciplinary and cross-layer trade-offs at design time as well as at run time. Moreover, applications exhibit incredible diversity across these dimensions, as shown in Table 2.1.

This naturally leads to the question of whether there are any common principles. As we see in the next two sections, *resource constraints* and *network autonomy* are the two pillars around which the key design principles for sensor networks have emerged.

2.3 Resource Constraints

Resource constraint is an issue due to the limited processing, storage, communication, and power resources that typical embedded wireless sensing platforms have. For example, even the high-end platforms (see Figure 2.1) have resources that are meager relative to modern smart phones, while at the low end the nodes

Table 2.1
Diverse Application Characteristics

	Ecological	Seismic	Structural	Machine	Security
Spatial density	10 m	km	m	cm	m
Spatial coverage	1 km^2	100 km^2	100 m^2	1 m^2	10 km^2
Sampling rate	< Hz	kHz	kHz	kHz	kHz
Lifetime	Months	Years	Years	Years	Weeks
Resolution	8 b	24 b	16 b	16 b	1 b–16 b
Latency	Months Hours	Minutes Days	Minutes	Seconds	Seconds
Access cost	Medium	Medium	Low	Low	High
Environment	Moderate	Benign	Benign	Benign	Dangerous
Nature of task	Field	Source	Source	Source	Source
Where the answer is needed	Centrally	Centrally	Centrally	Centrally	Distributed
Platform mobility	Some	No	No	Some	Yes

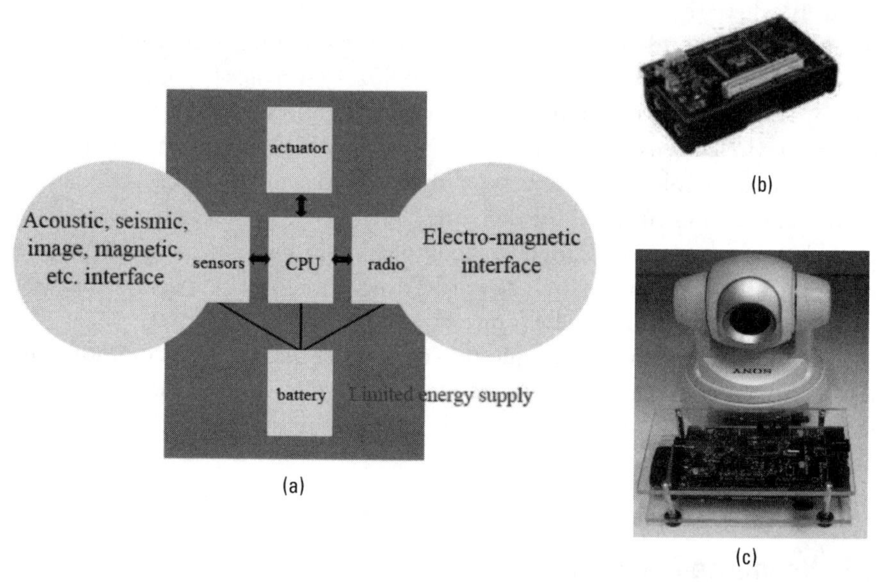

Figure 2.1 (a) Canonical wireless sensor node, (b) example of low-end sensor node: Berkeley/Crossbow's Mica Mote, and (c) example of high end sensor node: UCLA's LEAP.

only have 8-bit microcontrollers with few kilobytes of data RAM and few tens of kilobytes of Flash, short-range radios of few tens of kbps, and power consumption in milliwatts to tens of milliwatts range. While scaling of semiconductor technology will help, at the low end the resource limitations will continue as cost and size reasons drive increasing integration of node components on a single system on chip.

One manifestation of resource constraints is the limited battery energy, which has driven development of low-power and energy-aware nodes [10–12], and technology for in situ replenishment of energy by environmental harvesting [13]. A second manifestation of resource constraints is the limited processing, storage, and bandwidth, and has driven development of lightweight software frameworks [8, 14], and data centric protocols [15]. Data-centric protocols recognize that the multiple levels of indirections prevalent in conventional networks are not acceptable in the presence of resource constraints and organize network addressing and routing in terms of data availability and values as opposed to identity of nodes providing the data. Another form of bandwidth constraint comes from the way per node network capacity of ad hoc networks reduces with the increasing number of nodes, as shown by Gupta and Kumar in their seminal work [16]. Fortunately, nodes in an ad hoc sensor network are not independent data sources, and intelligently exploiting correlation in measurements at nearby nodes can alleviate the capacity-scaling problem [17]. We now explore some of the key design principles and techniques that have emerged from consideration of resource constraints.

2.3.1 In-Network Processing for Computing, Storage, and Communication Energy Trade-Offs

With energy being such a limited resource in sensor networks, an end-to-end perspective on energy optimization becomes important. One of the early recognitions by researchers was that the marginal energy cost of communicating a bit over a wireless link even at short distances using radios far exceeds that of processing an instruction on embedded processors. Indeed, this observation is embodied in the following quote by Gregory Pottie (UCLA): "Every bit transmitted brings a sensor network closer to death." A way of characterizing the relative energy efficiency of communication and computing is in terms of the ops/bit metric that indicates the number of operations that a sensor node can process in the amount of energy used to send a bit. As Table 2.2 shows, the op/bit metric is very large, even for low-power short-range nodes.

The design principle that emerges from this observation is that in-network processing, either at the source node or among groups of nodes, is an important approach for optimizing energy usage. From an overall energy perspective a sensor network will come out ahead if it can avoid sending a bit even at the cost of doing hundreds and thousands of instructions at the node. Systems such as TinyDB [14] and Diffusion [15] exploit this. A similar observation holds between communication and local storage as well, and has led to software architectures that exploit in-network storage [18].

2.3.2 Time Is Energy: Sleep Optimization Via Time Uncertainty Management

A characteristic of the short-range radios used in sensor networks is that the power consumption of the receiver electronics is significant and often dominates the power consumption of the transmitter (see Figure 2.2). Moreover, this energy is relatively unchanged whether the radio is actively receiving data or just idly listening. This is unlike long-range radios where the power consumed by the

Table 2.2
Communication Versus Processing Costs

	Transmit	Receive	Processor	Ops/Bit
Low-end sensor node	2,950 nJ/bit	2,600 nJ/bit	4 nJ/op	~1,400
High-end sensor node	6,600 nJ/bit	3,300 nJ/bit	1.6 nJ/bit	~6,000

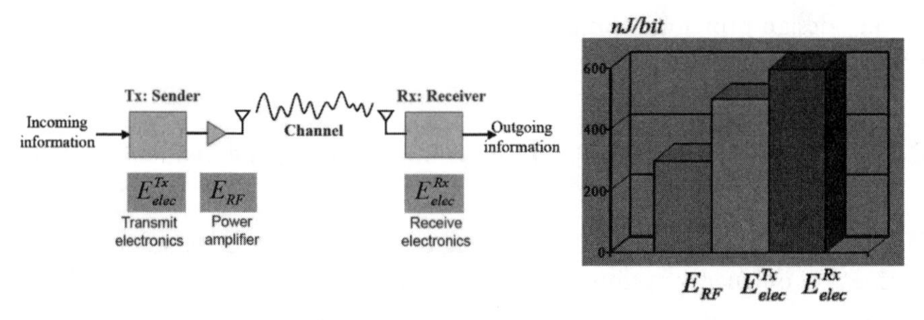

Figure 2.2 Dominance of receiver power.

RF amplifier at the transmitter is much larger than power consumed by receiver or transmitter electronics.

Therefore, conventional network protocols that assume data reception and idle listening are free are not a good match for sensor networks, which often sit idle for long intervals of time and perform a burst of activity upon detecting an event, as shown in Figure 2.3. Rather, the radio receiver is duty cycled whereby it is turned off most of the time but wakes up periodically to participate in potential network communication. While this has received significant attention in literature, the duty-cycling benefits achieved in theory and simulations do not translate to practice. This can be attributed mainly to the problem of time uncertainty between sensor nodes, which is illustrated in Figure 2.4. If the sleep–wakeup schedules of nodes do not intersect, they cannot communicate with each other. Note that each sensor node has its own notion of time governed by its local clock. The approaches used by medium access control (MAC) protocols to address this problem of time uncertainty determine their energy consumption. They can be classified into asynchronous or synchronous in nature.

Asynchronous protocols do not rely on any relative time synchronization information between the two communicating nodes. Instead, each packet is transmitted with a long enough preamble so that the receiver is guaranteed to wake up

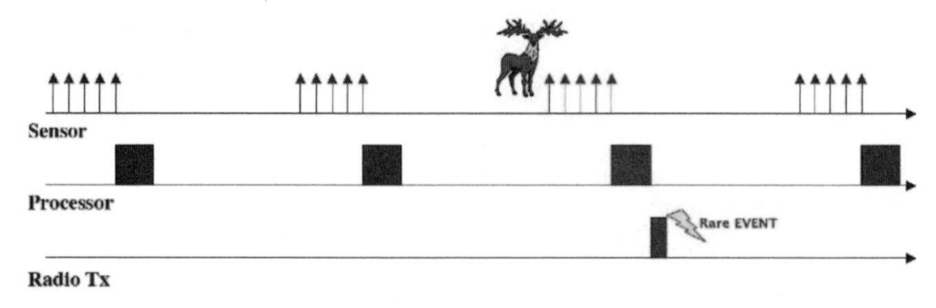

Figure 2.3 Ultra-low duty cycle operation.

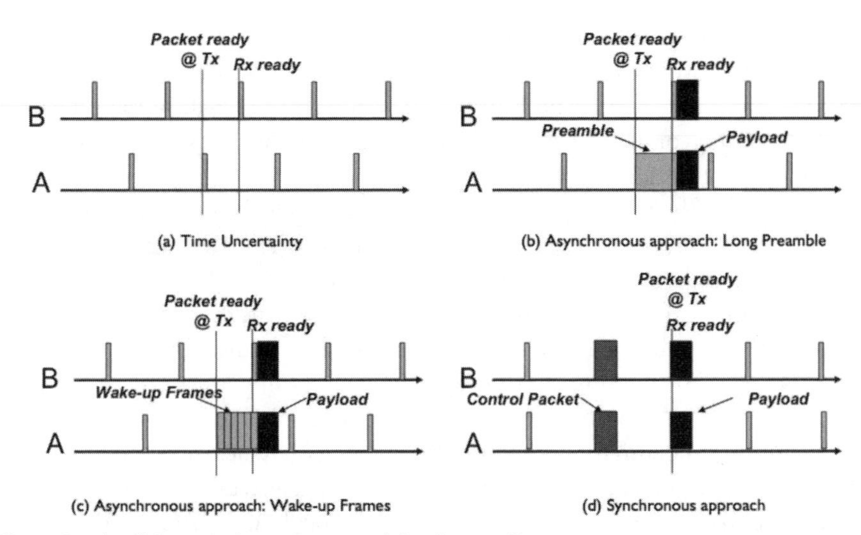

Figure 2.4 (a–d) Commonly used approach for duty cycling.

during the preamble transmission time, as shown in Figure 2.4(b). This implies that the preamble transmission time (preamble size divided by the radio's data rate) should be at least equal to the duty-cycling period. B-MAC [19], STEM-T [20], and WiseMAC [21] use slightly different variants of this approach. As an example, on the Mica2 Mote platform, B-MAC uses preambles of 250 bytes and 1212 bytes for 11.5% and 2.2% duty cycle, respectively, which is a significant energy overhead as the required preamble size for correct operation of the radio is just 4 bytes. Further, using a long preamble for every packet decreases the effective channel capacity as well as increases the receiving and listening overhead. All the overhearing nodes will have to be awake for half of the preamble transmission time, on average, before receiving the destination address information in the packet header and going back to sleep if they are not the intended receivers. A wakeup-frame–based scheme tries to reduce this overhead by transmitting multiple small wakeup frames instead of a single long preamble before the payload [refer to Figure 2.4(c)]. Each wakeup frame includes the destination address, so that overhearing nodes can go to sleep just after receiving a single wakeup frame. Furthermore, each wake-up frame also has a sequence number so that the destination node can calculate the start time of the data frame after receiving the wakeup frame. Using this information, the receiver can decide to go back to sleep for the time interval, equal to the transmission of the rest of the wake-up frames.

Synchronous approaches such as S-MAC [22] try to maintain time-synchronized duty cycling of nodes. This enables the transmitter to turn on the radio at the right moment and transmit the packet without incurring any extra overhead of a longer preamble or multiple wakeup frames [refer Figure 2.4(d)]. However, time synchronization is achieved by periodically exchanging control beacons. For instance, every node in S-MAC sends a time-synchronization packet after every 15 seconds, which results in a significant energy overhead, especially for low data rate applications.

Instead of being completely asynchronous or synchronous, an alternative is an uncertainty-driven approach to duty cycling, where a model of long-term clock drift is used to minimize the duty-cycling overhead. At the time of transmitting a packet, the node predicts its relative drift with the receiver node and chooses the preamble size accordingly instead of always choosing the worst-case preamble size. Further, instead of keeping the nodes in perfect synchrony at all times, just the right amount of synchronization is maintained between the nodes that minimizes the sum total overhead of achieving the synchronization precision and the extra preamble bytes that are added to every packet to overcome the time uncertainty. The uncertainty-driven MAC protocol (UBMAC) [23] is based on this concept and uses as its central building block a rate-adaptive, energy-efficient long-term time-synchronization algorithm that can adapt to changing clock drift and environmental conditions while achieving application-specified precision with very high probability. On Mica2 Motes, UBMAC achieves 3× energy improvement over B-MAC for higher duty cycles (>30%) and up to two orders of magnitude energy savings for lower duty cycles (<2%), simultaneously achieving the same packet loss rates as B-MAC. In contrast to using 100–2,500 bytes of preamble or transmitting a synchronization packet every 15 seconds, UBMAC achieves the same network performance by using just two extra bytes of preamble and one synchronization packet per hour. Increased effective channel capacity and reduced overhearing overhead are some of the auxiliary benefits that come as a result of using a smaller preamble and less control packets.

2.4 Network Autonomy

The sheer scale and deeply embedded nature of sensor networks make various configuration and maintenance tasks quite tedious and expensive to do manually. Configuration tasks include not only tasks such as address and cryptographic key assignments that are found in other wireless networks as well, but also tasks such as finding location and orientation of the nodes, synchronizing their clocks, and calibrating the sensors that are unique to sensor networks due to their physically coupled nature. Maintenance tasks include identifying and fixing or replacing malfunctioning nodes and sensors, recalibrating sensors, recalculating location and orientation nodes that may have moved, repositioning and reorienting nodes that become blocked by obstacles, and replenishing batteries.

2.4.1 Self-Configuration: Localizing Nodes in Space and Time

Knowledge of the location and local clock offset of each node in a sensor network is essential so as to put their measurements in the same coordinate reference frame in space and time for purposes of finding the time and location of occurrence of events being detected. Ad hoc deployment, demands of rapid deployment, and unknown obstacles in the environment make it difficult to configure the location and clocks manually before deployment, and in any case factors such as displacement of nodes and drifts in their clock due to subsequent ambient conditions make one time configuration insufficient. Self-configuration of location and time are therefore attractive, and commonly referred to as *node localization* and *time synchronization* problems.

These problems are specific instances of the more general class of calibration problem where some environment-dependent variable associated with each node needs to be estimated. Node localization involves estimating the <x, y, z> coordinate of each node, while time synchronization involves estimating <offset, drift> of each node's local clock source. Other examples of calibration problems include estimating orientation of sensor nodes, and estimating internal transducer parameters such as gain and offset of transducer electronics. Calibrating these parameters without having a human visit each node with a reference sensor (e.g., a GPS for location) and setting up the parameter being configured are clearly nonscalable. Recognizing this, much research has focused on distributed in situ approaches for these problems where sensor nodes collaborate among themselves and with the back-end to perform the self-configuration.

The time-synchronization problem, where the variable being calibrated is the local time at a node, arises because of imperfect clock sources at the nodes. These clock sources have different *initial offsets* (starting value of the clock C_0) and different drifts (time varying rate at which the clock $C(t)$ advances $\tau(t) = dC(t)/dt$. Without calibration, the local times reported by the clocks at each node $C(t)=C0+{}_{0_{\tau t}}\tau(\tau)d\tau$ would be different (see Figure 2.5), and make it impossible to combine measurements across sensors.

Representative of in situ approaches of solving calibration problems, solutions to time synchronization rely on a two-step process. First, pairs or small groups of nodes make measurements to establish constraints between their values of the calibration variable. Second, these constraints are combined in a distributed optimization framework to estimate the value of the calibration variables at each node.

For time synchronization the first step is done by pairs of nodes exchanging messages time stamped with local clocks, yielding instantaneous but noisy measurements of the relative clock offsets between the nodes $0_{ij}(t_k)=Ci(t_k)-C_j(t_k)$. Figure 2.6 shows two approaches of making such measurements, one based on a sender–receiver message exchange, while the other involves message exchange

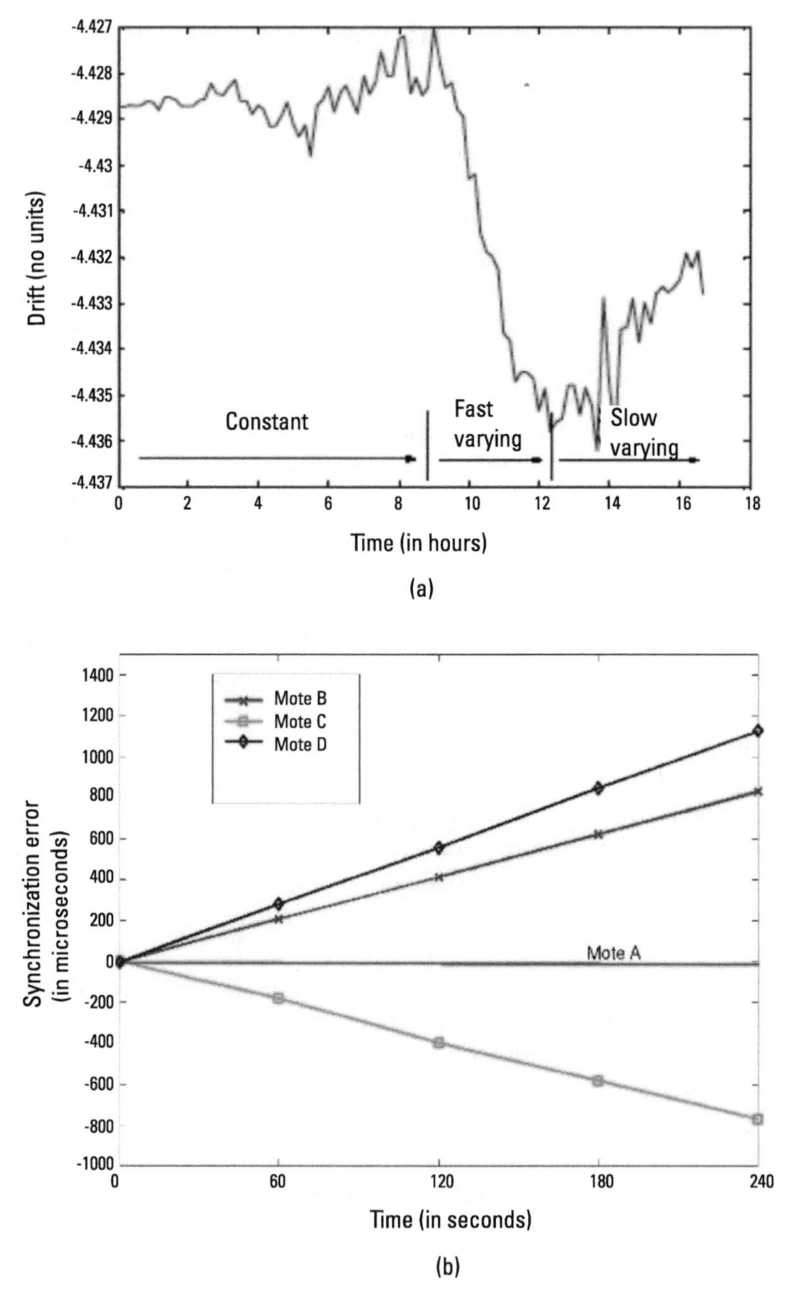

Figure 2.5 (a) Time varying clock drift between two sensor nodes, and (b) local time drifting apart in a group of sensor nodes.

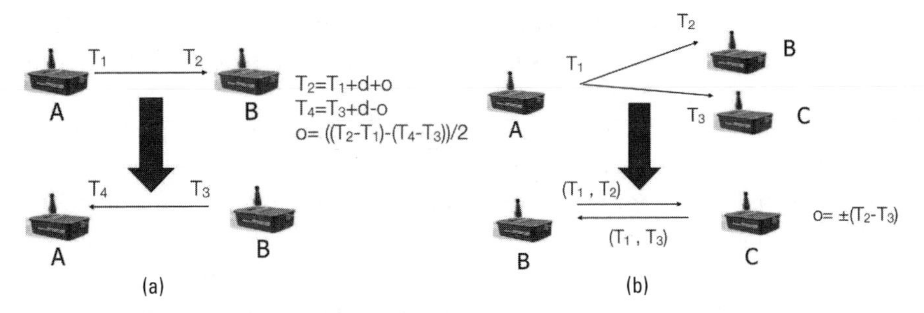

Figure 2.6 (a, b) Approaches for measuring clock offset.

between a beacon and a pair of receiver nodes. The receiver–receiver approach has the advantage that it totally eliminates any error due to delay uncertainty at the sender node (the beacon), including the queuing delay in the MAC layer, which could be rather large. The sender–receiver approach does not eliminate the sender side delay uncertainty, but reduces the impact of all the delay uncertainties in the end-to-end path by 2×. For sensor nodes where time stamping can be done below the MAC layer, the MAC uncertainty disappears, and the 2× error advantage of the sender–receiver approach makes it a superior choice. In other cases, such as when a sensor node uses a black box radio with an embedded MAC layer, the receiver–receiver approach offers large benefits.

The second step algorithmically combines these noisy measurements over time and across nodes to estimate the clock offset between arbitrary pairs of nodes in the network (or between an arbitrary node and a reference node) at arbitrary points in time. For example, some algorithms use prior noisy measurements of offset between the two nodes to develop a model of the relative clock drift as a low-order polynomial, and use it to estimate clock offset between the two nodes at an arbitrary time instant in the past or future [24]. Other algorithms [23] use the drift model to even control when the measurements themselves are made, thus optimizing the energy spent in time synchronization by minimizing how often nodes resynchronize. In the space dimension, many algorithms simply estimate offset between two nodes that are multiple hops away (and these have no direct measurements of their clock offsets) by estimating and aggregating hop-by-hop offsets along a path through the network [24, 25]. More sophisticated approaches, such as [26], formulate a global least square error optimization problem that assigns clock offsets such that within each loop in the network the constraint that offsets = 0 is satisfied. With low-end sensor node hardware such as Motes, instantaneous time synchronization accuracies of around 1–10 microseconds between neighboring nodes, and average accuracies between 10–1,000 microseconds across typical network sizes and resynchronization rates are easily achieved by the current algorithms. However, many applications of interest to military and homeland defense, such as RF-based tracking, require relative time-synchronization accuracies that are 3–4 orders of magnitude better. Such accuracies remain a challenge, and will require both algorithmic and platform advances.

Node localization, which is the other self-configuration problem, has proven to be much more challenging with robust solutions still elusive. The essence of the problem is for nodes to find their locations in one of two situations. In the first, nodes need to find their location in some absolute coordinate system given some geo-calibrated beacon nodes, which may be multiple hops away. This form of localization is useful in finding absolute location of a target. In the second, a group of nodes need to find their relative location in a local frame of reference, and is useful for tasks such as group motion of a team of robotic nodes. Figure 2.7 shows an example of the localization problem.

Akin to time synchronization, node localization approaches also involve a two-step process. First, nodes establish constraints on their location by measuring internode distances using time of flight or signal strength measurements, and direction using directional antenna or beam forming. These measurements usually employ RF, acoustic, or optical waveforms with different trade-offs and operating regimes. Second, these local internode distance and direction measurements are combined across space and time at the network scale to yield locations or trajectories of individual nodes. The typical approach is to solve a

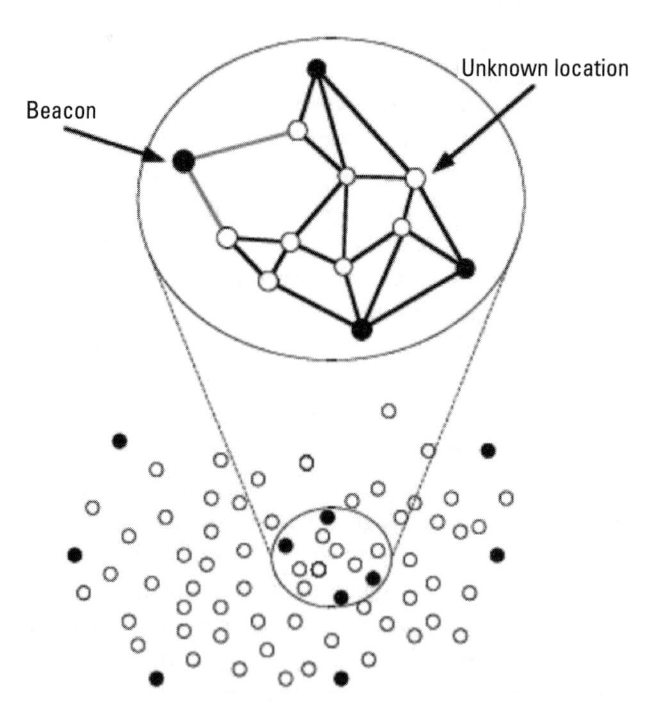

Figure 2.7 Node localization problem.

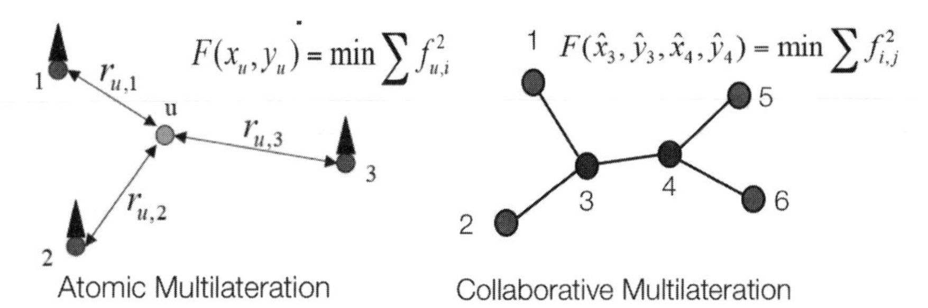

Figure 2.8 Algorithmically combining distances to obtain node locations in single-hop and multi-hop settings.

constrained nonlinear optimization problem under a cost metric that captures the estimation error. Figure 2.8 shows the formulation for both a local one-hop case as well as a multiple-hop case. Localization technology is considerably advanced and adequate for uncluttered indoor settings, but less so for cluttered outdoor environments with resource constrained nodes (thus ruling out GPS). However, emerging systems such as UCLA's self-calibrating acoustic array [27] and Vanderbilt's RIPS [28] are now able to localize the nodes to a few centimeters' location accuracy and few degree orientation accuracy in cluttered environments outdoors over a few tens of meter internode distances, albeit with trade-offs such as use of audible wideband acoustic chirps [27] and long convergence times [28].

2.4.2 Energy Neutral Operation

The need to replenish batteries is a major barrier to truly autonomous operation of sensor networks. Changing batteries in nodes deeply embedded in hard-to-access locations is costly due to the significant human effort involved. Techniques to improve energy efficiency by lowering the energy consumed by communications and computation, as discussed in Section 2.1, certainly help with this problem, but do not eliminate it even for the simplest and lower-power sensor nodes. For example, currently with typical battery sizes mote-class sensor nodes last for a year at very low duty cycles of around 1%, while more resource-rich sensor nodes last for a few days at best.

While techniques to lower energy consumption are essential and directly translate into increased network lifetime, they ignore a crucial component of the sensor node, namely the energy supply system, most commonly consisting of batteries. Batteries can only store a limited amount of energy, which places an upper bound on network lifetime. Environmental energy harvesting is a technique that helps circumvent this problem by exploiting energy sources that are already present in the operating environment of the sensor nodes, such as solar, wind,

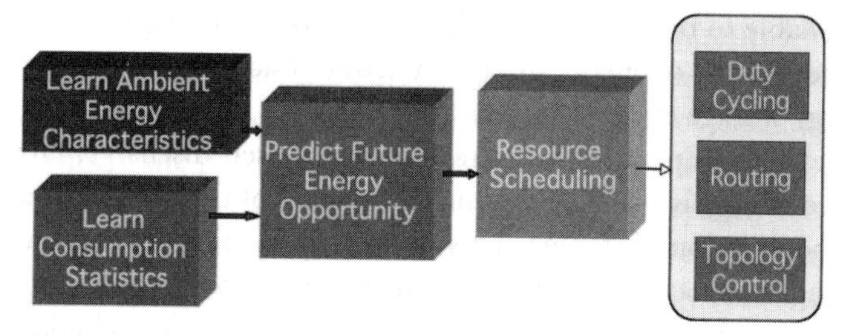

Figure 2.10 Harvesting-aware resource management.

via in-network collaborative processing. However, experience has proven this initial view of sensor networks as smart dust not appropriate for most application scenarios. Rather, sensor networks are rich and tiered ecologies consisting not only of the resource-constrained static nodes but also resource-rich sensor nodes and even nodes with actuation capabilities. These different types of nodes offer different information returns and trade-offs, and the structure of the network needs to be designed to achieve the desired sensing coverage, resolution, and latency while optimizing costs. Simple static and resource-constrained but low-power nodes may be deployed in large numbers over the physical space for making simultaneous measurements with high spatial density. However, their limited resources restrict sensing quality and sampling rate, and their static nature limits coverage in a cluttered environment despite large numbers. Static resource-rich sensor node arrays supporting complex modalities such as acoustic and image array sensors provide higher-quality sensing and higher sampling rate, but their cost and high power result in sparse deployments. Actuated sensor nodes with articulation and mobility can focus the sensor and communication resource at specific regions or directions in space. When coupled with smart sampling algorithms and networking protocols, they provide increased spatial density of measurements at the cost of reduced temporal density, and increased communication performance at the cost of reduced coverage.

2.5.2 Incorporating the Human Tier

Experience with deployments has indicated that sensor network architecture needs to be designed for the human tier as well since on-line interaction and tasking by humans continue to play a crucial role during deployment and management of a sensor network. The reason simply is that fully autonomous and robust engineering solutions to tasks such as localization, calibration, and sensor integrity checking continue to elude us. Intelligently incorporating human feedback can yield practical though not fully autonomous solutions to difficult deployment and run-time tasks. The human and back-end interaction has also become important as the envisioned nature of sensor network deployments has evolved.

While the early focus was largely on application-specific long-term deployments, in recent years alternate deployment models have emerged. One is the model of a sensor network as a long-term but multiuse instrument or observatory where over the lifetime of the system different applications for different missions are deployed after the sensor network infrastructure has been embedded in a space. A different model is where an application-specific but short-term deployment is done in a space for a particular mission or study. The human feedback plays a particularly important role under the second model where the system needs to be rapidly deployed for short durations in different locations without the luxury of a planned deployment in a well-characterized locale. Tasks such as proper emplacement of sensors, frequent calibration, and maintenance of nodes are simplified by the presence of users resulting in increased functionality and data quality.

2.5.3 To Sense or Not to Sense: Optimizing the Sampling Energy

As discussed in the previous section, the early design principles in sensor networks have been by energy resource constraints. Moreover, the focus of the design techniques has been on minimizing energy consumed in wireless communication, and to a lesser extent on storage and computing. The energy spent in acquiring the sensor data, consisting of energy used by the transducer, the A/D converter, and the front-end electronics and signal processing, was usually considered to be a relatively minor contributor to the overall energy budget. While certainly true for low-quality and low-rate sensors, and for simple sensing modalities such as temperature and light, which may be handled with passive transducers, this is not the case when one considers some other commonly used sensing modalities. For example, often sensors such as acoustic and seismic require high-rate and high-resolution A/D converters that can be tremendously power hungry. In describing their PASTA (power-aware sensing tracking and analysis) energy-aware sensor node for acoustic beam-forming–based tracking, authors of [11] say that as a result of optimizing the computation and communication aspects of their node the energy bottleneck is the four-channel 12-bit A/D converter that takes up more than half of their energy budget. CCD (charge-coupled device) or CMOS (complementary metal oxide semiconductor) image sensors are also characterized by higher power consumption. Indeed, the power consumption of the UCLA-Agilent Cyclops image sensor [35] is more than that of the Mica2 Mote radio and processor board that hosts the sensor. Active transducers, such as sonar, radar, and laser ranger, that need energy to send out a signal to probe the object being observed, and sensors used for detecting or identifying biological and chemical agents, also tend to be quite power hungry. In short, due to a variety of reasons, a common wisdom that the energy required to acquire sensor data can be ignored relative to the energy required to process and transport it is not true.

The solution to effective management of sensor power lies in applying to sensors the analogues of the two principles that also underlie much of the work on power management of processors and radios: *scalable fidelity* and *shutdown*. A key design principle is *energy-aware sensing* so that a sensor acquires a measurement sample only if needed, when needed, where needed, and with the right level of fidelity. Doing this not only reduces the energy consumed by the sensor, but in general reduces the processing and communication load as well. This goal is typically achieved using mechanisms such as changing the bit resolution of measurement samples (e.g., the number of bits per pixel or the number of pixels per frame in an image sensor); changing the sampling rate (e.g., the frame rate in an image sensor); adaptive spatiotemporal sampling (as opposed to uniform sampling) that focuses the sampling effort in interesting regions of space and intervals of time; exploiting redundancy, models, and correlations to predict a measurement instead of actually making it; and hierarchical sensing. In particular, three main design techniques that have emerged are the following.

Adaptive Sampling Instead of uniform sampling in time and space with a period dependent on the Nyquist frequency, adaptive sampling takes advantage of the variations in the characteristics of the physical phenomenon over the entire temporal and spatial region of interest. One could therefore sample more often in time or space in regions where there is more variation and unpredictability, or where the probability of event occurrence is higher, and less often elsewhere, without affecting the quality metrics of the final result such as fidelity of signal reconstruction, probability of event detection, and so forth. Where signal reconstruction is a goal, a possible strategy would be to do an initial coarse sampling followed by a finer-grained sampling in the region where the signal variations are higher. Of course such a strategy cannot work in time since the arrow of time prevents one from going back in time and resampling the signal at a finer granularity. However, this strategy is well suited for the spatial dimension where one can always acquire samples at additional points in space by activating static sensors located there or by making a robotic sensor node visit those points. The adaptive sampling concept can also be applied along the time dimension by changing the sampling rate as a function of the signal history, as long as transient reductions in fidelity when the signal changes rapidly are acceptable.

Triggered and Multiscale Sensing Often a physical event or phenomenon can be sensed with different sensors with different performance-power characteristics. For example, consider the task of detecting a vehicular target [11]. A low-power magnetometer or a simple acoustic energy detector can certainly detect such targets, but likely with a higher error rate and without an ability to identify the type of vehicle or track it. On the other hand, an acoustic beamformer with higher-quality acoustic sensors will consume higher power but will be more accurate in its detection accuracy and also provide the ability to identify and track the vehicle. Instead of having a field of only always-on acoustic beamformers, it is much better from an energy perspective to let the acoustic beamformer "trackers" sleep in the quiescent state and be woken up when a set of highly vigilant low-power "tripwire" nodes with magnetometer or simple acoustic energy detector detects a potentially interesting event. Such triggered wakeup across a hierarchy of nodes is a powerful mechanism for field-level energy management in sensor networks [36]. A closely related mechanism for field-level sensor energy management is that of multiscale sensing. The basic concept here is that measurements from a low-resolution wide-area sensor can be used to identify regions of interest and then higher-resolution sensors located in that region are woken up or tasked for measurement, or an actuated robotic sensor is tasked to visit or observe those regions. Significant energy savings can be realized by judicious tasking of higher-quality but higher-power sensors and relying on lower-quality but lower-power sensors for the long interval of time that the network spends waiting for interesting events to happen.

Model-Based Active Sampling Going beyond adaptive sampling and triggered sensing, we are now seeing the emergence of the more general notion of model-based active sampling to optimize the sensing process. The key idea is that the system learns spatiotemporal relationships among the measurements made by sensor nodes and events observed, and uses this knowledge to optimize the sensing for energy (i.e., whether, when, where, and at what fidelity level a sensor measurement should be made) for a required quality of end result. The process of learning the spatiotemporal relationships can be based on a variety of approaches, for example, modeling sensors as Gaussian processes and capturing the relationships among them in terms of covariances, or modeling the relationships among sensor values using nonparametric statistical models. Common to the different approaches is the ability to predict to some confidence level the value of a sensor measurement based on sensor measurements at other points in the <space, time, transducer type>. Now, faced with a user query that requires a specific sensor measurement, the sensor network can use the system and phenomenon model to decide whether making a new measurement is warranted, or whether it could be estimated with desired accuracy by other measurements already made (either at the same node in the past, or at other nodes, or with other sensor types) without spending additional sensing energy [37]. Likewise, faced with the need to estimate the spatial map at a certain time, the system can use the model to decide the minimal optimum set of measurements that can return a map of the phenomenon to desired fidelity, thus providing a generalization of adaptive sampling. Models of the phenomenon can also be used to predict the large-scale state of the physical system being observed, and the sensors can accordingly be activated or put in different energy-performance modes. Such model-based approaches to sensor data acquisition are in relative infancy, but promise a general framework for sensor power management while exploiting learned and prior information about the physical world.

2.5.4 Mobility as a Performance Amplifier

While networks of a large number of static nodes are good for dense spatial sampling, due to environmental clutter the likelihood of communication disconnections and sensing holes is surprisingly high at practically feasible node densities. Moreover, static sensors need to be placed on supporting surfaces, and thus sampling 3D volumes is hard. Mobility of actuated sensor nodes, while previously dismissed as too expensive and power hungry, has in fact emerged as a critical amplifier of sensing and communication coverage. Sensor nodes may be mobile at many different scales and with different degrees of system control over mobility. The scale dimension includes low-complexity small-scale constrained mobility (also called motility), large-scale mobility assisted by infrastructure as cables or other pathways to simple navigation, and autonomous infrastructureless large-scale mobility. The degree of control over mobility includes uncontrolled mobility that is opportunistically exploited, statistically predictable mobility, deterministic mobility, and fully controlled mobility.

Examples of large-scale controlled mobility sensing elements in sensor networks include networked UAVs with imagers being used in surveillance applications, USC's RoboDuck system [38] used for scientific investigations of aquatic systems, and UCLA's NIMS system [39], which uses a cable infrastructure to permit fine-grained sampling of 3D environmental phenomena. Large-scale controlled mobility is also useful for communication as well, such as using such nodes as roving base stations [40], as part of a dynamically formed network backbone [41], and as data mules [42] that can ferry bulk delay-tolerant data in on-board buffers between disconnected parts of the network. For example, the system in [42] uses a robotic data mule to extract delay-tolerant or bulk data from a sparse network of resource-constrained wireless sensors while minimizing energy consumption at the nodes. This is achieved by having the nodes send data at low transmit power and minimal hops when the robot is in the proximity. The trajectory and speed of the data mule are dictated by feedback from the network and a scheduling policy that seeks to minimize packet loss due to buffer overflows at the sensor nodes. In data systems a critical role is played by delay-tolerant protocols [43] that provide end-to-end transport between the sensor nodes and the back-end even though a complete network path between the two may never exist at a single point in time.

However, for mobility to be useful it does not need to be at large scale. Low-complexity small-scale constrained mobility, often called motility, is also very useful for improving sensing [44] and communication performance. Such mobility includes reorienting a directional antenna to improve network capacity and connectivity, and reorienting a narrow field-of-view sensor to focus sensing in the desired region of space. For example directional antennas at nodes in an ad hoc network may be mechanically steered after deployment to establish an energy-efficient topology tailored to the particular environment. Once oriented appropriately with a one-time energy investment, the network benefits in subsequent operation without having to spend higher energy for every bit exchanged as in signal processing–intensive smart antenna approaches.

Another example of the effectiveness of small-scale mobility is cameras or image sensors with pan, tilt, and zoom capability. Image sensor optics result in a field of view that is narrow, and a very large number of nodes would be needed for complete coverage, particularly in cluttered environments. As Figure 2.11 shows, the pan, tilt, and zoom capability of a commercial camera from Sony amplifies the volumetric coverage at a desired sensing resolution by a large factor of 226,940. In a network of distributed image sensors, coordinating the pan, tilt, and zoom settings of different cameras can be used for network-level benefits. For example, in a camera network monitoring urban spaces, in the quiescent state the cameras may be zoomed out for maximal field of view and their pan and tilt set to maximize coverage in the presence of clutter (see Figure 2.12).

Later, upon detection of an event of interest, one of the cameras may be tasked to observe the event at high resolution by zooming in with pan and tilt focusing the attention towards the event, as shown in Figure 2.13.

Meanwhile, the remaining cameras can reconfigure to maximize coverage of the remaining space. In effect, one gets a network with virtual high resolution while using low-resolution sensors at the cost of increased latency [45]. As shown in Figure 2.14, knowledge of medium characteristics (e.g., a map of clutter in the environment, represented as H) and prior knowledge of the phenomenon (e.g., a model of spatiotemporal distribution of the event, represented as q) may be used by a distributed coordination algorithm (f) that issues appropriate actuation

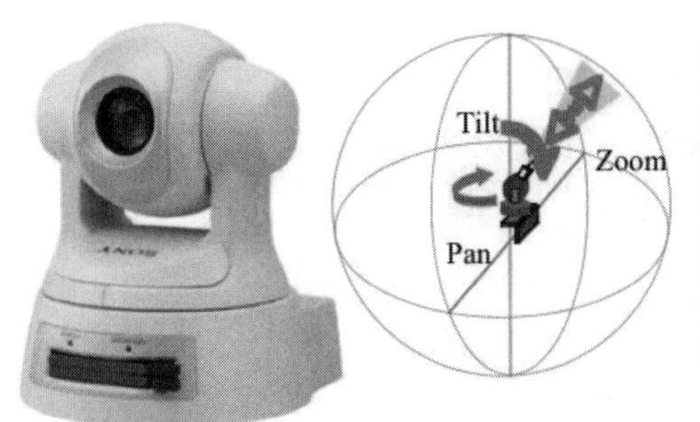

Pan	7.74
Tilt	4.04
Zoom	73
Pan and Zoom	6361
Pan, Tilt and Zoom	226940

Figure 2.11 Coverage gain from pan, tilt, and zoom (specified as the factor by which the volume of space that is sensed by the camera at some resolution is increased).

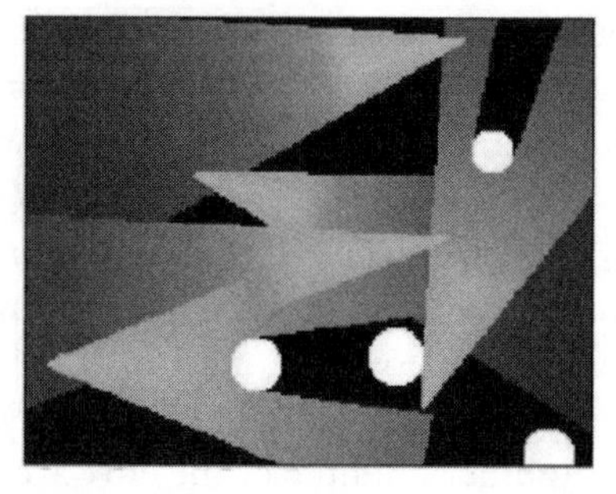

Figure 2.12 Orienting sensors for maximal coverage in the presence of clutter. On the left the sensors, whose field of view is represented by triangular shapes, are oriented randomly. On the right they are oriented judiciously using pan and tilt to maximize coverage. The clutter obstructing the sensors is represented by the circles.

commands (dx) to update the current pose of the sensors (X) to eventually arrive at an optimal configuration for the network under some cost function (C).

2.5.5 Monitoring the Monitors: Sensor Data Integrity

As noted earlier in the chapter, sensor networks represent a powerful new type of instrument for monitoring and studying our world. However, for an instrument to be useful, the information that it provides to its user must be of high integrity. In typical applications a sensor network detects sources, reconstructs physical fields, estimates statistical aggregates, and triggers other events, and communicates the results to client entities that make decisions based upon that information. The decision-making process can go awry if the sensor network provides a misleading picture of the physical phenomenon, some causes of which are shown in Figure 2.15: noise in the environment and transducer electronics; drift or other faults in the sensors; unfavorable channels from source to detectors (e.g., obstacles, multipath, and so forth); insufficient sampling in time or space; bugs and other faults in the computation hardware and software; noise, faults,

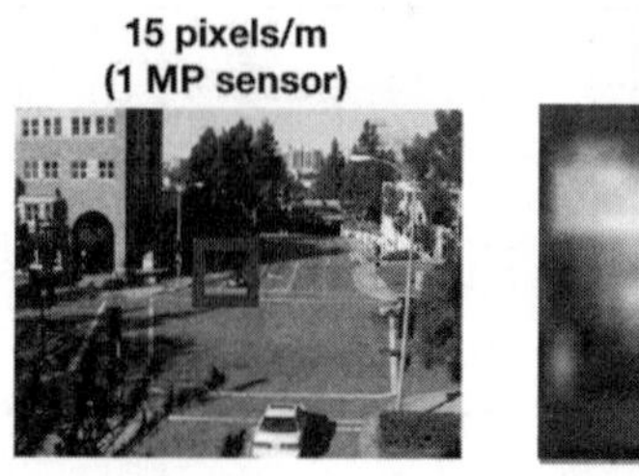

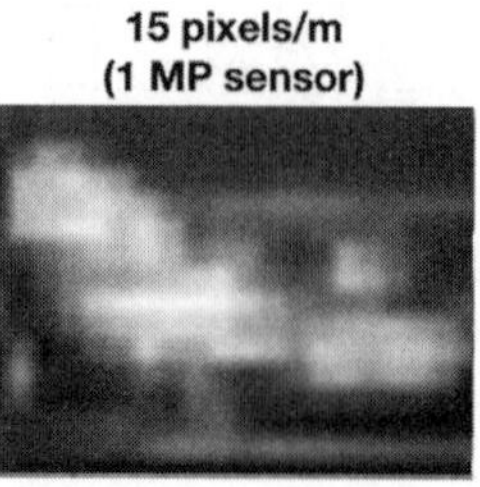

Figure 2.13 Virtual high resolution using a network of low-resolution imagers. Upon detection of the car arrival event, one of the cameras monitoring the scene at a wide angle is zoomed in.

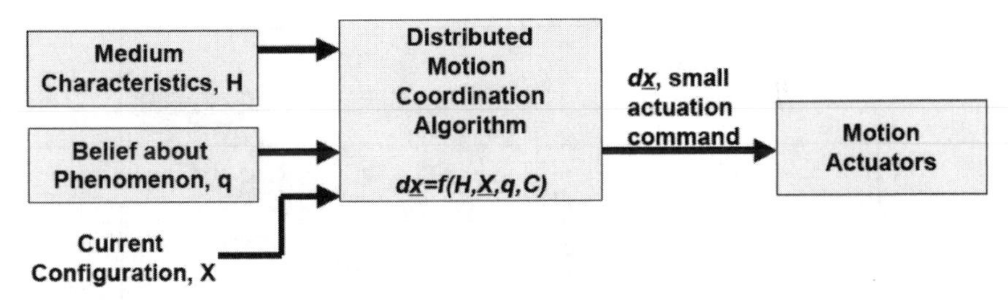

Figure 2.14 Framework for distributed coordinated actuation.

congestion, shadowing, multipath, compression loss, and other impairments in the networking subsystem; and attacks by an adversary during sensing, computing, or communication phases.

Data integrity represents a barrier challenge in designing and deploying sensor networks, albeit one, which unlike other challenges such as resource constraints and autonomy, has received far less attention. Current wireless sensor network designs are quite fragile, and there is still little systematic understanding of the multifaceted causes of how the quality of information returned by a sensor network is compromised, and a paucity of algorithms and software to tolerate or remediate them. In other information technology systems one pays for robustness by way of higher-quality hardware, infrastructure, and expert system administrators. In sensor networks, designers still focus on scale of the network and miniaturization of individual nodes. Consequently, the burden is on algorithms and software to provide robustness of the system and integrity of returned information. Specifically, algorithm support is needed for detecting misbehaving sensors, diagnosing the cause or locations of misbehavior, repairing or cleansing the corrupted measurements, and designing higher-level decision processes that make use of sensor measurements to be tolerant of data corruption. In addition, models play a crucial role since a fault is recognized when measurements do not correspond to what one expects according to some model that is based on external knowledge, previous readings at the same sensor taking into account diurnal and other cycles, readings from peer sensors, readings from wide-area sensors, readings from other correlated sensing modalities, and other variables that exhibit strong correlations. Figure 2.16 shows how temperature readings at a sensor

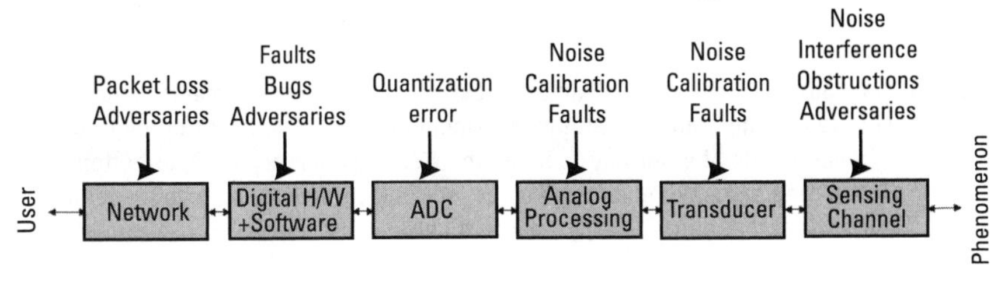

Figure 2.15 Sources of data integrity problems in sensor networks.

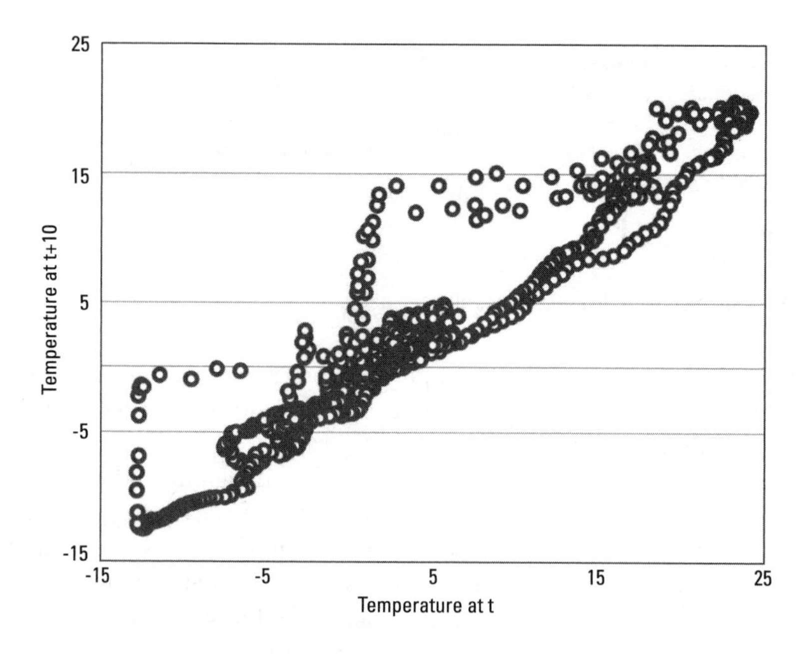

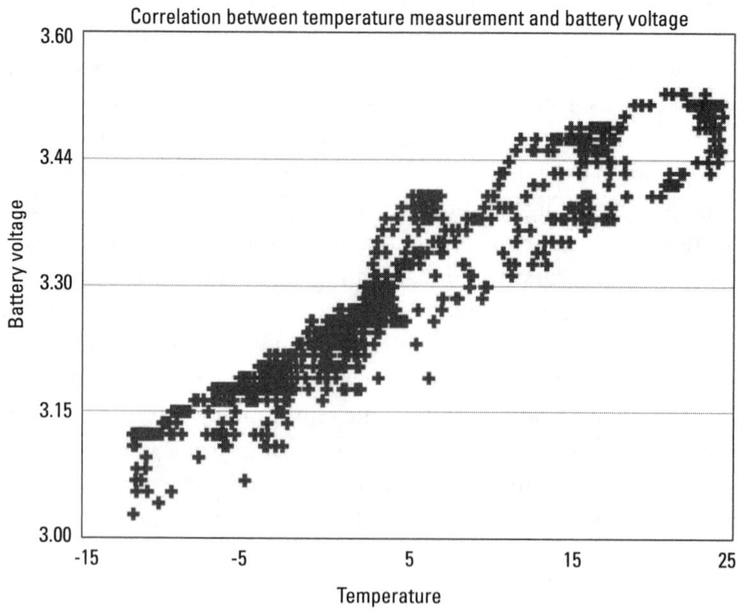

Figure 2.16 Fault detection corrections can exploit correlation of measurements made by sensors (e.g., temperature) with measurements made in the past (top plot), or with measurements of other physical variables such as battery voltage (bottom plot). The spread of measurements shown over the diagrams represents the statistical variation across different experiments, thus indicating the sensors' expected measurement range.

node exhibit a strong correlation not only with historical data, as expected, but also unexpectedly with the battery voltage.

In addition to benign causes, sensors may also misbehave due to malicious compromise of their integrity. For example, an adversary may interfere with the sensing process at the channel or the transducer, or may displace the sensor in space or time (or cause the sensor to believe that it is at a different location or time), and thus cause the sensor to report incorrect measurements. Such attacks are particularly tricky since conventional security techniques, which perform node authentication but not measurement authentication, are inadequate. They can ensure message integrity, confidentiality, and secure relaying, but not authenticate sensor data since compromised nodes have access to valid keys. Reputation-based approaches, adopted from social networks, are a potential way to provide data integrity resilience in such cases. Sensor nodes maintain "reputation" metrics for other nodes in the network. A sensor node continuously builds these reputation metrics for other nodes by monitoring their behavior and rating them as being good (expected behavior of the nodes in the network) or bad (unexpected behavior that is most likely the result of a system fault or node compromise). These reputation metrics, both directly observed as well as those obtained from other nodes via a chain of trust, are then used to evaluate the trustworthiness of other nodes and the data they provide. For example, a node can weigh the data provided by the other nodes with their corresponding reputation metrics while calculating the aggregates, or these reputation metrics can be used to identify misbehaving nodes in real time. Centralized versions of this model are in use in systems such as eBay and Amazon.com. Recently, reputation systems have been applied to security problems in ad hoc routing [46] and to sensor networks in [47] where the whole framework is implemented in a distributed fashion, with the function at each node being shown in Figure 2.17. The block labeled *Watchdog* is responsible for monitoring the actions of other nodes and characterizing these actions as cooperative or noncooperative. The actions of interest are the measurements made by the sensors. However, in general, the Watchdog can also encompass multiple layers such as routing, time synchronization, localization, etc. Instead of classifying a sensor reading as either cooperative or noncooperative, one can associate a level of confidence or probability (any real number between (0,1)) with it. A one-fit-all solution does not work for the Watchdog mechanism, and different applications might have different criteria for associating this level of confidence. The *Reputation* block maintains the reputation of a node by managing reputation representation, updating reputation based upon the new observations made by the Watchdog, integrating the reputation information based on other available information, aging the reputation, and creating an output metric of trust. Reputation-based approaches are lightweight and can be implemented even on low-end sensor platforms such as motes [48].

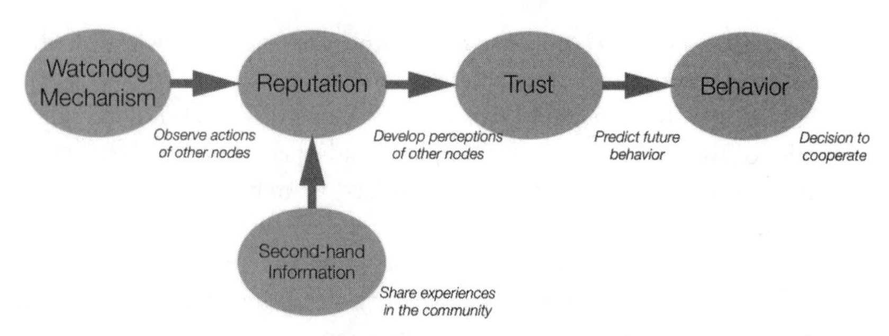

Figure 2.17 A reputation-based framework for assessing the trustworthiness of sensor nodes and the integrity of data provided by them.

2.6 Future Directions: Participatory Sensing

Although, as noted in Section 3.2, humans play an important role in deployment and management of current sensor networks, by and large they are not involved in the critical run-time sampling, processing, and communication of sensor measurements—tasks that are done solely by deeply embedded and unattended physical sensor nodes. However, information from human sources (human intelligence, or "humint") is also very important in many applications. This is certainly true for homeland defense and asymmetric and urban warfare situations. Unfortunately our understanding of humint information sources—their behaviors and ways in which their integrity is compromised—is very sparse. This is mostly because, unlike physical sensors, a formal engineering basis of humint information sources does not yet exist, although models of human coalitions and communications, human factors in context of interpreting information from sensor systems, and reputation-based systems have begun to emerge. However, in broader civilian application contexts frameworks for using humans as sensors have begun to emerge [49] for data collection and documentation vital in public health, urban planning, natural resource management, and culture. These frameworks marshal as sensor information sources the billions of people carrying cell phones equipped with image, acoustic, and various Bluetooth-based sensing modalities. Such frameworks provide searchable sensor feeds and blogs with geotime tags for context and in-network processing for privacy and control. In essence, these frameworks are creating an Internet of observatories for urban, social, and personal spaces by enabling people carrying cell phones to be part of a participatory sensing system. As these frameworks and the technology behind them mature, they could be used as valuable information resources for noncivilian applications as well.

There are two distinctive models of participatory sensing that are emerging. One is the *citizen-initiated* sensing, publishing, and sharing model where everyday users gather and share knowledge about their locale. Web services such as Flickr for images [50], and GasPriceWatch.com for gasoline prices [51] are based on this model. A second and more powerful model is that of *sensing campaigns* where professionally authored observation campaigns are conducted by automatically recruiting from the pool of participants based on their capabilities, reputation, and mobility profiles. Subsequently, the system automatically triggers the participants to make measurements (either automatically by the phone or by putting the human carrying the phone in the loop) based on their current location and the fidelity requirements and sensing uncertainty at that location taking into account previous measurements. Key challenges in such sensing include how to trigger data collection at the right time and place, how to ensure high quality and reliability, and how to provide incentives to foster participation. As an example, UCLA's Campaignr software framework [52] provides the ability for sensing campaigns to be conducted using a pool of users who volunteered by letting special software to be installed a priori on their Symbian cell phones.

Participatory sensing frameworks such as these raise serious questions of data quality, and, in civilian contexts, of privacy as well. From the perspective of participants, exercising control over the resolution and sharing rules of data they contribute are important. For example, a participant might be willing to reveal his or her location to different data consumers at different resolution in terms of location granularity and group size (individual vs. large group). From the perspective of the person conducting the sensing campaign, quality of data is important. In particular, data and their context are more valuable if verified. Subscribers want to know when and where the measurements were taken, and whether they can be corroborated. The system needs to verify, audit, and attest sensor measurements and their location and time stamps at a resolution the contributor is comfortable with. As an example, location information about a sensor measurement is prone to manipulation by either the contributing untrusted users or by an external adversary using attacks such as pulse delay, signal amplification, and wormhole. Recent approaches such as verified multilateration with cryptographic distance bounding [53] and covert listen-only base stations [54] seek to address this.

2.7 Conclusions

In nearly a decade of research and development sensor networking has yielded many interesting applications in science, the military, homeland defense, enterprises, etc. The experience thus far has also yielded insights that are transforming some of the initial design principles. In particular, the original smart dust model of sensor networks has yielded to system architectures that involve multiple modalities, multiple scales, diverse forms of mobility, and heterogeneous resource capabilities. Moreover, the early focus on resource constraints and autonomy is giving way to a focus on information quality and integrity, which have emerged as barrier challenges.

The involvement of humans, not just in configuration and management, but also in sensing, is a new direction leading to participatory sensing frameworks that make use of the billions of cell phones already deployed as sensors. Further changes are on the horizon as sensor networks become more integrated with the Internet by using a variant of IPv6 for low-power networking [55] and through the emergence of services supporting publishing and sharing of sensor data with architectural support for verification, privacy, and selective sharing.

Acknowledgments

The author would like to acknowledge the contributions of his current and former students and collaborators for the material on which this chapter is based, in particular Deborah Estrin, Saurabh Ganeriwal, Deepak Ganesan, William Kaiser, Aman Kansal, Gregory Pottie, Vijay Raghunathan, and Andreas Savvides. The research support of various funding agencies including ARL, DARPA, NSF, and ONR is also gratefully acknowledged.

References

[1] D. Culler, "Toward the Sensor Network Macroscope," Keynote talk, *Proceedings of the 6th ACM International Symposium on Mobile Ad Hoc Networking and Computing (MobiHoc 2005),* May 2005.

[2] G. Tolle, et al., "A Macroscope in the Redwoods," *Proceedings of the 3rd International Conference on Embedded Networked Sensor Systems (SenSys 2005)*, November 2005.

[3] P. Anthony, *Macroscope*, Mundania Press, 2003.

[4] J. de Rosnay, *The Macroscope: A New World Scientific System*, New York: Harper & Row, 1979.

[5] R. Braybrook, "Urban Rush Hour," Armada International, pp. 1–24, April 2006, www.armadainternational.com/06-4/complete_06-4.pdf.

[6] K. Bult, et al., "Low Power Systems for Wireless Microsensors," *Proceedings of the 1996 International Symposium on Low Power Electronics and Design*, pp. 17–21, 1996.

[7] J.M. Kahn, R.H. Katz, and K.S.J. Pister, "Next Century Challenges: Mobile Networking for 'Smart Dust'," *Proceedings of ACM MOBICOM*, pp. 271–278, 1999.

[8] J. Hill, et al., "System Architecture Directions for Networked Sensors," *Proceedings of Architectural Support for Programming Languages and Operating Systems (ASPLOS)*, 2000.

[9] IEEE Computer Society—LAN/MAN Standards Committee, "Part 15.4: Wireless Medium Access Control (MAC) and Physical Layer (PHY) Specifications for Low-rate Wireless Personal Area Networks (WPANs)," *IEEE*, 2006.

[10] J. Polastre, R. Szewczyk, and D. Culler, "Telos: Enabling Ultra-low Power Wireless Research," *Proceedings of the 4th International Conference on Information Processing in Sensor Networks: Special Track on Platform Tools and Design Methods for Network Embedded Sensors (IPSN/SPOTS)*, April 25–27, 2005.

[11] B. Schott, et al., "A Modular Power-Aware Microsensor with >1000X Dynamic Power Range," *Proceedings of the 4th International Symposium on Information Processing in Sensor Networks (IPSN 2005)*, April 2005.

[12] D. McIntire, et al., "The Low Power Energy Aware Processing (LEAP) Embedded Networked Sensor System," *Proceedings of the 5th International Conference on Information Processing in Sensor Networks (IPSN 2006)*, April 2006.

[13] V. Raghunathan, et al., "Design Considerations for Solar Energy Harvesting Wireless Embedded Systems," *Proceedings of the 4th International Symposium on Information Processing in Sensor Networks (IPSN 2005)*, April 2005.

[14] S. Madden, et al., "TinyDB: An Acquisitional Query Processing System for Sensor Networks," *ACM Transactions on Database Systems*, vol. 30, no. 1, March 2005.

[15] C. Intanagonwiwat, et al., "Directed Diffusion for Wireless Sensor Networking," *ACM/IEEE Transactions on Networking*, vol. 11, no. 1, February 2002.

[16] P. Gupta and P.R. Kumar, "The Capacity of Wireless Networks," *IEEE Transactions on Information Theory*, vol. 46, no. 2, March 2000.

[17] A. Scaglione and S. Servetto, "On the Interdependence of Routing and Data Compression in Multi-Hop Sensor Networks," *Proceedings of the 8th Annual International Conference on Mobile Computing and Networking (MobiCom 2002)*, September 2002.

[18] D. Ganesan, et al., "Multiresolution Storage and Search in Sensor Networks," *ACM Transactions on Storage*, vol. 1, no. 3, August 2005.

[19] J. Polastre, J. Hill, and D. Culler, "Versatile Low Power Media Access for Wireless Sensor Networks," *Proceedings of the 2nd ACM Conference on Embedded Networked Sensor Systems (SenSys '04)*, November 2004.

[20] C. Schurgers, et al., "Optimizing Sensor Networks in the Energy-latency-density Design Space," *IEEE Transactions on Mobile Computing*, vol. 1, January–March 2002.

[21] C. Enz, et al., "WiseNET: An Ultralow-Power Wireless Sensor Network Solution," *IEEE Computer*, vol. 37, no. 8, August 2004.

[22] W. Ye, J. Heidemann, and D. Estrin, "An Energy-Efficient MAC Protocol for Wireless Sensor Networks," *Proceedings of the IEEE Infocom*, June 2002.

[23] S. Ganeriwal, et al., "Estimating Clock Uncertainty for Efficient Duty-Cycling in Sensor Networks," *Proceedings of the 3rd ACM Conference on Embedded Networked Sensor Systems (SenSys '05)*, November 2005.

[24] J. Elson, L. Girod, and D. Estrin, "Fine-Grained Network Time Synchronization Using Reference Broadcasts," *Proceedings of the 5th Symposium on Operating Systems Design and Implementation (OSDI 2002)*, December 2002.

[25] S. Ganeriwal, R. Kumar, and M.B. Srivastava, "Timing-Sync Protocol for Sensor Networks," *Proceedings of the 1st International Conference on Embedded Networked Sensor Systems (SenSys '03)*, November 2003.

[26] R. Solis, V. Borkar, and P.R. Kumar, "A New Distributed Time Synchronization Protocol for Multihop Wireless Networks," *Proceedings of the 45th IEEE Conference on Decision and Control*, pp. 2734–2739, December 2006.

[27] L. Girod, et al., "The Design and Implementation of a Self-Calibrating Acoustic Sensing Platform," *Proceedings of the 4th ACM Conference on Embedded Networked Sensor Systems (SenSys '06)*, November 2006.

[28] M. Maroti, et al., "Radio Interferometric Geolocation," *Proceedings of the 3rd ACM Conference on Embedded Networked Sensor Systems (SenSys '05)*, November 2005.

[29] A. Kansal, et al., "Harvesting Aware Power Management for Sensor Networks," *Proceedings of the 43rd Design Automation Conference (DAC)*, July 2006.

[30] A. Kansal, D. J. Potter, and M.B. Srivastava, "Performance Aware Tasking for Environmentally Powered Sensor Networks," *ACM Joint International Conference on Measurement and Modeling of Computer Systems (SIGMETRICS)*, June 2004.

[31] A. Kansal, et al., "Power Management in Energy Harvesting Sensor Networks," *ACM Transactions on Embedded Computing Systems*, vol. 6, no. 4, December 2007.

[32] J. Hsu, et al., "Adaptive Duty Cycling for Energy Harvesting Systems," *ACM-IEEE International Symposium on Low Power Electronics and Design (ISLPED)*, October 4–6, 2006, Tegernsee, Germany.

[33] C. Vigorito, D. Ganesan, and A. Barto, "Adaptive Control for Duty-Cycling in Energy Harvesting-Based Wireless Sensor Networks," *Proceedings of the 4th Annual IEEE Communications Society Conference on Sensor, Mesh, and Ad Hoc Communications and Networks (SECON 2007)*, June 2007.

[34] A. Kansal and M.B. Srivastava, "An Environmental Energy Harvesting Framework for Sensor Networks," *ACM/IEEE Intl. Symposium on Low Power Electronics and Design (ISLPED)*, August 2003.

[35] M. Rahimi, et al., "Cyclops: In Situ Image Sensing and Interpretation in Wireless Sensor Networks," *Proceedings of the 3rd ACM Conference on Embedded Networked Sensor Systems (SenSys '05)*, November 2005.

[36] R. Kumar, V. Tsiatsis, and M. Srivastava, "Computation Hierarchy for In-Network Processing," *Mobile Networks and Applications (MONET) Journal*, vol. 10, no. 4, August 2005.

[37] A. Deshpande, et al., "Model-Driven Data Acquisition in Sensor Networks," *Proceedings of the 30th International Conference on Very Large Data Bases (VLDB '04)*, September 2004.

[38] B. Zhang and G.S. Sukhatme, "Adaptive Sampling for Estimating a Scalar Field Using a Robotic Boat and a Sensor Network," *Proceedings of IEEE International Conference on Robotics and Automation*, pp. 3673–3680, 2007.

[39] R. Pon, et al., "Networked Infomechanical Systems: A Mobile Embedded Networked Sensor Platform," *4th ACM/IEEE International Symposium on Information Processing in Sensor Networks*, April 2005.

[40] J. Luo, et al., "MobiRoute: Routing towards a Mobile Sink for Improving Lifetime in Sensor Networks," *Proceedings of the 2nd IEEE/ACM International Conference on Distributed Computing in Sensor Systems (DCOSS '06)*, June 2006.

[41] I. Rubin, "UV Aided Ad Hoc Wireless Networking Using Mobile Backbones," *Information Theory and Applications Workshop, UCSD*, 2006.

[42] A.A. Somasundara, et al., "Controllably Mobile Infrastructure for Low Energy Embedded Networks," *IEEE Transactions on Mobile Computing*, August 2006.

[43] K. Fall, "A Delay-Tolerant Network Architecture for Challenged Internets," *Proceedings of the 2003 Conference on Applications, Technologies, Architectures, and Protocols for Computer Communications (Sigcomm '03)*, August 2003.

[44] A. Kansal, et al., "Sensing Uncertainty Reduction Using Low Complexity Actuation," *ACM 3rd International Symposium on Information Processing in Sensor Networks (IPSN)*, April 2004, pp. 388–395.

[45] A. Kansal, et al., "Virtual High Resolution for Sensor Networks," *Proceedings of the 4th ACM Conference on Embedded Networked Sensor Systems (SenSys '06)*, November 2006.

[46] S. Buchegger and J.L. Boudec, "Performance Analysis of the CONFIDANT Protocol," *Proceedings of the 3rd ACM International Symposium on Mobile Ad Hoc Networking and Computing (MobiHoc '02)*, June 2002.

[47] S. Ganeriwal and M.B. Srivastava, "Reputation-Based Framework for High Integrity Sensor Networks," *The 2004 ACM Workshop on Security of Ad Hoc and Sensor Networks (SASN '04)*, October 2004.

[48] S. Ganeriwal, "Trustworthy Sensor Networks," Ph.D. Thesis, UCLA Electrical Engineering Department—Networked and Embedded Systems Laboratory, June 2006.

[49] T. Abdelzaher, et al., "Mobiscopes for Human Spaces," *IEEE Pervasive Computing*, 2007.

[50] http://www.flickr.com.

[51] http://www.gaspricewatch.com.

[52] http://wiki.urban.cens.ucla.edu/index.php?title=Campaignr.

[53] S. Capkun and J.P. Hubaux, "Secure Positioning in Wireless Networks," *IEEE Journal on Selected Areas in Communications: Special Issue on Security in Wireless Ad Hoc Networks*, February 2006.

[54] S. Capkun, M. Cagalj, and M. Srivastava, "Securing Localization with Hidden and Mobile Base Stations" *Proceedings of IEEE INFOCOM*, 2006.

[55] D.E. Culler, "6LoWPAN: Low-Power IP Connectivity," *Network World*, May 10, 2007, www.networkworld.com/news/tech/2007/051007-tech-update.html.

3

Visual Detection and Classification of Humans, Their Pose, and Their Motion

Stefano Soatto and Alessandro Bissacco

3.1 Introduction

Automated analysis of images and video plays an important role in modern surveillance systems, a role that is destined to further increase as the need to survey larger areas and critical strategic assets grows. In this chapter we review the state of the art concerning the automated detection and analysis of human motion. Many of the techniques described are viable for both visible and infrared or multispectral data.

The problem of understanding human motion has occupied psychologists, physiologists, mathematicians, engineers, and most recently computer scientists for decades. What is surprising to the uninitiated is the difficulty of performing even the most seemingly simple tasks, such as determining the presence and locating a person in a photograph, especially when compared with the apparent effortlessness with which humans can perform the same task. Obviously, the human visual system has evolved to be particularly attuned to human and animal motion, and a significant portion of the neo-cortex is devoted to processing visual information [1].

What makes the problem difficult is that the same event, for instance, the presence of a person, or the execution of an action such as walking or running, can manifest itself in a myriad of different ways. A person can be walking in different portions of the scene, at different distances from the viewer, under different illuminations, wearing different clothes, moving at different speeds and in different directions, and so forth. The goal is to tease out such "nuisance factors" from the data so that one can infer the object of interest, say the presence of a person, regardless of pose, vantage point, illumination, and clothing. This would be ideally achieved by an *invariant statistic* that is a function of the data that is independent of the nuisance factors that are of no direct interest but that nevertheless affect the data. While one can always construct statistics that are invariant to nuisance factors (the function that maps any datum to a constant is trivially invariant), the goal is for this statistic to be *discriminative*, that is to not be constant with respect to the object of interest. Ideally, one would like to be able to extract from the data the *sufficient statistics* for the task at hand, be it classification, categorization, recognition. This would be a statistic that is at the same time invariant with respect to the nuisance factors, and contains all and only the information necessary for the task. Unfortunately, it is rather straightforward to show that *there exist no discriminative illumination-invariant statistics* [2]. Similarly, if one factors in occlusions, it is possible to show that *there exist no discriminative viewpoint-invariant statistics* [3]. Therefore one is faced with the question: How can humans be so effective at detecting people, or recognizing them and their actions regardless of pose and illumination, when no discriminative invariants exist? Clearly the human visual system makes extensive use of prior information, and it is a fascinating scientific question to explore how such priors come to be in the first place, whether they are learned from other sensory modalities, or hard-wired. However, even the performance of the human visual system is often overrated, as it is very simple to conceive scenes that, from certain viewpoints and certain illumination, produce erroneous interpretations, or *visual illusions*. Even macroscopic changes in the image, such as the presence or absence of an object, or the presence of a large cast shadow, are often undetected by humans under the right circumstances [4]. So, given that the goal of devising optimal decision systems that are invariant to nuisance factors in the strict sense, and that produce sufficient statistics or "signatures" for the events of interest, is unattainable, we will have to be content with systems that *learn* the natural statistics from images, and that trade off discriminative power invariance, *but no error-free operation can be guaranteed in general.*

Large-scale competitive studies of simple test cases, such as the detection of a person in a static image, have been conducted in the past, for instance, by the U.S. Defense Advanced Research Project Agency (DARPA), and have exposed discouragingly low performance, both in terms of detection rate and false alarms. However, the field is evolving rapidly, partly under pressure from the applications, and performance in certain tasks improves by the month, with some of the tasks, such as the detection of faces, being rather mature with systems working in real time on commercial hardware. It should be mentioned that humans are among the most difficult "objects" commonly encountered in natural everyday scenes. They are subject to dramatic pose and shape variations, and exhibit a wide variety of appearance due to clothing. For simple objects, such as vehicles or products on a supermarket shelf, commercial systems are already in place that can automatically detect the presence and track the motion of an object.[1] The equal-error rate for object categories in static image has been steadily increasing, and benchmark datasets exist, such as the Caltech 101. Current performance rates are around 70% (compared with chance performance at 1%), but what's more important is that there are certain categories that are rather easy to classify (for instance, cars and motorcycles), whereas others are significantly more problematic. Humans in their natural habitat are among those. Indeed, detection and classification of human actions often hinge on *context*, that is, on what surrounds the person rather than on the area of the image that contains the person itself. For instance, an elongated blob at a corner of a street near a bus stop is classified as a person by human observers, but the same blob, rotated by 90 degrees and placed in the middle of the road, is classified as a car [5].

In the rest of this chapter we review each element of an ideal pipeline that can detect humans, track them, classify, categorize, and recognize their motion. This could then be fed to a module that extracts higher-level semantic information to "reason" about actions, or simply be fed to a human operator that is so spared the attention-intensive process of watching surveillance video, and could instead simply respond to "flagged events" from image and video processing.

3.2 Detection of Humans

Automatically detecting the presence of a human in a photograph can be trivial or impossible depending on a variety of conditions that have to do with scale and resolution, pose, occlusion status, etc. Most algorithms recently proposed can reliably detect a human that occupies a significant portion of the image (the so-called "300-pixel man"), is unoccluded, sports a familiar and benign pose, and wears clothing that contrasts with the background. On the other hand, most algorithms would fail to detect a person occupying a small portion of the image (say the "30-pixel man," or even the "3-pixel man" [6]), is partially occluded by other objects or by itself, is in an unfamiliar pose, and wears clothes that blend with the background. It is impossible to measure the performance of an algorithm in absolute terms, and often evaluation hinges on the choice of dataset. We start with the case of faces, where the most significant success stories have been recorded.

3.2.1 Faces

Early face detection algorithms were inspired by the theory of deformable templates, championed by Grenander [7]. The basic idea is to devise (by learning

1. See, for instance, the product Lane Hawk, developed by Evolution Robotics, www.evolution.com.

or by design) a "template" of a face, and then scan the image by correlation and report the local maxima as putative locations where a face might be. Because faces can appear in different sizes, one would have to test templates of all scales. This causes the computational complexity of algorithms to explode, especially if one wants to account for nuisance factors other than scale. Furthermore, partial occlusions cause missed detection, so one could break down the full-face template into "parts," including templates for eyes, noses, and mouths. Again, these have to be designed for all scales, and tested at all positions and orientation in the image. Unfortunately, images are rather ambiguous in the local, so cross-correlation based on, say, a "nose" template finds hundreds of noses in an image, most being false alarms. The same goes for eyes and mouths. In [8], the idea was set forth that, while local detectors fire a large number of false alarms, only in the presence of a face do their locations appear in the characteristic configuration of a face (with two eyes nearby, and a nose and a mouth below them). This "constellation model" [9] used procrustean statistics [10] to perform decisions based on the mutual configuration of parts, and was later extended to account for partial occlusions.

While elegant and principled, the constellation model was costly both to learn—as the computational complexity explodes with the number of parts—and to test, as one would have to perform template matching or feature selection for multiple complex templates at multiple scales and orientations. The basic ideas, however, remain valid and indeed can be found in the current approach to face detection based on *boosting*. The basic idea is to devise a so-called "weak classifier" that performs only slightly above chance, so it can reject locations where most likely there is no face, thus generating a large number of false alarms. If this step is computationally efficient, it can be run iteratively in a "cascade" manner so that eventually a combination of weak classifier produces the desired result. The most popular scheme is based on AdaBoost [11], a simple cascade of classifiers, and uses primitive templates, also called *Haar features*, that are binary masks that respond to coarse-oriented structures in the image. The resulting algorithm, known as Viola-Jones, has been in use since 2000 and it still constitutes the basis for the most effective face detectors. We refer the reader to [11], and to the follow-up journal version [12], for details.

Other successful alternatives for face recognition include work [13] based on the use of neural networks. However, since these approaches do not scale to the full body we will not further investigate them here and refer the reader to their original publication instead.

3.2.2 Full Body

While the use of neural networks has been successful on face detection, including some pose variation (frontal, partial out-of-plane rotation), the extension to full body has proven prohibitive because of the significant pose variation exhibited by the human body, and the impossibility of training for all possible variants. Similarly, clothing causes significant appearance variations that are more severe than in the case of faces. In order to extend successful approaches to face detection to the case of full body, one can seek more powerful classifiers and learning techniques, or for more sophisticated features.

On the side of a more powerful classifier, the work of [14] has demonstrated that the boosting approach can be useful even for the case of full body detection, especially if used as a preprocessing step to be followed by more accurate body localization and pose detection. On the side of more elaborate representation, histograms of oriented gradients have shown promising results [15].

The approach we advocate consists in a fast approximate detection stage that can return several putative locations and bounding boxes where humans can be present with high probability, following [14]. This can then be followed by more sophisticated processing to determine pose and, where desired, motion. This is an alternative to more accurate but computationally expensive dynamic programming approaches [16], which can find the optimal pose estimate of an articulated body model in a neighborhood of a proposed body configuration. Other alternatives have also used elaborate mixture models to resolve ambiguities in pose, especially when binary silhouettes are used instead of the full image [17–19].

The detector proposed in [15] consists on a support-vector machine trained on histograms of oriented gradients. Although considerably slower than the approach just described, there are recent attempts to speed up the classifier by using boosting [20].

3.2.2.1 Single Versus Multiple Images

Motion imagery (e.g., video) provides an increased volume of data that can, under certain conditions, be beneficial for the detection task. For instance, if one uses a boosting approach with weak learners [21], simple features can be employed in the spatial as well as temporal domain, with the latter providing added discriminative power that serves to more effectively prune false alarms. This is the basis of the approach that [14] followed to extend their AdaBoost to the case of full body. The task is further facilitated when the camera is known to be static, so that silhouettes can be inferred by simple *background subtraction* techniques [22–24], providing a representation that is insensitive to clothing and illumination. Unfortunately, many modern applications including surveillance from moving platforms (aerial or ground vehicles) do not allow for such a feature.

While a human detector has to operate despite pose, in the sense that one wants to be able to detect a person regardless of his/her pose, it is true that in many applications pose is not a nuisance factor, but instead contains significant information on the action or intention of that person. Therefore, most recently the attention has turned to the simultaneous detection and pose estimation, which we describe next.

3.2.2.2 Pose: Nuisance or Information?

Estimating pose from a single image without any prior knowledge is an extremely challenging problem. It has been cast as a deterministic optimization [16, 25], as inference over a generative model [26–29], as segmentation and grouping of image regions [6], or as a sampling problem [27]. Proposed solutions either assume very restrictive appearance models [16] or make use of cues, such as skin color [30] and face position [28], which are not reliable and can be found only in specific classes of images (e.g., athletes).

On the other hand, pose knowledge can simplify the detection problem, or rather the two can be solved jointly: a large body of work in pose estimation focuses on the simpler problem of estimating the 3D pose from human body silhouettes [18, 22–24]. It is possible to learn a map from silhouettes to poses, either directly [22], one-to-many [24], or as a probabilistic mixture [17, 18]. However, silhouettes are inherently ambiguous as very different poses can generate similar silhouettes so to obtain good results either we resort to complex mixture models [18] or restrict the set of poses [31], or use multiple views [23]. Shakhnarovich et al. [32] demonstrate that combining appearance with silhouette information greatly improves the quality of the estimates. Assuming segmented images, they propose a fast hashing function that allows matching edge-orientation histograms to a large set of synthetic examples. We experimented with a similar basic representation of the body appearance by masking out the background and computing a set of oriented filters on the resulting patch. However, likely due to the low quality of the silhouettes obtained from the dataset by simple background subtraction, the masking step does not yield sensible improvements.

In a combined detector/pose estimator, one starts with a video sequence with some putative locations and bounding boxes extracted by a coarse detector that processes all possible locations in the images and returns a sparse set of candidates for human presence (e.g., [14]). One does not require, nor enforce, continuity of the detector responses across frames at this stage. If available, one may also take advantage of binary silhouettes, which can be extracted from the sequence using any background subtraction or segmentation scheme. However, in practical real-time scenarios the quality of the extracted silhouettes is generally low and in our experiments we noticed that using bad silhouettes degrades the estimator performance.

In the following subsections we describe a sample embodiment of this problem, following in part [33]. We use a set of differential filters tailored to the human body to extract essential temporal and spatial statistics from the images. We create a large pool of features, which later will be used in a boosting scheme to learn a direct map from image frames to 3D joint angles.

3.2.2.3 Motion and Appearance Patches

The starting point, after preprocessing via the detector, is a collection of image patches that putatively portrays a person. Patches are normalized in intensity value and scaled to a default resolution (for instance, 64×64 is a trade-off between computational complexity and discriminative power).

We can use the silhouette of the human body (extracted by any background subtraction technique) to mask out the background pixels in order to improve learning speed and generalization performances. However, this step is by no means necessary: given a sufficient amount of data and training time, the boosting process automatically selects only the features whose support lies mostly in the foreground region. In our experiments we noticed that using low-quality silhouettes compromises performance, so we opted to omit this preprocessing step.

Motion information is represented using the absolute difference of image values between adjacent frames: $\Delta_i = \text{abs}(I_i - I_{i+1})$, where I is an array of positive numbers of size 64×64, with values in the positive integers between 0 and 255, and the subscript simply indicates the index of the frame. From the image difference Δ_i we compute the motion patches by extracting the detected patch. We could use the direction of motion as in [14] by taking the difference of the first image with a shifted version of the second, but in order to limit the number of features considered in the training stage we opt for not using this additional source of information. In Figure 3.1 we can see some sample appearance and motion patches. Normalized appearance I_i and motion Δ_i patches together form the vector input to our regression function: $\mathbf{x}_i = \{I_i, \Delta_i\}$.

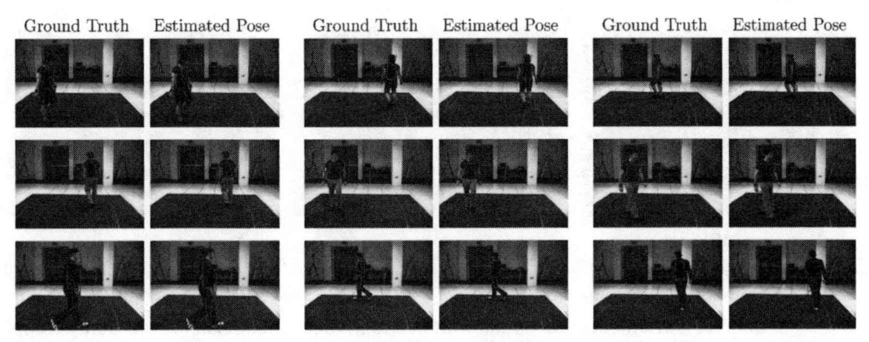

Figure 3.1 Sample frames from the dataset, and extracted appearance and motion patches. There are 3 subjects (S1, S2, S3), each performing a pair of cycles in a circular walking motion, for a total of 2,950 frames (1,180, 875, and 895, respectively, for each subject). In the first row we show a sample frame for each subject. In the second row, we display appearance and motion patches, scaled to the default resolution 64×64 and normalized in intensity.

3.2.2.4 Features for Body Parts

A simple human pose estimator is based on Haar-like features similar to the ones proposed by Viola and Jones in [14]. These filters measure the difference between rectangular areas in the image with any size, position, and aspect ratio. They can be computed very efficiently from the "integral image" [11]. However, in the context of this work a straightforward application of these filters to appearance and motion patches is not doable for computational reasons.

For detection of either faces or pedestrians, a small patch of about 20 pixels per side is enough for discriminating the object from the background. But if the goal is to extract full pose information, and if we were to use similar resolutions, we would have limbs with area of only a few pixels. This would cause their appearance to be very sensitive to noise and would make it extremely difficult to estimate pose. Augmenting the size beyond 64×64 greatly increases the number of basic features that fits in the patch (approximately squared in its area), therefore we need a strategy for selecting a good subset for training.

Another weakness of these basic features is that, by using vertical rectangles only, they are not suited to capture edges that are not parallel to the image axes. For pose estimation this is a serious shortcoming, since the goal is to localize limbs that can have arbitrary orientation. Therefore, one needs to extend the set of basic Haar features by introducing their rotated versions, computed at a few major orientations, as shown in Figure 3.2. Notice that these filters are very similar to oriented rectangular templates commonly used for detecting limbs in pose detection approaches [25, 34]. Oriented features can be extracted very efficiently from integral images computed on rotated versions of the image

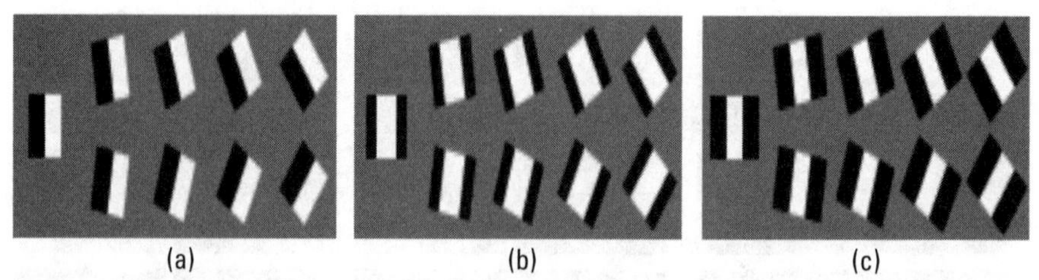

Figure 3.2 Basic types of Haar features (correlation templates) for body pose estimation: edges (a) and thick (b) and thin (c) lines. Each of these features can assume any position and scale within the estimation window [(although for scale some restrictions apply, see text for details). Each feature can assume a set of 18 equally spaced orientations in the range $[0,\pi]$. Here we show the 9 horizontal orientations and vertical ones are obtained by swapping axes. The value of the feature is computed by subtracting the sum of pixel values inside white regions from pixels in black regions, scaled by their area. It is intuitive to see how features (c) are suitable to match body limbs, while features (a) and (b) can be used to match trunk, head, and full body.

patch. Notice also that by introducing orientation in the features we further increase their number, so a good subset selection in the training process becomes crucial.

Among the possible configurations of features that one can select, we found that one type of edge feature shown in Figure 3.2(a) and two types of line features, Figure 3.2(b) and Figure 3.2(c), are the best performers. Each feature can assume any of 18 equally spaced orientations in the range $[0,\pi]$, and they can have any position inside the patch. To limit the number of candidates, we restrict each rectangle to have a minimum area of 80 pixels, do not come closer than 8 pixels from the border, and have even width and even height.

With this configuration, we obtain a pool of about 3 million filters for each of the motion and image patches. Since this number is still too high, we randomly select K of these features by uniform sampling. The result is a set of features $\left\{f^k\left(\mathbf{x}_i\right)\right\}_{k=1,\ldots,K}$ that map motion and appearance patches $\mathbf{x}_i = \left\{I_i, \Delta_i\right\}$ to real values.

3.2.2.5 Multidimensional Gradient Boosting

In order to learn the regression map from motion and appearance features to 3D body pose we start with the robust boosting of [35]. This algorithm provides an efficient way to automatically select from the large pool of filters the most informative ones to be used as basic elements for building the regression function. In order for this to be useful, however, the technique in [35] has to be extended to multidimensional maps, so that regressors for all joint angles can be learned jointly. This has been done in [33]. Joint learning is important because it exploits the high degree of correlation between joint angles to decrease the effective dimensionality of the problem. The resulting algorithm is sensibly faster than the collection of scalar counterparts; it is reported in detail in [33]: given a training set $\left\{y_i, \mathbf{x}_i\right\}_1^N$, with inputs $\mathbf{x}_i \in \mathbb{R}^n$ and outputs $y_i \in \mathbb{R}$ as independent samples from some underlying joint distribution, it finds a function $F^*(\mathbf{x})$ that maps $\mathbf{x}$ to y, such that the expected value of a loss function $E_{\mathbf{x},y}[\Psi(y, F(\mathbf{x}))]$ is minimized. Typically, the expected loss is approximated by its empirical estimate, thus the regression problem can be written as:

$$F^*(\mathbf{x}) = \arg\min_{F(\mathbf{x})} \sum_{i=1}^{N} \Psi(y_i, F(\mathbf{x}_i)) \tag{3.1}$$

Regularization can be imposed by assuming an additive expansion for $F(\mathbf{x})$ with basic functions h:

$$F\left(\mathbf{x}\right) = \sum_{m=0}^{M} h\left(\mathbf{x}; \mathcal{A}_m, \mathcal{R}_m\right) \tag{3.2}$$

where $h(\mathbf{x}; \mathcal{A}_m, \mathcal{R}_m) = \sum_{l=1}^{L} a_{lm} 1(\mathbf{x} \in R_{lm})$ are piecewise constant functions of $\mathbf{x}$ with values $\mathcal{A}_m = \{a_{1m},\ldots,a_{Lm}\}$ and input space partition $\mathcal{R}_m = \{R_{1m},\ldots,R_{Lm}\}$.[2] For $L = 2$ the basic functions are decision stumps, which assume one of two values according to the response of a feature $f^{k_m}(\mathbf{x})$ compared to a given threshold θ_m. In general h is an L-terminal node classification and regression tree (CART) [36], where each internal node splits the partition associated to the parent node by comparing a feature response to a threshold, and the leaves describe the final values $\mathcal{A}_m$.

We solve (3.1) by a greedy stagewise approach where at each step m we find the parameters of the basic learner $h(\mathbf{x}; \mathcal{A}_m, \mathcal{R}_m)$ that maximally decreases the loss function (3.1). The details can be found in [33].

One of the major limitations for experimental validation of any approach to human pose estimation from images is the availability of labeled training data. Ground truth for human motion is difficult to collect, since it requires expensive motion capture systems that must be synchronized and calibrated with the imaging device. Moreover, motion capture data can only be collected in controlled indoor environments, often requiring the performer to wear a special suit and having her appearance altered by the motion capture sensors attached to her body.

In our experiments we used the recently made public synchronized video and human motion dataset for human tracking and pose estimation [37]. The dataset consists of four views of people with motion capture makers attached to their body walking in a circle, see Figure 3.1 for some sample frames. In our experiments we use only the walking sequences for which both video and motion data are available, having a total of three subjects and 2,950 frames (first trial of subjects S1, S2, S3). Our goal is pose estimation, therefore we use only the images taken from a single camera (C1). In order to assess the performances of our pose estimator we trained the model using fivefold cross-validation.

Motion information for this data consists of 3D transformations from a global reference frame to the body part local coordinates. We have a total of 10 parts (head, torso, upper and lower arms, and upper and lower legs). We represent motion as the relative orientation of adjacent body parts expressed in the exponential map coordinates. By discarding coordinates that have constant value in the performed motions we reduce to 26 degrees of freedom.

The first step is to extract the human body patches and scale them to the default resolution of 64×64 pixels and normalized in intensity. As we have mentioned, one could extract the foreground pixels, but in our experiments such a preprocessing step actually degrades performance. We then extract motion patches as the differences between adjacent frames, scaled and normalized as just described. Although eventually in real applications the patches will be provided

2. Here we denote $1(c)$, the function that is 1 if condition c is true, 0 otherwise.

by a detector, we used the calibration information available in the datasets to draw the patches. Some sample outputs of this preprocessing stage are reported in Figure 3.3.

Given a set of motion and appearance patches $\mathbf{x}_i$ with associated normalized joint angles $\mathbf{y}_i$, we train a multidimensional boosting regressor for both least-squares (LS) and least-absolute-deviation (LAD) loss functions, using either decision stumps or CART as basic learner. At each iteration, for each patch we evaluated 10^5 features randomly sampled from the pool of oriented filters described previously. We experimentally found the optimal learning rate to be 0.5, and ran the boosting process until the improvement on the training residual was negligible. We also experimented with different basic functions, and obtained the best results using 5-node classification and regression trees. We believe that decision stumps do not perform as well for this kind of problem because body part configurations for articulated objects such as humans are highly dependent, and an approximation of the pose map as a sum of functions of single features cannot capture these dependencies. On the other hand, CART trees of n nodes can model functions having arbitrary interactions between $n-1$ variables. Table 3.1 shows mean and standard deviation of the L_1 and L_2 pose error norms on the entire dataset during validation. Rather surprisingly, the least-squares approach outperforms the least-absolute-deviation in all settings. Since we noticed similar results with other motion representations (i.e., 2D marker positions), we believe

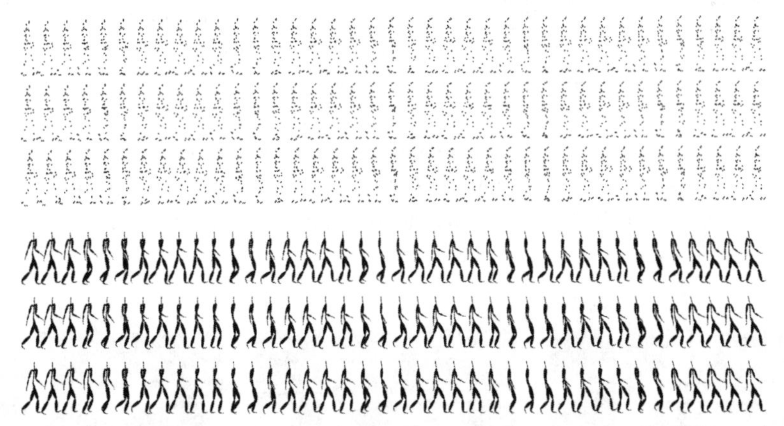

Figure 3.3 Synthesis results for models learned from motion capture data, both for marker positions (first three rows) and joint angle representations (last three rows). For each group, the first row shows the original sequence, the second row displays the synthesis obtained with the minimum phase system, while the last row shows the synthesis from the optimal nonminimum phase system. It is perceivable (see also the movies, downloadable from [http://www.cs.ucla.edu/~bissacco/dynamicICA]) the better fidelity of the nonminimum phase synthesis to the character of the original motion.

Table 3.1
Pose Estimation Errors

Algorithm	Mean	Standard Deviation	Number of Features	Time (s)
Zhou et al. [38]	0.3031 (0.3026)	0.0750 (0.0703)	52,000	40.55
LS stump	0.2818 (0.2777)	0.0931 (0.0885)	2,000	4.22
LAD stump	0.2922 (0.2757)	0.1020 (0.0909)	8,000	16.14
LS 5-CART	0.2736 (0.2701)	0.1158 (0.1067)	2,850*	3.28
LAD 5-CART	0.2996 (0.2863)	0.0972 (0.0929)	6,000*	5.94

Table showing mean and standard deviation of the relative L_2 error norm (i.e., $\|\hat{y}_i - y_i\|/\|y_i\|$) for the entire dataset during the validation phase. In brackets are the mean and standard deviation of relative L_1 error norm. We report results for the multidimensional boosting [38] multidimensional treeboost with both least-squares (LS) and least-absolute-deviation (LAD) loss [33], using stumps and classification and regression trees [36] with 5 internal nodes as basic functions. We see that the best performer (in bold) is the CART regressor with least-squares loss, while the approaches using LAD criterion do not score as well. For each classifier we also report the number of features used (for CART regressors the * denotes an upper bound; for each evaluation only a subset of features are computed) and the evaluation time on the entire dataset (2,950 frames).

that this is due to the approximate nature of the algorithm for nonsquared loss, and for a dataset with few outliers the benefits of the L_1 criterion are overcome by the error in the approximation. Our experiments also show that there is an advantage in selecting a smaller number of features over alternative approaches [38]. Compare this with approaches based on exemplar matching or kernel machines [17, 19], which often need to retain a large part of the training examples. In Figure 3.4 we show some sample motion and appearance patches together with the estimated pose represented as the outline of a cylinder-based human model superimposed onto the original images. Here we use a regressor trained using least-absolute-deviation and 5-node CARTs as basic functions. From these results it is clear that the lack of prior information adversely affects the estimations of occluded parts.

Figure 3.4 Sample estimation results. The first, third, and fifth columns show the provided ground truth, while the second, fourth, and last columns show the estimated pose. Samples in the last column show that, as we could expect, large estimation errors occur for poses characterized by prominent self-occlusions, visible from the overlay of the original image and the projected model.

3.3 Tracking Human Motion

Once a human has been detected in an image, and even better if his/her pose has been classified, one may attempt to track that person throughout a sequence. We discuss the issue in this section.

3.3.1 Tracking as Transition Between Key Frames

The simplest tracking methods consists of a sequence of pose classifiers. In this case the temporal model is simply a transition between poses, considered independently at each frame [34]. One can expand this to learn a transition map between poses [18], or build a likelihood function as a matching score between a configuration hypothesis and a given image [16]. These approaches can help maintain focus on a particular individual, but they lack temporal dynamics in the sense that the temporal discretization is often too coarse to enable fine-grain classification of motion, let alone of an individual. We refer the reader to [34] for details on the implementation for such a system as well as a review of recent progress using this approach to tracking human motion.

3.3.2 Continuous Temporal Models

Tracking by estimating the evolution of pose continuously has been historically the first approach to the problem [39, 40]. A common approach is to learn a statistical model of the human dynamics and to use it in a sampling scheme [41] where, given the body configuration in the current frame and the motion model, we can compute a probability distribution that allows us to make informed guesses on the limb positions in the next frame. We refer the reader to the survey in [42] for more extensive references.

A common approach is to assume a parametric model of the optical flow, which can be either designed [43] or learned from examples [44]. This can then be fitted to a model of the human body, designed from first principles or from simplified physical and photometric models of humans. An important issue in the design of these models is the explosion of the dimensionality if one is to learn the spatial, temporal, and photometric variability during tracking. In fact, both the geometry (pose) of an individual, his/her motion (dynamics), and appearance (photometry) can change drastically as the person moves towards or away from the camera, in and out of shadows, and so forth. One approach has been to strongly constrain these phenomena and their variability, for instance, by assuming that they live on linear subspaces of the infinite-dimensional space of configurations and appearance [45], or that a simple representation of appearance, for instance, the silhouette of a binary image sequence, is constrained to evolve on a linear space in some feature representation [46]. A prior model of motion can greatly help tracking by providing strong constraints on the allowable state space where predictions are to be performed.

Tracking human motion, when a proper initialization is available, is well studied, with early approaches such as [39, 40] providing satisfactory tracking for at least short sequences. Given recent progress in pose estimation, we advocate a hybrid approach, where a continuous model is initialized with a human detection and pose estimation, and then tracking is performed on a frame-by-frame basis. When the tracker fails, another detection step, followed by pose estimation, can be executed.

3.4 Classification and Recognition of Human Motion

The ability to classify and recognize events as they unfold is an important skill for biological as well as engineering systems to have. The intentions of a predator in the wild, or a suspicious individual at an airport, may not be obvious from its stance or appearance, but may become patent by observing its behavior over time. The pioneering experiments of Johansson [47] with moving dots illustrate eloquently just how much information is encoded in the observation of time series of data, as opposed to static snapshots.

Classification and recognition of events are very complex issues that depend on what kind of sensor data are available (e.g., optical, acoustic), and what representation is used to characterize the events of interest. Certainly a given event (e.g., a traffic accident) can manifest itself in many possible ways, and a suitable representation should exhibit some sort of invariance or insensitivity to nuisance factors (e.g., illumination and viewpoint for the case of optical images) since it is unlikely that one could "train them away" with extensive datasets. Also, many cues contribute to our perception of events. For instance, it is easy to tell that a person is running (as opposed to, say, walking) by a static image snapshot, and whether the classification task is trivial or impossible depends on the null set as well as the alternate hypothesis: it is easy to tell a walking person from a banana even from a static image, but it is not so easy to tell whether he/she is limping.

But whatever the sensors, whatever the representation, and whatever the null set, in the end one will need the ability to *compare time series of data.* This problem has been approached in a variety of ways that we discuss in the next sections.

3.4.1 Integral Representations

In order to compare functions of time, one has to devise a way to measure a distance (a positive number) given two (generally infinite-dimensional) functions of time. The simplest way to achieve this is to "kill time" by extracting some statistic that somehow summarizes the temporal aspect of the data. For instance, given a sequence of images of finite duration, one can simply take the average of these images, or of binary images obtained from them via background subtraction. A variety of different statistics have been proposed in the past. The difficulty with this approach is that these statistics are usually rather coarse representations of the temporal dynamics, so that subtle variations in the temporal execution of an action cannot be appreciated. In the remaining subsections we review ways to compare time series directly without resorting to static signatures.

3.4.2 Global Measures of Similarity

The simplest form of comparison between two time series is to treat them as elements of some standard function space such as L^1 or L^2 , and measure the norm of their difference. Unfortunately, this is not very effective because of the intrinsic variability in the data due to the nuisance factors. Other discrepancy measures that are somewhat less sensitive to small perturbations include various forms of correlation [48]. However, again one has to rely on the data being "close enough" in the data space itself, and does not model the intrinsic variability explicitly.

A different approach to compare functions is via their likelihood. The basic idea is that, in order to compare one signal, say, "A," to another, say, "B," one learns a generative model for "A," and then measures the likelihood that this model generates the data "B." Again, the generalization model implicit in this comparison is that each class of data is a perturbation ball around "A." Paradoxically, the longer the time series (hence the more accurate the model inferred from "A"), the more peaked the likelihoods, and the worse the classification is, a phenomenon that is treated with palliative measures such as likelihood rescalings or partitioning of the temporal axis.

One step towards modeling the intrinsic variability is done in dynamic time warping (DTW), where comparison is performed by minimizing over all possible time domain deformations the difference between two time series. The problem with this is at the opposite end of the spectrum: many different actions will look similar because the time warping does not take into account physical constraints. This issue is described in [49], where a modification of DTW is proposed that takes into account dynamic constraints.

3.4.3 Classification of Models

When an action is represented by a dynamical model, for instance, for the case of quasi-periodic motion, where the physical constraints such as masses and inertias are represented by a dynamical system, comparison of action is posed in terms of comparison between dynamical models.

When the models are discrete, both in time and in the granularity of the states, such as hidden Markov models (HMM), then it is customary to compute—or approximate—the Kullback-Liebler divergence between the outputs.

This is computationally expensive and is subject to significant drawbacks in that the divergence between output distributions is affected by nuisance factors that cannot be easily teased out.

If we think of a discrete-time series $\{y(t) \in \mathbb{R}^N\}_{t=1,\ldots,T}$ as a function $y: \mathbb{N}_+ \rightarrow \mathbb{R}^N$, then comparison between any two sets of data can be performed with any functional norm. However, again it will be difficult to do so while discounting simple nuisances such as reparametrizations of the spatial and temporal scale, or the initial time of the experiment: for instance, we may want to recognize a person from her gait regardless of speed, or detect a ball bouncing regardless of height. For this reason, we find it more helpful to think of the time series as *the output of a dynamical model* driven by some stochastic process.

Under mild assumptions [50] $y(t)$ can be expressed as an instantaneous function of some "state" vector $x(t) \in \mathbb{R}^n$ that evolves in time according to an ordinary differential equation (ODE) driven by some deterministic or stochastic "input" $v(t)$ with measurement noise $w(t)$, where these two processes are jointly described by a density $q(\cdot)$, which can be degenerate for the limiting case of deterministic inputs. They can be thought of as errors that compound the effects of unmodeled dynamics, linearization residuals, calibration errors, and sensor noise. For this reason they are often collectively called input (state) and output (measurement) "noises." For reasons that are discussed in detail in [51], we assume that the noise process is temporally independent, or *strongly white*. In general, $q(\cdot)$ may or may not be normal. For the case of human gaits, one can think of limit cycles generating nominal input trajectories,[3] stable dynamics governing muscle masses and activations, initial conditions characterizing the spatial distribution of joints, and the input depending on the actual gait, the terrain, and the neuromuscular characteristics of the individual.

So, comparing time series entails *endowing the space of dynamical models with a metric structure* so we can measure the distance between models. Such a distance should include all elements of the model: the input, the state and its dynamics, the output map, and the initial condition, but allow the possibility of discounting some depending on the task at hand.

The simplest conceivable class of dynamical models is linear ones, where the time series $\{y(t)\}$ is generated via the model

$$\begin{cases} x(t+1) = Ax(t) + v(t) & x(t_0) = x_0 \\ y(t) = Cx(t) + w(t) & \{v(t), w(t)\} \overset{IID}{\sim} q(\cdot) \end{cases} \tag{3.3}$$

that is determined by the matrices $A \in \mathbb{R}^{n \times n}$, $C \in \mathbb{R}^{m \times n}$, and the density $q(\cdot)$. If the latter is Gaussian with zero-mean, this is determined by its covariance $Q \in \mathbb{R}^{k \times k}$.

3. This input can itself be considered as the output of an "exo-system," for instance, the central nervous system, that is not explicitly modeled.

Consider the simplest model (3.3), driven by a Gaussian input. It can be thought of as a point in a space that is embedded in $\mathbb{R}^{n^2+mn+k^2}$. Unfortunately, even for such linear-Gaussian models, this space is nonlinear, due to the constraints on the parameters and the fact that there are equivalence classes of models that yield the same output statistics [52]. Thus, in order to define a proper distance that takes into account the geometry of the space, one would have to define a Riemannian metric in the (homogeneous) space of models and integrate it to find geodesic distances between any two points on the space.

A simpler approach is to define a *cord distance*, one that does not come from a Riemannian metric, between any two points. Many such distances have been proposed recently, from subspace angles [53] to spectral norms [54], to kernel-induced distances [55].

Linear-Gaussian models capture the second-order statistics of a stationary time series. In fact, there is an entire equivalence class of models that have the same second-order statistics, so it is common to choose a realization that is stable and has minimum phase as a representative of the class [56–58]. That is often sufficient to classify coarse classes of events [59].

However, the assumptions of stability and phase minimality are often violated in important classes of data: for instance, humans are a collection of inverted pendulums, the prototypical example of nonminimum phase mechanical systems [50], and their gaits are often quasi-periodic, or marginally stable. So, we need to broaden our attention to nonminimum phase, marginally stable models.

So, the class of models we are honing into is a linear one, possibly of unknown order, with non-Gaussian input. Since one can interpret a white non-Gaussian independent and identically distributed (IID) process as a Gaussian one filtered through a static nonlinearity, we are left with considering so-called *Hammerstein models* that are linear models with static input nonlinearities [60]. In [51] we have shown how to perform identification (learning) of such models. To compare such models, a good starting point is the work of Vishwanathan et al. [55] that introduces an inner product in the embedding space of an output time series and uses it to define a cord distance between dynamical models. Unfortunately, in order to compare two models M_1, M_2, the method proposed in [55] requires knowledge of the *joint density* of the noises, i.e., $p(v_1,w_1,v_2,w_2)$, which is seldom available. An extension of this approach that does not require knowledge of such a distribution has been presented in [60]. The main idea is to *identify a model that generates the same output statistics (of all orders) of the original system, but that has a canonical input* that is strongly white and with independent components. Then all the information content of the input is transferred to the model that becomes nonlinear (Hammerstein) even if the original one was linear. One can then proceed to define a kernel in a manner similar to [55], but extended to take into account the nonlinearity. This can be done by solving an *optimal transport* problem which, given a finite amount of data, can be done in closed form.

The details are described in [51], but the overall outcome is the endowment of the space of dynamical models with a metric and a probabilistic structure that consider all of their components: the states and their *dynamics*, the *measurement maps*, the *initial conditions*, and the *inputs* or *noise* distributions. Different components may play different roles depending on the application; for instance, one may want to discard the transient behavior or the input distribution, but it is important to have machinery to account for all if needed.[4]

One can test the phenomenology of these models by using them to synthesize novel sequences. The idea is to use training data to learn a model, and then use the model to "hallucinate" novel sequences, and then display them to see if they capture the basic phenomenology of the event of interest, for instance, a human walking or running or limping.

In these experiments we use two publicly available datasets, the CMU motion capture [61] and the CMU Mobo dataset [62]. We show results for:

- *Video*: fast walking sequence (340 frames), directory moboJpg/04002/fastWalk/vr03_7 in the CMU Mobo dataset. The images have been scaled to 128×128 pixels and converted to 8-bit grayscale.
- *Markers*: walking sequence 02_01.c3d from CMU motion capture, consisting of 343 frames describing the 3D positions of 41 markers attached to the subject body.
- *Angles*: sequence 02_01.amc from CMU mocap (same as above); this time motion is represented as 52 joint angles of a skeletal model plus 6 DOFs for the reference frame.

The first step is to remove mean and linear trends from the original data. This is necessary given that our models represent zero-mean stationary processes. The best linear fit is removed from 3D position data, while in body joint angles and image data no significant linear trend is observed and so only the mean is subtracted. For synthesis, mean and linear trends are added back to the output sequence produced by the learned model to give the final motion.

A second preprocessing step consists in applying principal component analysis (PCA) to reduce the dimensionality of the observations. This step is not required theoretically but necessary in practice since we are dealing with short realizations (a few hundred frames) of high-dimensional processes. For motion capture data, we verified experimentally that 8 PCA bases are sufficient to synthesize sequences that are perceptually indistinguishable from the original. For

4. For the case of human gaits, one can think of limit cycles generating nominal input trajectories, the stable dynamics governing muscle masses and activations, initial conditions characterizing the spatial distribution of joints, and the input depending on the actual gait, the terrain, and the neuromuscular characteristics of the individual.

Figure 3.5 Comparison between our non-Gaussian linear models and standard Gaussian ARMA models [64]. We set the dimension of the state in both systems to 16, assigning in our model n_d=8 components to the periodic part and n_s=8 components to the stochastic part. The first row shows the original sequence after PCA projection, the second row displays corresponding frames produced by a dynamic texture model, and the last row is the output of our model. These few frames are sufficient to show the ability of our model to deliver better quality in the synthesis (notice how images are less blurry and more synchronized with the original compared to dynamic textures), thus validating the importance of explicitly representing periodic modes and high-order statistics. See also the entire movies at http://www.cs.ucla.edu/~bissacco/dynamICA.

image data, the required dimensionality is higher. In the sequence shown here we use 16 components.

In Figure 3.3 we show some sample frames of synthesis for motion capture data (both marker positions and joint angles) from minimum-phase and optimal nonminimum phase systems. As expected, motion produced by the nonminimum phase model is perceptually closer to the original.

For video sequences, the low quality of the synthetic images due to the linear PCA projection does not allow us to visually appreciate the difference between outputs of minimum and nonminimum phase models. However, as we show in Figure 3.5, correctly representing periodic modes and non-Gaussian statistics allows us to largely outperform the results obtained with standard Gaussian linear models such as "dynamic textures" [63].

3.4.4 Gait Recognition

Directly modeling gait video sequences using linear systems is a viable approach to synthesis but does not yield satisfactory results for recognition problems [59]. It is necessary to derive a representation insensitive to changes in background, illumination, and appearance of the subject performing the motion, and a simple choice is binary silhouettes of the moving body. We used the ones provided in the Mobo dataset, obtained by simple differencing with a background image followed by thresholding. Given that the extracted silhouettes are rather noisy (in particular, in the inclined walk sequences large parts of the treadmill are labeled as foreground), we derive a set of coarse features providing a robust description of the shape.

In Figure 3.6 we show a sample image from background subtraction and the corresponding representation with the projection features. Given a binary silhouette, the projection features encode the distance of the points on the silhouette from lines passing through its center of mass. The bounding box of the silhouette is divided uniformly in the $2n$ region, n to each side of the projection line, and for each region the average distance from the line is computed. In our experiments we used 2 lines (horizontal and vertical) and $n=8$ features on both sides for a total of 32 components.

In Figure 3.7 we illustrate the effect of using kernels between dynamical models to recognize various stationary actions (gaits), such as walking, running, walking with a ball, and up an incline. The figure shows the confusion matrices (matrices of pairwise distances) of the distances between learned models defined by initial state trace kernels (left) and the full trace kernels, including input distributions (right). It is evident that the inclusion of the stochastic part modeled by the input statistics improves the gait discrimination performances, visible by the block diagonal structure of the corresponding confusion matrix and the higher number of same-gait nearest-neighbor matches.

Stationary gaits are a special case, in that their temporal statistics are constant, so one can easily extract a model and compare different models to recognize actions. The same cannot be said for "transient actions" that require different representations, different models, and different methods for comparison. This is

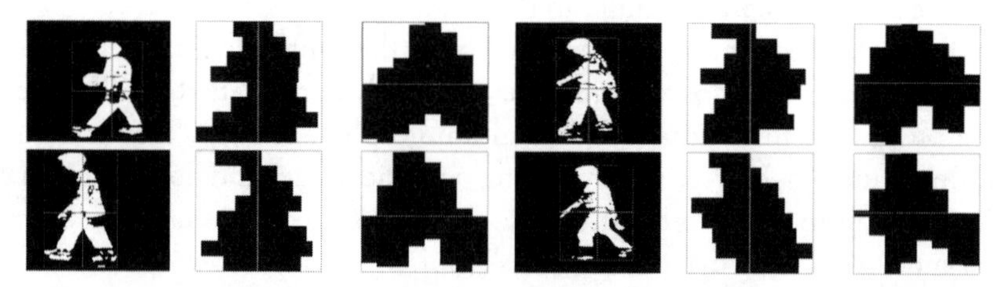

Figure 3.6 Sample silhouettes and associated shape features. First and fourth columns show some sample background subtraction frames from the gait dataset [63]: walking with ball, normal walk, fast walk, and inclined walk. Superimposed to the binary silhouette we plot the bounding box and the horizontal and vertical lines passing through the center of mass used to extract the features. On columns (2,5) and (3,6) we show the features obtained by computing the distance of the points on the two sides of the silhouette to the vertical and horizontal lines, respectively, discretized to $n_f = 8$ values.

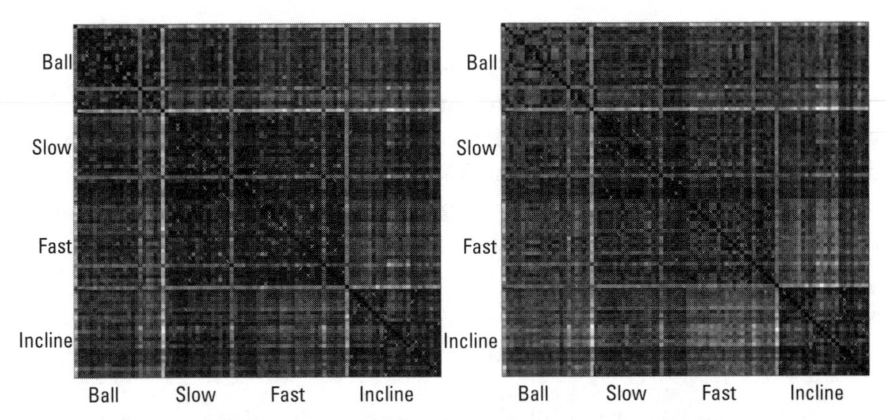

Figure 3.7 State and input kernel distances. We show the confusion matrices representing trace kernel distances between non-Gaussian linear models learned from walking sequences in the CMU Mobo dataset. Each element in row i, column j, represents the distance between the motion indexed with i and the motion indexed with j. The brightness of the element (i, j) indicates the distance (darker is small; brighter is larger). The minimum is indicated with a cross. There are 4 motion classes and 24 individuals performing these motions, for a total of 96 sequences. For each sequence we learn a linear model and then measure the distance between models by the trace kernels. Because motions are organized by groups, the confusion matrix should be block-diagonal, and the minima should be within the corresponding block. On the left we show results using kernels on initial states only; on the right we display the confusion matrix obtained from the trace kernels that include the effect of the input. It is clear how the additional information provided by the input statistics results in improved gait classification performances: we have 17 (17.7%) nearest-neighbor mismatches (i.e., closest models that do not belong to the same gait class) using the state-only distance, while only 9 (9.3%) with the complete trace kernel distance.

a subject of ongoing activity in the research community, and some future directions are outlined in [49].

3.5 Outlook

Automating surveillance is not an easy task, because human operators exploit a large amount of prior information and experience that is not easily transferred to engineering systems. While tremendous progress has been made in the past few years on at least some of the most critical tasks, such as detecting the presence of a human in an image, we are still far from being able to match or surpass human performance; that is, assuming that a fully attentive person is performing a task. It is more often the case that one human operator is in charge of tens if not hundreds of security monitors, and that attention issues play an important role as the need for live, online monitoring increases. Therefore, we expect that automatic detection of events will gain increasing importance in years to come.

There are important issues of user interface that will play an important role in the diffusion of automatic systems, as the cost of false alarms may become rather high if it causes an operator to switch off an automated module. In this chapter we have given an expository overview of the state of the art in terms of human detection, pose estimation, and motion estimation for the purpose of human motion analysis. Considerable work is yet to be done, but the field is evolving rapidly under the pressure of applications in industry, entertainment, and national security.

References

[1] D. J. Felleman and D. C. van Essen. "Distributed hierarchical processing in the primate cerebral cortex." *Cerebral Cortex*, 1:1–47, 1991.

[2] H. F. Chen, P. N. Belhumeur, and D. W. Jacobs. "In search of illumination invariants." *Proceedings of the Conference on Computer Vision and Pattern Recognition*, 2000.

[3] A. Vedaldi and S. Soatto. "Features for recognition: viewpoint invariance for non-planar scenes." *Proc. Int. Conf. Computer Vision (ICCV)*, October 2005, pp. 1474–1481.

[4] J. T. Enns and R. A. Rensink. "Influence of scene-based properties on visual search." *Science*, 247:721–723, 1990.

[5] A. Torralba, K. P. Murphy, W. T. Freeman, and M. A. Rubin. "Context-based vision system for place and object recognition." *ICCV*, 2003.

[6] G. Mori, X. Ren, A. A. Efros, and J. Malik. "Recovering human body configurations: combining segmentation and recognition." *CVPR*, 2:326–333, 2004.

[7] U. Grenander. *General Pattern Theory*. Oxford University Press, 1993.

[8] M. Fischler and R. Elschlager. "The representation and matching of pictorial structures." *Transactions on Computer*, 22(1):67–92, 1973.

[9] T. K. Leung, M. C. Burl, and P. Perona. "Probabilistic affine invariants for recognition." *Computer Society Conference on Computer Vision and Pattern Recognition*, June 1998.

[10] D. G. Kendall. "The diffusion of shape." *Advance in Applied Probability*, 9:428–430, 1977.

[11] P. Viola and M. Jones. "Robust real-time object detection." *2nd Int. Workshop Stat. Comput. Theories Vis.*, Vancouver, Canada, 2001.

[12] P. Viola and M. Jones. "Robust real-time face detection." *International Journal of Computer Vision*, 57(2):137–154, 2004.

[13] H. Schneiderman and T. Kanade. "Object detection using the statistics of parts." *IJCV*, 2004.

[14] P. Viola, M. Jones, and D. Snow. "Detecting pedestrians using patterns of motion and appearance." *ICCV*, pp. 734–741, 2003.

[15] N. Dalal and B. Triggs. "Histograms of oriented gradients for human detection." *Proceedings of the Conference on Computer Vision and Pattern Recognition*, 2005.

[16] P. F. Felzenszwalb and D. P. Huttenlocher. "Efficient matching of pictorial structures." *CVPR*, 2:2066–2074, 2000.

[17] A. Agarwal and B. Triggs. "Monocular human motion capture with a mixture of regressors." *CVPR*, 2005.

[18] C. Smichisescu. "Learning to reconstruct 3D human motion from Bayesian mixture of experts." *CVPR*, 2005.

[19] A. Thayananthan, R. Navaratnam, B. Stenger, P. H. S. Torr, and R. Cipolla. "Multivariate relevance vector machines for tracking." *ECCV*, 2006.

[20] Q. Zhu, S. Avidan, M.-C. Yeh, and K.-W. Cheng. "Fast human detection using a cascade of histograms of oriented gradients." *Proceedings of the Computer Vision and Pattern Recognition*, pp. 1491–1498, 2006.

[21] A. Levin, P. Viola, and Y. Freund. "Unsupervised improvement of visual dectors using co-training." *Proceedings of the Ninth IEEE International Conference on Computer Vision*, 2003.

[22] A. Agarwal and B. L. Triggs. "3D human pose from silhouettes by relevance vector regression." *CVPR*, 2:882–888, 2004.

[23] K. Grauman, G. Shakhnarovich, and T. Darrell. "Inferring 3D structure with a statistical image-based shape model." *ICCV*, 2003.

[24] R. Rosales and S. Sclaroff. "Learning body pose via specialized maps." *NIPS*, 2002.

[25] R. Ronfard, C. Schmid, and B. Triggs. "Learning to parse pictures of people." *ECCV*, 4:700–714, 2002.

[26] G. Hua, M. H. Yang, and Y. Wu. "Learning to estimate human pose with data driven belief propagation." *CVPR*, 2:747–754, 2005.

[27] S. Ioffe and D. A. Forsyth. "Probabilistic methods for finding people." *International Journal of Computer Vision*, 43,(1):45–68, 2001.

[28] M. W. Lee and I. Cohen. "Proposal driven MCMC for estimating human body pose in static images." *CVPR*, 2004.

[29] L. Sigal, M. Isard, B. H. Sigelman, and M. J. Black. "Attractive people: assembling loose-limbed models using non-parametric belief propagation." *NIPS*, pp. 1539–1546, 2003.

[30] C. Wren, A. Azarbayejani, T. Darrell, and A. Pentland. "Pfinder: real-time tracking of the human body." *Transactions Pattern Analysis and Machine Intelligence*, 19:780–785, 1997.

[31] A. Elgammal and C. S. Lee. "Inferring 3D body pose from silhouettes using activity manifold learning." *CVPR*, 2004.

[32] G. Shakhnarovich, P. Viola, and T. Darrell. "Fast pose estimation with parameter-sensitive hashing." *ICCV*, 2:750–757, 2003.

[33] A. Bissacco, M.-H. Yang, and S. Soatto. "Fast pose estimation via multi-dimensional boosting regression." *Proc. IEEE Conf. Comput. Vis. Pattern, Recog.*, June 2007 pp. 1–8.

[34] D. Ramanan, D. A. Forsyth, and A. Zisserman. "Strike a pose: tracking people by finding stylized poses." *CVPR*, 2005.

[35] J. H. Friedman. "Stochastic gradient boosting." *Computational Statistics and Data Analysis*, 38:367–378, 2002.

[36] L. Brieman, et al., *Classification and Regression Trees*. Wadsworth & Brooks, 1984.

[37] L. Sigal and M. J. Black. "Humaneva: Synchronized video and motion capture dataset for evaluation of articulated human motion." *Brown University TR*, 2006.

[38] S. K. Zhou, B. Georgescu, X. S. Zhou, and D. Comaniciu. "Image based regression using boosting method." *ICCV*, 2005.

[39] C. Bregler and J. Malik. "Tracking people with twists and exponential maps." *Proc. of Int. Conf. Comput. Vis. Pattern Recog.*, pp. 141–151, 1998.

[40] H. Sidenbladh, M. Black, and D. Fleet. "Stochastic tracking of 3D human figures using 2D image motion." *Proceedings on European Conference on Computer Vision*, pages II: 307–323, 2000.

[41] M. Isard and A. Blake. "Condensation—conditional density propagation for visual tracking. *International Journal of Computer Vision*, 1:5–28, 1998.

[42] D. Gavrila. "The visual analysis of human movement: a survey." *CVIU*, 73,(1):82–98, 1999.

[43] Y. Yacoob and M. J. Black. "Parameterized modeling and recognition of activities." *Computer Vision Image Understanding*, 1999.

[44] R. Fablet and M. J. Black. "Automatic detection and tracking of human motion with a view-based representation." *ECCV*, 2002.

[45] G. Doretto and S. Soatto. "Dynamic shape and appearance models." *IEEE Transactions on Pattern Analysis and Machine Intelligence*, 28(12):2006–2019, December 2006.

[46] D. Cremers and S. Soatto. "Motion competition: a variational approach to piecewise parametric motion segmentation." *International Journal of Computer Vision*, pp. 249–265, May 2005.

[47] G. Johansson. "Visual perception of biological motion and a model for its analysis." *Perception Psychophysics*, 14:201–211, 1973.

[48] L. Z. Manor and M. Irani. "Event-based analysis of video." *CVPR*, 2001.

[49] S. Soatto. *On the Distance Between Non-Stationary Time Series. Modeling, Estimation and Control.* New York: Springer Verlag, 2007.

[50] T. Kailath, A. H. Sayed, and B. Hassibi. *Linear Estimation*. Prentice Hall, 2000.

[51] A. Bissacco, A. Chiuso, and S. Soatto. "Classification and recognition of dynamical models: the role of phase, independent components, kernels and optimal transport." *IEEE Transactions on Pattern Analysis and Machine Intelligence*, Vol. 29, No. 11, November 2007, pp. 1958–1972.

[52] A. Lindquist and G. Picci. "A geometric approach to modelling and estimation of linear stochastic systems." *J. of Math. Sys., Est. and Cont.*, 1:241–333, 1991.

[53] K. De Coch and B. De Moor. "Subspace angles and distances between arma models." *Proc. Int. Symp. Math. Theory of Netw. Syst.*, 2000.

[54] R. Martin. "A metric for arma processes." *Transactions Signal Processing*, 48(4):1164–1170, 2000.

[55] S. V. N. Vishwanathan, R. Vidal, and A. J. Smola. "Binet-Cauchy kernels on dynamical systems and its application to the analysis of dynamic scenes." *International Journal of Computer Vision*, 2005.

[56] L. Ljung. *System Identification; Theory for the User*. Prentice Hall, 1997.

[57] P. Van Overschee and B. De Moor. *Subspace Identification for Linear Systems: Theory, Implementation, Applications*. Norwell, MA: Kluwer, 1996.

[58] T. Söderström and P. Stoica. *System Identification*. Upper Saddle River, NJ: Prentice-Hall, 1989.

[59] A. Bissacco, A. Chiuso, Y. Ma, and S. Soatto. "Recognition of human gaits." *Proceedings of the Conference on Computer Vision and Pattern Recognition*, pp. II 52–58, December 2001.

[60] I. Goethals, K. Pelckmans, J. A. K. Suykens, and B. D. Moor. "Subspace identification of Hammerstein systems using least squares support vector machines." *Transactions in Automatic Control*, 50(10):1509–1519, October 2005.

[61] CMU. Carnegie-Mellon MOCAP Database, 2003. http://mocap.cs.cmu.edu.

[62] R. Gross and J. Shi. "The CMU motion of body (Mobo) database." Technical report, Robotics Institute, Carnegie Mellon University, 2001.

[63] G. Doretto, A. Chiuso, Y. N. Wu, and S. Soatto. "Dynamic textures." *International Journal of Computer Vision*, 51(2):91–109, 2003.

4

Cyber Security Basic Defenses and Attack Trends

Alvaro A. Cárdenas, Tanya Roosta, Gelareh Taban, and Shankar Sastry

4.1 Introduction

Our society, economy, and critical infrastructures have become largely dependent on computer networks and other information technology solutions. As our dependence on information technology increases, cyber attacks become more attractive, and potentially more disastrous.

Cyber attacks are cheaper, more convenient, and less risky than physical attacks: they require few expenses beyond a computer and an Internet connection, they are unconstrained by geography and distance, they are not physically dangerous for the attacker, and it is more difficult to identify and prosecute the culprits of a cyber attack. Furthermore, cyber attacks are easy to replicate. Once a single attacker writes a malicious program, several other people in any part of the world can reuse this program to attack other systems.

Given that attacks against information technology systems are very attractive, and that their numbers and sophistication are expected to keep increasing, we need to have the knowledge and tools for a successful defense.

Cyber security is the branch of security dealing with digital or information technology.[1] This chapter presents a selected overview on topics in cyber security.

1. Throughout the chapter, we use the terms *security*, *information security*, and *computer security* interchangeably.

Following the subject of this book, we also explore the role of cyber security as part of the strategies for homeland security. Cyber security is an essential component in the protection of any nation. For example, the Department of Homeland Security (DHS) in the United States has a National Cyber Security Division (NCSD) with the following two objectives: (1) to build and maintain a cyberspace response system, and (2) to implement a cyber-risk management program for the protection of critical infrastructure.

The cyberspace response system unit is in charge of determining the actions that need to be taken when a cyber incident arises. The cyber-risk management program is in charge of assessing risks, prioritizing resources, and building a national awareness program in an effort to build stronger defenses against cyber attacks.

As we explore in Section 4.6.3 (with the recent attacks in Greece and Estonia), similar cyber security response programs can greatly benefit other countries, providing assistance and guidance for responding and handling cyber security incidents.

This chapter is divided into five sections:

- In Section 4.2 we describe the basic goals and terminologies used in information security to provide the necessary background for the subsequent sections.
- In Section 4.3 we give a brief overview of cryptography. Cryptography is an essential tool for securing communication by providing integrity, authentication, and confidentiality.
- In Section 4.4 we turn to network security. We show examples of widely used cryptographic protocols in Internet communication, such as IPsec and SSL. Cryptographic protocols, however, are only one link for achieving Internet security: firewalls are a complementary means for preventing attacks; intrusion detection systems are useful for detecting attacks; and honeypots provide security researchers the ability to study and understand the attacks.
- In Section 4.5, we focus on software security. We first summarize some of the most common problems and then give an overview of the ways security researchers attempt to prevent and limit the problems arising from malicious codes.
- Finally, in Section 4.6, we discuss some of the trends and threats of cyber security, and their relation to homeland security. We first discuss the growing cyber crime and botnet problem. Then, we discuss some of the threats to the communication infrastructure, such as worms and distributed denial of service (DDoS) attacks. We then finalize by discussing

some of the threats that may have a wider impact for homeland security: cyber espionage and attacks to the computer systems supporting and controlling critical infrastructures, such as electric power distribution, or water supply systems.

Cyber security is a very large field, therefore there are several topics that we are not able to cover in this short survey. Some topics missing in our survey include access control, usability, information flow control, security policies, secure operating systems, trusted computing, and advanced topics in cryptography, network security, and software security. We refer the reader interested in details on these topics to one of several computer security books [1–6].

4.2 Basic Concepts

A systematic study of the security of any system requires the description of three concepts: the security goals we want to achieve, the threats we expect to face, and the mechanisms and tools we can use to protect the system.

4.2.1 Common Security Goals

When a system is said to be "secure," it usually means that it has one or more of the following properties [7]:

- *Confidentiality* (or secrecy) refers to the concealment of information or resources from all but those who are authorized. A violation of confidentiality results in *disclosure*: a situation where an unauthorized party gets access to secret information.
- *Integrity* refers to the trustworthiness of data or resources. The goal of integrity is to prevent an attacker from tampering or corrupting the system's data or resources. A violation of integrity results in *deception*: a situation where a legitimate party receives false information and believes it to be true.
- *Availability* refers to the ability to use the information or resource desired. A violation of availability results in *denial of service* (DoS): the prevention (or "noticeable" delay) of authorized access to the information or resource.

Some security policies, such as preventing unauthorized users from using a resource (free-riding), are not directly covered by these three goals. However, most other security properties, such as privacy—the ability of a person to choose which personal details are to be kept confidential—or authentication—the verification of an identity (a subset of integrity)—rely on these three goals.

Integrity and availability are properties that we would like to have in almost every system. Even if we are not faced with a malicious entity, there are systems that can fail, corrupt data, and become unavailable because of design failures or accidents. Under these types of failures, integrity and availability become part of *reliability*.

What differentiates the fields of security and reliability is that in security we need to provide these goals under the presence of a malicious entity attacking the system of interest.

4.2.2 Threat Modeling

When people say "my system is secure" they usually mean "my system is secure as long as my threat model is satisfied in practice." Incorrect threat models (such as focusing on security against *outsiders* and doing little to prevent attacks by *insiders*) are common causes for breaches of security.

An essential part of threat modeling is identifying the entities or systems that we *trust*. Trusted systems are systems we rely on (i.e., trust is accepted dependence).

System designers should try to minimize the number of entities they trust; however, trust is essential for the security of any system. There is little chance of building a secure system if everyone is an enemy. A system, computer program, or security protocol is *trustworthy* if we have *evidence* to believe it can be trusted.

4.2.3 Security Analysis

Once the security goals and policies are defined, and the threat to the system has been assessed, we begin an iterative process of designing secure mechanisms and assessing new attack vectors against our mechanisms. Secure mechanisms in this context refer to the technology employed to enforce the intended security policy.

There are several ways to analyze the security of computer systems, some of which are software testing, red teams, certifications, and formal analysis.

In this chapter we focus on describing some traditional security mechanisms. We do not describe the different approaches for analyzing the security of a system. Instead, we intend to give a high-level view of security and the most common mechanisms used to enforce it.

4.3 Cryptography

Cryptographic protocols are a fundamental building block for securing computer systems. While cryptography is not a panacea—in practice, most security problems occur because of software or design bugs, human errors, or a bad security policy—it can be argued that without cryptographic algorithms, such as hash functions, digital signatures, key distribution, and encryption schemes, we would have very little chance of securing any distributed system.

Cryptography attempts to achieve *confidentiality*, *message integrity*, and *authentication* in an insecure network of computers, devices, and resources. A standard means to model the vulnerabilities associated with such a large network is the threat model proposed by Dolev and Yao [8]. In this threat model, all communications go through an attacker who can *eavesdrop*, *intercept*, *relay*, *modify*, *forge*, or *inject* any message. The attacker can also try to impersonate other parties in the system and send messages on their behalf.

Modern cryptography gives us the basic tools to achieve confidentiality with *encryption* algorithms, message integrity with *digital signatures* and *message authentication codes*, and authentication with protocols for authentication and key establishment.

In this section, we start by describing two fundamental concepts: hash functions and the differences between secret-key and public-key cryptography. Then, we give an overview of the techniques used to provide confidentiality, integrity, and authentication.

4.3.1 Hash Functions

An important primitive in cryptography is a hash function. The basic goal of a cryptographic hash function is to provide a *seemingly random* and *compact* representation (the hash value) of an arbitrary-length input string (which can be a document or a message). A hash function has two properties: (1) it is *one way* (it is *hard* to invert, where *hard* means it is *computationally infeasible*), and (2) it is *collision resistant* (it is *hard* to find two inputs that map to the same output).

Because the input to a hash function is any arbitrary-length string, and its output is a 160-bit binary string, there must be several collisions; however, in practice we should not be able to find any collisions of a well-designed hash function.

Hash functions have a variety of applications from integrity verification to randomization functions. Network administrators can store in a database the hash of the passwords instead of the raw passwords themselves. The *one-wayness* property of hash functions prevents an attacker from obtaining the passwords if the database is compromised [9]. Furthermore, hash functions are used in several cryptographic algorithms, such as message integrity codes, digital signatures, and encryption schemes (as we explain in the next sections).

Hash functions in practice are most susceptible to collision attacks. In this attack, the adversary tries to find two inputs to the hash function that map to the same output. If hash functions are used for signature schemes, a collision attack can allow an adversary to forge a signed message [10]. Most recently SHA-1, the most popular hash function at the moment, has been successfully attacked [11, 12]. Although these attacks have not yielded a practical use for the two inputs mapping to the same hash output, the potential for finding useful attacks is increasing. In order to avoid practical attacks on hash functions, in January 2007 NIST announced [13] a public competition for a new cryptographic hash function that would become the new federal information processing standard.

4.3.2 Secret-Key and Public-Key Cryptography

In an analogy to the way locks are opened with keys, cryptographers have used the idea of a *key* to refer to the information necessary to access cryptographically protected data. Modern cryptography follows Kerckhoffs' principle, which states that "the security of a system should depend on its key, not on its design remaining obscure"[14]. In short, the common practice in cryptography is not to *rely* on the secrecy of the algorithms, only on the secrecy of the *secret keys*.

There are two types of encryption algorithms, one which uses a secret key shared between the two communicating parties, and another one in which only one party knows the secret key, and everyone else (even the adversary) knows the public key (known as public-key cryptography).

Secret-key cryptography, also known as symmetric cryptography, involves the use of a single key shared between a pair of users. The fact that you need to share a secret key with every other party that you wish to communicate with makes secret-key cryptography cumbersome for several applications.

There are two main problems with the key management in symmetric key systems. First, since secrets are shared between *pairs* of users, a large system will contain a large number of secrets, which is hard to manage. The second problem is related to the initial sharing of secrets between users. In particular, the difficulty of establishing an initial secret key between two communicating parties, when a secure channel does not already exist between them, presents a chicken-and-egg problem. This problem is most commonly solved using key distribution centers (KDC), which are trusted intermediaries between communicating parties. By having trusted intermediaries, a party only needs to share a secret key with the KDC. Whenever two new parties need to communicate, they establish a secret key with the help of the KDC. Figure 4.1 provides a simplified overview of Kerberos, one of the most common key distribution protocols.

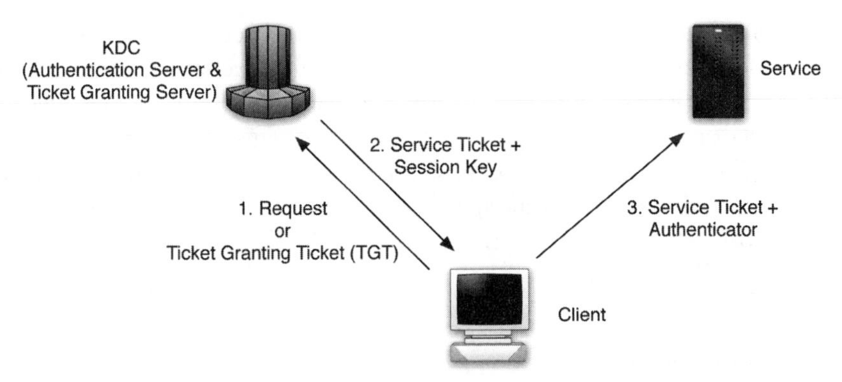

Figure 4.1 A simplified version of the Kerberos authentication system: both the client and a service share their own secret key with the KDC. To communicate securely with the service, the client sends a request to the KDC, which first authenticates the user. The KDC then selects a secret key to be shared between the client and the server (called a *session key*), and distributes it to the client and the service via a service ticket. Subsequent authentication can be streamlined by the use of ticket-granting tickets, issued by the KDC.

Using a KDC has two important shortcomings: first, the KDC introduces a single point of failure and if it is unavailable, the whole system fails. Second, the security of the entire system breaks if the KDC is compromised. These two problems are solved through the use of public-key cryptography.

In *public-key cryptography*, also known as asymmetric cryptography, parties *do not share any secrets* and different keys are used for encrypting and decrypting. This is a particularly powerful primitive as it enables two parties to communicate secretly without having agreed on any secret information in advance.

In this setting, one party (the receiver) generates a pair of keys, called the public key and the secret key. The public key can then be made openly available so anyone can (for example) encrypt a message for the receiver. The receiver then uses its secret key to recover the message.

Public-key algorithms are less efficient than their secret-key counterparts; therefore, in practice public-key cryptography is often used in combination with secret-key cryptography. For example, in the pretty good privacy (PGP) set of algorithms for encrypting emails, a public key is used to encrypt a symmetric key. The symmetric key, in turn, is used to encrypt the bulk of the message.

Although public-key cryptography is computationally more intensive than secret-key cryptography, it requires simpler key management. However, a central problem in public-key cryptography is ensuring that a public-key is *authentic*; that is, we need to make sure that the public key we have was created by the party with whom we wish to communicate, and that it has not been modified or fabricated by a malicious party.

A public key infrastructure (PKI) provides the necessary services to distribute and manage *authentic* public keys. A trusted server called a certification authority (CA) issues a public-key certificate to each user in its system, certifying the user's public-key information. The main benefits of a CA are that it can operate *offline* and that its compromise does not lead to the compromise of the secrets of the existing users of the system. There are two main approaches to PKI: centralized (such as the X.509 model) and decentralized (such as the "web of trust" used in PGP) [15].

4.3.3 Confidentiality

Encryption schemes are used to protect messages against eavesdroppers, and therefore achieve message confidentiality. Specifically, the information we want to protect (commonly called the *plaintext*) is transformed with an encryption algorithm (also called a *cipher*) into mangled data (commonly called the *ciphertext*) such that it is unintelligible to anyone not possessing the secret key.

In general, the security of an encryption scheme depends on the difficulty of breaking the secrecy of the plaintext. The security level of an encryption scheme is defined with respect to the knowledge and the capability of the attacker. The weakest level of security is achieved against a ciphertext-only attacker who can only see the random ciphertexts. The strongest attacker has the capability to decrypt any ciphertext it wishes (except for the ciphertext the attacker is interested in breaking). For detailed definitions of these and other security notions refer to books on theoretical cryptography [5].

In practice, an encryption scheme needs to be only as secure as the system requires. In some cases, such as closed systems, a chosen ciphertext attacker cannot be realized and weaker notions of security are sufficient. Smart card systems, however, might require stronger notions of security as the attacker has more control over the input of the cryptosystem. We refer the interested reader to books on theoretical cryptography for detailed definitions of these and other security notions.

The most common examples of public-key encryption schemes are RSA [16, 17], and ElGamal [18]. Both cryptosystems rely on exponentiation, which is a fairly expensive operation. More efficient schemes have been introduced that rely on elliptic curve cryptography.

Secret-key encryption algorithms can be divided into stream ciphers and block ciphers. Stream ciphers encrypt the bits of the message one at a time, and block ciphers take a number of bits and encrypt them as a single unit. The most common examples of secret-key algorithms include the data encryption standard (DES) and advanced encryption standard (AES) block ciphers, and the RC4 stream cipher.

4.3.4 Integrity

Data integrity techniques are used against unauthorized modification of messages. Specifically, the sender generates a code based on the message and transmits both the message and the code. The receiver then uses a verification algorithm that checks if the message has been altered in an unauthorized way during the transmission. The receiver can also verify that the message has indeed come from the claimed source.

4.3.4.1 Secret-Key Cryptography

Data integrity in secret-key cryptography is achieved using message authentication codes (MACs). Given a key and a message, a MAC value is generated that protects the integrity of the message by allowing verifiers (who also possess the secret key) to detect any changes to the message content. MACs can also be used to provide authentication.

In general, there are two types of MAC schemes. An HMAC is based on keyed hash functions and is characterized by its efficiency. HMAC-SHA-1 and HMAC-MD5 are used within the IPsec and SSL protocols, respectively (see Section 4.4). Another type of MAC is generated based on block ciphers, such as CBC-MAC and OMAC [19].

4.3.4.2 Public-Key Cryptography

Unlike secret-key cryptography, a ciphertext generated by a public-key encryption, accompanied by its associated plaintext, can provide data integrity for the plaintext and authentication of its origin. The integrity code of a message can only be generated by the owner, while the verification of the integrity check can be done by anybody (both properties of a signature). Therefore, the integrity check in public-key cryptography is called a *digital signature*. These characteristics allow for the provision of *non-repudiation*. Non-repudiation means that the owner cannot deny a connection with the message and is a necessary requirement for services such as electronic commerce. Examples of signature schemes include the digital signature standard (DSS) and RSA-PSS.

4.3.5 Authentication

There are two classes of authentication: data origin authentication and entity authentication.

Data origin authentication, also called message authentication, is the procedure whereby a message is transmitted from a purported transmitter (or origin) to a receiver who will validate the message upon reception. Specifically the receiver is concerned with establishing the identity of the message transmitter as well as the data integrity of the message subsequent to its transmission by the sender.

In entity authentication, which is concerned with validating a claimed identity of a transmitter, a "lively" correspondence is established between two parties, and a claimed identity of one of the parties is verified. Important properties of authentication include the establishment of message *freshness*: verifying whether data has been sent sufficiently recently, and user *liveness*: the lively correspondence of the communicating parties. The main techniques that handle user liveness include challenge–response mechanisms, time stamps, or freshness identifiers such as nonces.

User authentication can be divided into three categories [20]:

1. Knowledge-based authenticators ("what you know")—characterized by secrecy or obscurity (e.g., passwords, security questions such as mother's maiden name, and so forth).
2. Object-based authenticators ("what you have")—characterized by physical possession (e.g., security tokens, smart cards, and so forth).
3. ID-based authenticators ("who you are")—characterized by uniqueness to one person (e.g., a biometric such as a fingerprint or iris scan).

Different types of authenticators can be combined to enhance security. This is called *multifactor authentication*. For example, the combination of a bank card plus a password (two-factor authentication) provides better security than either factors alone.

4.4 Network Security

In this section we give a brief overview of network security, focusing on cryptographic protocols used in the Internet for achieving secure services (such as online banking) and other security paradigms such as firewalls, intrusion detection systems, and honeypots.

The networking protocols used in the Internet can be viewed as a set of layers or protocol stack. Each layer is responsible for solving a different set of problems related to networking. The open systems interconnection (OSI) model describes a seven-layered network stack: physical layer, data link layer, network layer, transport layer, session layer, presentation layer, and application layer (shown in Figure 4.2).

In general, we are interested in providing end-to-end security, that is, ensuring that the communication between a client and a server in the Internet has been authenticated (both parties know who they are communicating with), is confidential, and has not been tampered with. In order to achieve these goals, protocols such as IPsec, SSL, and DNSSEC have been proposed.

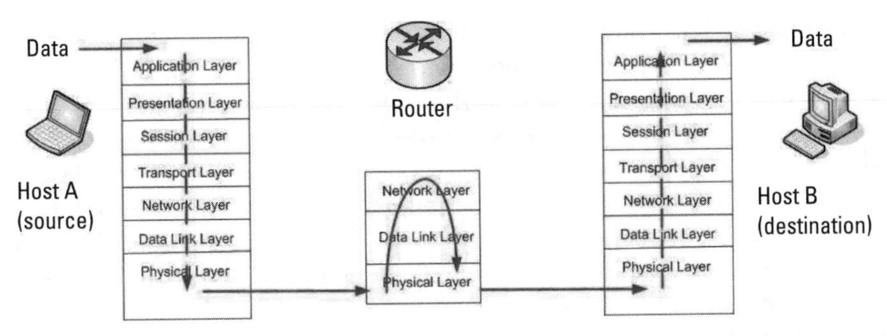

Figure 4.2 The network layer stack and routing of the data from a source to a destination.

IP Security (IPsec) The Internet protocol (IP) is the main protocol of the network layer: it provides the information needed for routing packets among routers and computers of the network. One of the shortcomings of the original IP protocol was that it lacked any kind of general-purpose mechanism for ensuring the authenticity and privacy of IP data while in transmission. Since IP data packets are generally routed between two devices, over unknown networks, any information included in the packets is subject to being intercepted or possibly changed. With the increased use of the Internet for critical applications, security enhancements for IP were needed. To this end, IPsec was developed to provide transparent end-to-end encryption of IP traffic and user data. IPsec is predominantly used in the commercial sector.

IPsec consists of two parts: the Internet key exchange (IKE) protocol, which provides mutual entity authentication and establishes a shared symmetric key, and the encapsulating security payload and authentication header (ESP/AH), which provides end-to-end confidentiality and authentication.

One of the main uses of IPsec today is for the creation of a virtual private network (VPN). VPNs are generally used by enterprises to connect remote offices across the Internet. IPsec is used in VPNs for creating a secure channel across the Internet between a remote computer and a trusted network.

Secure Socket Layer (SSL) SSL was originally developed to provide end-to-end confidentiality, integrity, and authentication between two computers over the transmission control protocol (TCP), a protocol running at the transport layer. SSL and its successor, transport layer security (TLS), have been very popular, receiving the endorsement of several credit card companies and other financial institutions for commerce over the Internet.

SSL/TLS is commonly used with http to form https, a protocol used to secure Web pages. (Https is the protocol used when your browser shows a closed lock in one corner of the browser window, to indicate a secure connection.)

SSL is composed of a set of protocols: the protocol to ensure data security and integrity, called the SSL record protocol, and the protocols that are used for establishing an SSL connection. The latter consists of three sub-protocols: the SSL handshake protocol, the SSL change cipher protocol, and the SSL alert protocol.

The main difference between IPsec and SSL is the layer where they are implemented. The main motivation for IPsec is to avoid the modification of any layer on top of the network layer. IPsec is implemented in the operating systems, so no modifications are required from applications. The main motivation for SSL is to create a secure channel by creating (or modifying) the applications (as long as the application runs over TCP) without changing the infrastructure of the Internet or the operating system of the user.

Domain Name Service Security Extensions (DNSSEC) The domain name server (DNS) is the protocol that translates the human-readable host names into 32-bit Internet protocol (IP) addresses: it is essentially a "yellow book" for the Internet, telling routers to which IP address to direct packets when the user gives a name such as http://www.google.com.

Because DNS replies are not authenticated, an attacker may be able to send malicious DNS messages to impersonate an Internet server [21]. Although SSL can try to prevent this impersonation (the attacker's Web site should not have the secret key of the real Web site), there are many Web sites and other services that run without SSL, but that still need to be reached in a trustworthy manner.

In order to ensure the integrity of the DNS service, the IETF is currently working on DNSSEC.

Another major concern about DNS is its availability. Because a successful attack against the DNS service would create a significant communication disruption in the Internet, DNS has been the target of several DoS attacks. We explore these attacks in Section 4.6.3.

4.4.1 Firewalls, IDSs, and Honeypots

In this section we describe some very popular *noncryptographic* tools in network security.

Firewalls A firewall is a software program or a hardware device sitting between the Internet and a private network that filters the incoming traffic to the private network [22]. If there are suspicious packets coming through the connection, the firewall prevents them from entering the private network. The concept of a firewall as an entity monitoring traffic between a local network and the Internet is shown in Figure 4.3.

A firewall can use three different approaches to filter traffic:

- *Packet filtering (inspection):* In this method, packets are analyzed against a set of filters (rules). The packet filter is implemented at the network layer, and operates on the information available at this layer, such as

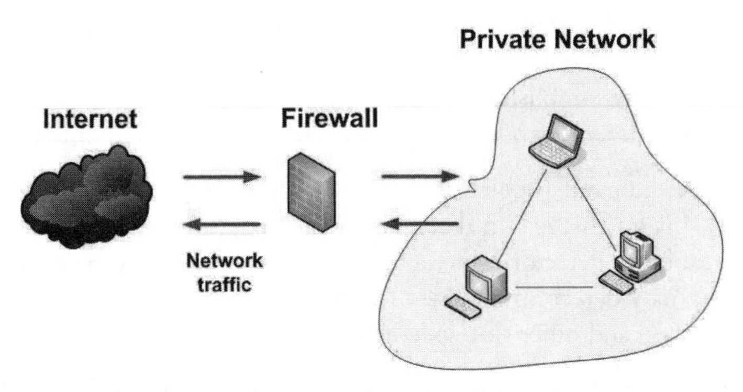

Figure 4.3 A firewall filters the traffic entering a private network from the Internet.

source address, destination address, and port information contained in each packet.

- *Application proxy:* The firewall is implemented at the application layer and processes the incoming packets all the way up to this layer. It acts as a proxy between the Internet and the private network, and verifies that the packets are clean in terms of both their origin and content. Application proxy firewall has a complete view of the network connections and application data, and as a result, it can filter bad data at different layers of the network.
- *Stateful packet filtering (inspection):* This type of firewall adds states to the packet filter firewall. Instead of examining the contents of each packet on an individual basis, it compares certain key parts of the packet to a database of trusted information. A stateful packet filter is implemented at the transport layer. Therefore, it is not capable of monitoring the packets at the network or application layer. The advantage of this type of packet filtering is that it keeps a history of the transport layer connections, such as TCP or UDP connections. As a result, this type of firewall can help prevent attacks that exploit existing connections, or certain DoS attacks.

Intrusion Detection System (IDSs) Generally, an IDS consists of: sensors, analysis center, databases, and response center. IDSs tend to be classified in two types:

- *Signature-based detection:*[2] here, the system looks for the known attack patterns and tries to match the so-called *signature* of the attack to identify these intrusions. For example, failed login attempts could indicate a password-cracking attack. Signatures are essentially a *black-list* of activities that are not allowed in the network.

2. This is also known as a misuse detection system.

- *Anomaly detection*: in this type of intrusion detection, the normal user behavior is defined and the system looks for any significant deviation from these established normal uses. The goal of anomaly detection is to create a *white-list* of the only activities that are allowed in the network.

One of the main problems with IDSs is their relatively large false alarm rates [23]. It is for this reason that most of the practical systems, such as *Snort*, are signature-based detectors, since anomaly detectors tend to generate more false alarms. Anomaly detection schemes are, however, a popular tool for detecting credit card fraud and other very specific scenarios.

Honeypots A honeypot is a network decoy set as a trap to detect and deflect attempts at unauthorized use of information systems. A honeypot is composed of a network site that is isolated, protected, and monitored. The network site appears to contain information or a resource that would be of value to attackers. Honeypots are used to distract adversaries from more valuable devices on a network, to provide early warning regarding new attacks and exploitation trends, and to allow in-depth examination of attackers' behavior. Most antivirus companies use honeypots to capture and study malware.

4.5 Software Security

Programming errors in computer programs can create software vulnerabilities. An attacker can compromise or obtain illegal access to databases, or cause a denial of service to computers running vulnerable software. To achieve these goals, attackers make use of an *exploit*, a computer code that takes advantage of the software vulnerability.

Software vulnerabilities are the most persistent problem in information security. It does not matter if the network security protocols or cryptographic algorithms are correct and theoretically secure if the software implementing them has errors. Ironically, even security software such as firewalls, IDSs, or implementations of security protocols (such as SSH) can open the door to attackers if they have software vulnerabilities. Therefore, preventing, detecting, and reacting to software vulnerabilities are the most active fields of information security.

In this section we give a small survey of *software security*: the study of the problems arising from programming errors and malicious software.

4.5.1 Software Vulnerabilities

In this section we briefly describe some of the main vulnerabilities in software. We start by describing some of the first discovered exploits, such as *buffer overflows* and *race conditions*. Then, we summarize two vulnerabilities that were discovered in the early years of this decade and led to a wave of exploitable bugs being discovered in all kinds of programs: *format string vulnerabilities* and *integer overflows*. Finally, we describe *SQL injection* and *cross site scripting*, the most commonly reported vulnerabilities.

- **Buffer overflows**: A buffer overflow occurs when a computer instruction tries to store data that needs a larger computer memory space than what the program had allocated. The data that does not fit into the preallocated buffer overwrites something else in the computer memory.

  Buffer overflows have consistently been among the top reported vulnerabilities in the last two decades, and are the cause of most of the widely spread worms in the Internet. The most common type of buffer overflow is "smashing the stack" [24]. Pincus and Baker [25] provide a recent summary of this vulnerability and its variations.
- **Race conditions**: Race conditions occur when two processes compete to access the same resource before the other. One of the most common examples is the time-of-check-to-time-of-use (TOCTTOU) flaw. A TOCTTOU flaw occurs when a program checks for a particular characteristic of an object and then takes some action that assumes the characteristic still holds when in fact it does not [26].
- **Format string vulnerabilities**: Format string vulnerabilities were discovered in 2000 and immediately led to a wave of exploitable bugs being discovered in all kinds of programs [27]. These vulnerabilities arise from errors in the use of format functions (such as printf() in C). When an attacker is able to specify the format string, it can force the function to output values from parts of the computer memory that should not be available to the attacker.
- **Integer overflows**: Integer overflows occur because most programming languages give integer types a fixed maximum upper bound. When an attempt is made to store a value greater than this maximum value, an integer overflow occurs. One of the first and most significant integer overflow vulnerabilities was discovered in 2001, and allowed remote attackers to obtain full administrative privileges on an affected computer without any credentials [28].
- **Command injection**: This vulnerability occurs when data is interpreted as control data by a program. Structured Query Language (SQL) injection attacks (a particular type of command injection) allow attackers to bypass access control mechanisms of databases that support many Web applications [29]. The increase of Web applications has made SQL injection attacks widespread in the last couple of years. They currently rank

among the top three vulnerabilities reported, along with buffer overflows and cross site scripting attacks [30] (another attack to Web security).

- **Cross site scripting (XSS)**: Because Web browsers usually download and run (potentially untrusted) programs each time a user visits a Web site, browsers have implemented a *same-origin policy*, where the code downloaded from a Web site can only access the user's credentials of the Web site from where it was downloaded. In an XSS attack, an honest user is fooled to click on a malicious link that downloads a malicious program created by the attacker but that appears to have been generated from the correct Web site [31].

4.5.2 Malicious Software

Malicious software, or *malware*, is a generic term to identify computer programs with a harmful purpose. Malware usually exploits a program vulnerability to achieve its harmful purpose. There are several types of malware [1, 7, 32]:

- **Trojan**: (or Trojan horse) a program that appears to have a useful function, but also has a hidden and potentially malicious effect.
- **Computer virus**: a program that inserts itself into another program or file, and then performs a malicious action. A virus cannot run by itself, and requires someone to run its host program.
- **Worm**: a computer program that can run independently, and copies itself from one computer to another.
- **Spyware**: software that gathers information of the users of a system without their knowledge.
- **Rootkit**: a set of tools that attempts to make malware invisible to the users of the system.

Although these definitions are generally accepted, sometimes there is no clear distinction between these concepts. For example, computer programs, such as e-mail viruses, or e-mail Trojans, depend on the user action to abet their propagation and are commonly referred to as worms in the media.

4.5.3 Defenses

Attempts to mitigate and solve the effects of vulnerabilities and malware can be classified as; (1) detection of malicious code; (2) detection of software flaws; (3) reducing programming errors, which requires support by programming languages; (4) reducing the impact of programming errors by confining untrusted code; and (5) correcting software flaws by patches.

Patching is a relatively straightforward technical problem. The main problems for patching systems are usually legal or economic or due to human factors. For example, patching may void certification of certain software used in critical applications, which is a legal concern. Therefore, we focus on the first four approaches in our discussion.

4.5.3.1 Detecting Malicious Code

IDSs and antivirus scanners are the most common tools for detecting attacks. IDSs focus on the detection of an exploit, or an interactive attacker, while antivirus scanners focus on the detection of malware.

Antivirus scanners are currently the most popular defense against malware. These malware detectors often look for known sequences of code (*signatures*). In order to obstruct code analysis, malware writers sometimes use *polymorphism*, *code obfuscation*, and *encryption* [33].

One of the fundamental results in the theory of malware asserts that it is undecidable whether an arbitrary program contains malware [34, 35]. Therefore, it is impossible to create an algorithm that can recognize all malicious logic instructions from normal program instructions. Any defense will be imprecise, allowing false negatives (not recognizing malware) and/or false positives (labeling nonmalicious code as malware).

4.5.3.2 Software Testing

Finding software errors is very difficult. In his Turing Award lecture, Edsger W. Dijkstra stated, "Program testing can be quite effective for showing the presence of bugs, but it is hopelessly inadequate for showing their absence" [36]. A practical example given more than 10 years later by Thompson [37] shows that no amount of source-level verification or scrutiny can stop the use of untrusted code.

Moreover, Rice's theorem, a basic theoretical result, states that there is no general computer algorithm that can classify code as safe or malicious with perfect accuracy. Software testing tools, therefore, produce false negatives (the program contains bugs that the tool does not report) and/or false positives (the tool reports bugs that the program does not contain).

Despite these negative results, approximate software testing algorithms are very successful in finding software flaws.

Automatic software testing can be divided into *static analysis* (analysis of programs without their execution [26, 38–40]) and *dynamic analysis* (analysis of programs by executing them [41–45]).

4.5.3.3 Type-Safe Languages

Many vulnerabilities are caused because programmers have to manage the memory used by their programs. Type-safe programming languages avoid these vulnerabilities. A data type, such as the data type int in C, assigns meaning to values and variables in a program. Type safety is the property of a programming language that does not allow the programmer to treat a value as a type it does not belong to. Since a data type defines specific memory requirements for values and variables, type safety implies memory safety. A type-safe language prevents several memory management errors, such as buffer overflows. Java is a type-safe language, whereas C and C++ are not type-safe languages.

4.5.3.4 Confining Code

Most users and operating systems allow programs to read, modify, or delete any file in the computer. This allows an attacker to get full control of the system using any vulnerabilities of a program. Therefore, it is of interest to find ways to limit the privileges that a program uses when running.

The *principle of least privilege*, by Saltzer and Schroeder [46], is an old design principle. The idea is to give a program only the privileges it needs to accomplish its task. By limiting the privileges, the damage is limited when the program is compromised.

Sandboxing is a way to limit the ability of untrusted code for doing harmful things. A sandbox executes code in a heavily restricted environment, limiting access to the file system or the ability to open network connections. Janus [47] is a research example of sandboxing: it monitors untrusted applications and disallows system calls that the untrusted code is not permitted to execute. A practical example of sandboxing is *Systrace* [48], a utility available in BSD, Linux, and Mac OS X that can limit how a computer program accesses the operating system. Another popular example is the Java applet. Applets are typically executed in a sandbox, preventing the untrusted code that is usually downloaded from Web sites from accessing information on the executing computer.

Java also uses *code-signing*, in which a code producer signs its code to provide a proof of trustworthiness. Based on the reputation of the code producer, Java (or the user) can decide whether the code can be trusted to be executed.

Proof carrying code is another technique to avoid trusting the producer of the code. Here, the user specifies a safety requirement, and the creator of the code must generate a proof that the code meets the desired safety properties and integrates them with the code [49].

4.6 Cyber Attack Trends, Threats, and Homeland Security

The Internet and information technology at large have become ingrained in our lives. This assimilation of information technology in our lives has made cyber attacks more attractive. There is a clear shift from cyber attacks carried out by a few technically savvy and curious hackers, to cyber attacks that can be mounted by a wide range of people or groups. The latter has clearly defined economic, political, religious, or national motivations.

In this section we discuss these trends by studying some recent examples. We start with cyber crime and botnets that are the driving force behind most of the Internet-based attacks. Worms and distributed denial of service (DDoS) attacks that threaten the communication infrastructure of the Internet are explained next. Finally, we describe cyber espionage and cyber attacks against the critical infrastructures, which are of paramount importance in homeland security.

4.6.1 Cyber Crime and Botnets

There is a large underground market based on a wide variety of criminal activities. Theft of intellectual property, extortion based on the threat of DDoS attacks, fraud based on identity theft, credit card fraud, spamming, phishing, sales of *bots* (i.e., compromised computers), sales or rentals of *botnets* (i.e., networks of compromised computers under the control of the botnet owner), sales of stolen source code from software companies, and sales of malware or tools to create attacks (the vendor even provides technical support) are just a few examples of ways that they are making it easier for unskilled attackers to commit cyber crimes.

The prevalence of these activities and the large amount of compromised computers in the world can be recognized by the fact that the average compromised computer is being sold at an estimated average of 4 cents [50].

Botnets are one of the most coveted electronic goods. With several million messages sent per day, sending spam is the primary source of income for botnet operators. Botnets can be rented and sold for performing attacks, such as DDoS attacks (see Section 4.6.3). Other forms of attacks are possible. For example, a botnet attack, targeting user accounts in eBay and making fraudulent transactions, was detected in September 2007 [51].

Botnet operators are also very active in protecting their networks. They buy and steal botnets from each other, protect their compromised computers against compromise by other people, and actively attack security companies focusing on defeating spam. For example, in 2006, botnets launched DDoS attacks against BlueSecurity (forcing the company to abandon its antispam efforts), and other antispam and antimalware groups such as CastleCops, Spamhaus, and URIBL were victims of DDoS attacks in 2007. From the standpoint of those defending against cyber war, the problem is that the existence of lucrative spamming and underground criminal business keeps the botnets in operation and motivated to stick around and fight security companies [52].

At the time of this writing, the Storm botnet (first detected in January 2007) was considered to be the largest botnet identified by security researchers. The number of compromised computers it controls has been estimated to be from several hundred thousands up to 50 million [53]. Although estimating the size of a botnet is a very difficult task, and estimates themselves may not be very reliable [54], it is clear that the size of botnets is very large.

The Storm botnet is also highly resilient to efforts to take it down. Its command and control architecture is based on a peer-to-peer (P2P) network, with several redundant hosts spread among 384 providers in more than 50 countries. Furthermore, in order to hide infected machines and malicious Web sites, the Storm botnet uses a technique called *fast flux*, a method in which the DNS records of a Web site's domain name are constantly rotating and changing their IP addresses. Additionally, the infection code is polymorphic, so antivirus designers have a harder time crafting signatures against it. Storm's malware also tries to avoid infecting virtual machines and, therefore, security researchers have less opportunities to study the malicious code—virtual machines are typically used in honeypots; see Section 4.4.1.

4.6.2 Widely Spread Malware

In this section we summarize the recent worms, email viruses, and email Trojans, focusing on their side effects and their impact in different infrastructures.

The timeline of widespread worms and viruses did not see any major contenders to the original Internet worm of 1988 until 1999, when the Melissa email virus was released. Although the virus did not carry a malicious payload, it saturated email servers, creating an unintentional DoS attack. A similar email virus, the ILOVEYOU virus, was released a year later.

In 2001, three major Internet worms were released [55]: Code Red 1, Code Red 2, and Nimda. Code Red 1 infected approximately 360,000 computers in just six days, creating routing disruptions in the Internet. From the 20th until the 27th of each month it attempted to use all the compromised computers to launch a DDoS attack against the IP address of the White House. The White House responded to this threat by changing its IP address. Code Red 2 and Nimda showed improved spreading techniques over Code Red 1. Computers infected by Code Red 2 searched for vulnerable computers in their vicinity (local area networks), while Nimda had different methods to spread itself (IIS vulnerability, emails, and scanning for backdoors). Code Red 2 left a backdoor in infected computers, allowing an attacker to connect and control the computer at a later stage, thus having the potential of creating a botnet. Interestingly enough, Nimda used these backdoors to help its propagation.

The fastest spreading worm to date (Slammer) was released in 2003. Slammer compromised over 75,000 computers in just 10 minutes by exploiting (via UDP) a buffer overflow in Microsoft's SQL Server software. Due to this fast spread, the worm disrupted parts of the Internet, the phone service in Finland, airline reservation systems, credit card networks, 13,000 ATMs on the Bank of America network [56], and it shot down display systems at the Davis-Besse power plant in Ohio [57]. A couple of months later, the Blaster worm was released, and it was widely suspected of contributing to a power loss at a plant providing electricity to parts of New York state.

Although these worms created widespread damage, this damage was mostly a side effect of the spreading behavior of the worm. These attacks could have been much more damaging if they had included a malicious payload. In 2004, Witty carried such a destructive payload: it corrupted the hard drive in compromised computers. Witty also showcased the increased sophistication of worms, since it started compromising hosts with a list of known vulnerable computers, and it was released just one day after the ISS firewall vulnerability was made public [58].

The current most widely spread computer infection is due to the Storm Trojan/virus. Storm has been used to construct the largest botnet to date. At the time of this writing, the botnet is still active and increasing. It spreads via e-mails directing users to a Web page hosting the malicious code, or to open a malicious Trojan that attempts to exploit several vulnerabilities in Windows computers. Storm also reflects the trend of creating worms and e-mail Trojans for profit. The size of the Storm botnet should also be a concern for any nation, since the owner of the botnet can not only use it to commit fraud against hundreds of citizens in many countries, but it can potentially launch DDoS attacks several times larger than the ones against Estonia in 2006.

4.6.3 DDoS Attacks, Estonia, and Hacktivism

A DoS attack can be defined as an attack designed to disrupt, or completely deny, legitimate users' access to networks, servers, services, or other resources. In a distributed denial of service (DDoS) attack a botnet (or several botnets) launches a coordinated attack onto a single computer by flooding it with massive volumes of useless traffic. The flood of incoming messages to the target can occupy all the resources that the target could use to service legitimate traffic. Figure 4.4 illustrates this concept.

The first widely publicized DDoS incident occurred in 2000, when several of the most popular Web sites—Yahoo, Amazon.com, E-Trade, eBay, and CNN.com—fell victim to DDoS attacks.

Since then, DDoS attacks have become commonplace. The vast majority of DDoS attacks are not publicized and include a wide range of global victims, from small commercial sites to educational institutions, public chat servers, and government organizations. In a recent study, Moore et al. [59] estimated that the vast majority of victims of the attacks were home users and small businesses rather than large corporations. Over three years, the researchers witnessed over 68,000 attacks, and yet this number is an underestimate, since they only monitored attacks where the bots used spoofed IP addresses (and spoofing IP addresses is not necessary in DDoS attacks).

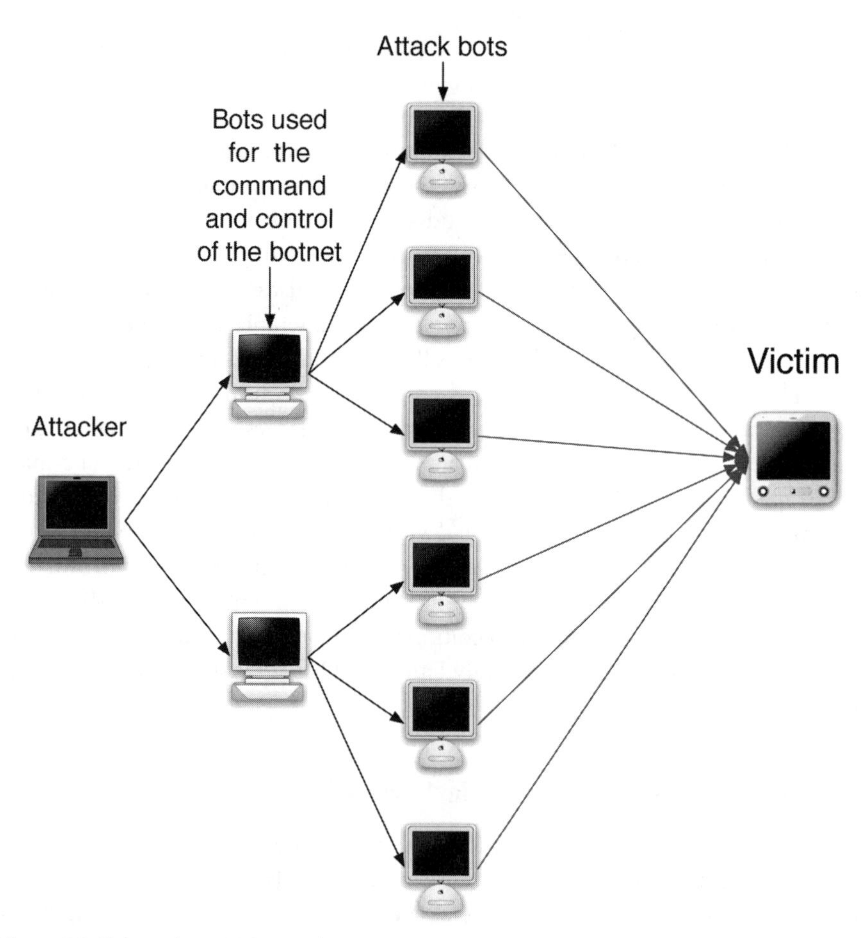

Figure 4.4 Using a (possibly rented) botnet, an attacker can launch a DDoS attack on a desired victim by flooding it with useless connections.

Two of the most relevant DDoS attacks for the scope of this book are the DNS attacks in 2002 and 2007, and the attacks against Estonia in 2006.

Although the DDoS attacks against the DNS infrastructure in 2002 and 2007 had little impact on Internet users, they are relevant because of the potential consequences of a successful attack: if the attacks had been successful, the Internet communication infrastructure would have suffered a major disruption in service. However, since the DNS service is one of the most resilient services on the Internet, the amount of resources required to mount a successful attack is extraordinarily large. On the other hand, the attacks against Estonia were extremely successful by using even less resources than those required for the DNS attacks and much less than the resources used by botnets in their battles against antispam companies [52].

The cyber attacks against Estonia began as a retaliation to Estonia's plans for moving Soviet-era war memorials. The attacks were incited via Russian-language chat rooms: some people volunteered their computers by following the instructions on how to launch a ping attack and some other attackers defaced several Estonian Web sites. However, most of the damage came as DDoS attacks from botnets. The DDoS attacks blocked online access to banks and government Web sites, including the government ministries.

Estonia initially blamed Russia for the attacks, but the Kremlin denied the accusations. Most experts found the evidence of official Russian involvement weak, and concluded that the attacks in Estonia were the work of *hacktivists*, activists that use the Internet as a tool to advance their cause. At the end, Estonia accepted the idea that it was the work of individuals, but hopes Russia will cooperate in tracking the perpetrators.

Hacktivist activities are nothing new [60]. Some of the activities sparked by international tensions include the Web defacements between China and Taiwan in 1999, between Israel and Palestine in 2000, between China and the US after the bombing of the Chinese embassy in Belgrade in 1999, and in 2001 when one of the US spy planes collided with a Chinese airplane. In the recent Iraq war, there were a series of DDoS attacks against Al Jazeera's Web site, and the defacement of the Iraqi government Web site [61]. Although in several cases there is speculation about government involvement in these attacks, most of them are assumed to be part of independent groups.

However, even if these attacks were the product of independent groups, they should serve as a reminder of the fragility of the security of several Internet Web sites, and of the threat posed by large botnets. Based on estimated prices for renting botnets, the attacks on Estonia could have been carried by $100,000 [52], a sum within the reach of independent groups.

4.6.4 Cyber Espionage and the Athens Affair

Cyber espionage is another major threat for the security of a nation. It is natural to assume that several nations (or even independent groups) perform some level of espionage or intelligence gathering on each other by complex computer network attacks [62]. However, the details of these attacks are generally kept confidential.

One of the cases that researchers have been able to study and disclose publicly was the "Athens affair" [63], a case in which hackers broke into Vodafone's telephone network in Athens and subverted its built-in wiretapping features for their own purposes.

The cell phone bugging began sometime during the run-up to the August 2004 Olympic Games in Athens and remained undetected until January 24, 2005. The tapped cell phones included the prime minister of Greece, the defense and foreign affair ministers, top military and law enforcement officials, the Greek

EU commissioner, activists, journalists, the mayor of Athens, and at least 100 other high-ranking dignitaries.

In 2005, while looking into what appeared to be a glitch, Ericsson found unauthorized software had been installed in two of Vodafone's central offices. The main reason the attack succeeded was because the rogue software used the lawful wiretapping mechanisms of Vodafone's digital switches to tap about 100 phones, a wiretapping mechanism that is supposed to be available only to the law enforcement agencies in Greece and with the appropriate warrants.

After the forensic investigation was performed on the rogue software, it was concluded that the developers who created this malware were experts. The intruders included a backdoor to install, operate, and update their wiretapping software without being detected by Vodafone or Ericsson. The software also included a rootkit that made it invisible to network operators and deleted all logs of its activities. Finally, the program was written in the PLEX language, a language where most of the information is available to only very trained individuals. Due to these facts, people have speculated that the culprits behind the attacks required the resources available only to insiders, or to a foreign spy agency.

The implications of this incident are far reaching and very relevant due to current developments in the US. First, the recent warrantless wiretapping advances by the US government on calls within the US have raised concerns because these procedures can in fact make the job of a hacker (or foreign spy agency) easier for listening to these conversations [64].

Additionally, the recently released information about DCSNet [65], the wiretapping network that makes it easy for FBI operatives to tap into conversations, has raised similar concerns. Critics argue that the lack of proper security measures for DCSNet means that there is no trustworthy way to prevent an attacker from exploiting the system; in other words, creating an infrastructure for wiretapping (in addition to recent warrantless wiretapping laws under the "Protect America Act") just makes the job of a malicious wiretapper easier [66].

4.6.5 Critical Infrastructure and Cyber Security

All of the critical infrastructures (energy, telecommunications, transportation, banking and finance, continuity of government services, water supply systems, gas and oil production, and emergency services) are dependent on the computer communication infrastructures. Moreover, the computer information infrastructures are themselves dependent on many of the critical infrastructures, such as electric power grid and telecommunications systems. A successful cyber attack on the supervisory control and data acquisition (SCADA) and other control systems for the critical infrastructures could have a significant impact on public health, economic losses, and potential loss of lives. Securing control systems in critical infrastructures is thus a national priority for the department of homeland security [67].

To date, there have been relatively few attacks on the critical infrastructures. However, security studies from the U.S. Department of Energy (DOE) and commercial security consultants have demonstrated the cyber vulnerabilities of control systems [67–69]. In one of the most recent demonstrations of the vulnerability of the critical infrastructure, a security researcher was able to break into a nuclear power station and within a week take over the control plant [70]. This is not the first time that a critical infrastructure has been penetrated. In 2000, a 48-year-old Australian man, who was fired from his job at a sewage-treatment plant, remotely accessed his workplace computers and poured toxic sludge into parks and rivers [70].

In January 2003, computers infected with the Slammer worm (SQL Server worm) shut down safety display systems at the Davis-Besse power plant in Oak Harbor, Ohio. A few months later, another computer virus was widely suspected by security researchers of leading to a power loss at a plant providing electricity to parts of New York state. A third incident was the power outage of August 2003 in the Midwest and Northeast of the United States, and Canada. Even though the incident was not an act of terrorism, it demonstrates the vulnerability of the electric power grid. In fact, some of the documents gathered from Al Qaeda in 2002, suggested that they were considering a cyber attack on the power grid.

Cyber security is one of the most fundamental aspects of critical infrastructure protection. We need to study and assess the threats and risks of possible cyber attacks, and implement suitable defenses.

4.7 Conclusions

There has been a lot of speculation about state-funded cyber militias and espionage, but without identifying the clear culprits of several of the recent attacks, there is no conclusive evidence for most claims. However, three things are clear: (1) most governments are investing in cyber warfare activities (in defense and offense), given that any conflict in the twenty-first century will necessarily involve the use of information technology, (2) any conflict, or international tension in the physical world, will have its counterpart effect in the Internet, and (3) there are large economic incentives for crime in the Internet. If we leave these activities unchecked, they will increase.

Although cyber war, cyber terrorism, and hacktivism pose real threats, they tend to attract most of the media attention and public opinion, leaving the thriving criminal activities on the Internet relatively unnoticed. Cyber crime is also expected to expand to new technologies: cell phones, personal digital assistants, music players, and embedded hardware can give rise to new vulnerabilities and risks.

Even though the threats and concerns about cyber security seem daunting, we believe that efforts to raise public awareness, new policy rules, investment in research for cyber security technologies, and an expansion in diplomacy and international cooperation for tracking and prosecuting criminals across different countries will greatly help in securing cyberspace. By providing a better cyber security, governments will be one step closer to providing homeland security to their citizens.

Acknowledgments

This work was supported in part by TRUST (Team for Research in Ubiquitous Secure Technology), which receives support from the National Science Foundation (NSF award number CCF-0424422) and the following organizations: AFOSR (#FA9550-06-1-0244) Cisco, British Telecom, ESCHER, HP, IBM, iCAST, Intel, Microsoft, ORNL, Pirelli, Qualcomm, Sun, Symantec, Telecom Italia, and United Technologies.

References

[1] M. Bishop. *Computer Security, Art and Science*, Reading, MA: Addison-Wesley, 2003.

[2] C. Kaufman, R. Perlman, and M. Speciner. *Network Security: Private Communication in a Public World*. Prentice-Hall, 2nd ed, 2002.

[3] R. Anderson. *Security Engineering*, New York: Wiley, 2001.

[4] D. Gollmann. *Computer Security*, 2nd ed., New York: Wiley, 2006.

[5] J. Katz and Y. Lindell. *Introduction to Modern Cryptography*. Chapman & Hall/CRC, 2007.

[6] C. Boyd and A. Mathuria. *Protocols for Authentication and Key Establishment*. Springer, 2003.

[7] R. Shirey. *RFC 2828—Internet Security Glossary*. http://www.faqs.org/rfcs/rfc2828.html, 2000.

[8] D. Dolev and A. C. Yao. "On the security of public key protocols." *IEEE 22nd Annual Symposium on Foundations of Computer Science*, 1981, pp. 350–357.

[9] X. Boyen. "Halting password puzzles—hard-to-break encryption from human-memorable keys." *16th USENIX Security Symposium—SECURITY 2007*, pp. 119–134, August 2007.

[10] A. Lenstra, X. Wang, and B. de Weger. Colliding X.509 certificates. Report 067, Cryptology ePrint Archive, 2005.

[11] C. De Cannière and C. Rechberger. "Finding SHA-1 characteristics: General results and applications." *Advances in Cryptology— Asiacrypt*, pp. 1–20, 2006.

[12] X. Wang, Y. L. Yin, and H. Yu. "Finding collisions in the full SHA-1." *Advances in Cryptology, CRYPTO 2005*, Santa Barbara, CA, 2005.

[13] *NIST's Plan for New Cryptographic Hash Functions*. http://www.csrc.nist.gov/pki/HashWorkshop/index.html, January 24, 2007.

[14] A. Kerckhoffs. "La cryptographie militaire." *Journal des Sciences Militaires*, pp. 5–38, January 1883.

[15] R. Perlman. "An overview of PKI trust models." *IEEE Network*, 13(6):38–43, November/December 1999.

[16] R. Rivest, A. Shamir, and L. Adleman. "A method for obtaining digital signatures and public-key cryptosystems." *Communications of the ACM*, 21(2):120–126, 1978.

[17] M. Bellare and P. Rogaway. "Optimal asymmetric encryption—How to encrypt with RSA." *Advances in Cryptology—Eurocrypt '94*, 1995.

[18] T. El-Gamal. "A public-key cryptosystem and a signature scheme based on discrete logarithms." *IEEE Transactions on Information Theory*, 31(4):469–472, 1985.

[19] W. Mao. *Modern Cryptography: Theory and Practice*. Prentice-Hall, 2004.

[20] L. O'Gorman. "Comparing passwords, tokens, and biometrics for user authentication." *Proceedings of the IEEE*, 91(12):2021–2040, December 2003.

[21] S. M. Bellovin. "Using the domain name system for system break-ins." *Proceedings of the 5th USENIX Security Symposium*, 1995.

[22] W. Cheswick, S. M. Bellovin, and A. D. Rubin. *Firewalls and Internet Security: Repelling the Wily Hacker*. Addison-Wesley Professional Computing Series, 2002.

[23] A. A. Cárdenas, J. S. Baras, and K. Seamon. "A framework for the evaluation of intrusion detection systems." *Proceedings of the 2006 IEEE Symposium on Security and Privacy*, pp. 63–77, Oakland, May 2006.

[24] Aleph One. Smashing the stack for fun and profit. *Phrack Magazine*, 7(49), 1996.

[25] J. Pincus and B. Baker. "Beyond stack smashing: Recent advances in exploiting buffer overruns." *IEEE Security & Privacy Magazine*, 2(4):20–27, July–August 2004.

[26] M. Bishop and M. Dilger. "Checking for race conditions in file accesses." *Computing Systems*, 9(2):131–152, 1996.

[27] Scut/Team Teso. *Exploiting Format String Vulnerabilities.* http://julianor.tripod.com/teso-fs1-1.pdf, March 2001.

[28] D. Ahmand. "The rising threat of vulnerabilities due to integer errors." *IEEE Security & Privacy Magazine*, 1(4), July–August 2003.

[29] Z. Su and G. Wassermann. "The essence of command injection attacks in Web applications." *33rd ACM SIGPLAN-SIGACT Symposium on Principles of Programming Languages*, pages 372–382, 2006.

[30] S. Christey and R. A. Martin. *Vulnerability Type Distributions in CVE.* http://cwe.mitre.org/documents/vuln-trends/index.html, May 22, 2007.

[31] A. Klein. "Cross site scripting explained." *Sanctum White Paper*, May 2002.

[32] E. H. Spafford. "A failure to learn from the past." *Proceedings of the 19th Annual Computer Security Applications Conference*, pp. 217–231, December 8–12, 2003.

[33] A. Shevchenko. *The Evolution of Self-Defense Technologies in Malware.* http://www.viruslist.com/analysis?pubid=204791949, July 2007.

[34] L. Adleman. "An abstract theory of computer viruses." *Advances in Cryptology—Proceedings of CRYPTO '88*, pp. 354–374.

[35] F. Cohen. "Computational aspects of computer viruses." *Computers and Security*, 8(4):325–344, November 1989.

[36] E. W. Dijkstra. *The Humble Programmer*. ACM Turing Lecture, 1972.

[37] K. Thompson. "Reflections on trusting trust." *Communications of the ACM*, 27(8):761–763, August 1984.

[38] B. Chess and G. McGraw. "Static analysis for security." *IEEE Security & Privacy Magazine*, 2(6):76–79, November 2004.

[39] D. Wagner, et al. "A first step towards automated detection of buffer overrun vulnerabilities." *Proceedings of the 7th Networking and Distributed System Security Symposium (NDSS)*, pp. 3–17, February 2000.

[40] J. Yang, et al. "MECA: an extensible expressive system and language for statically checking security properties." *Proceedings of the 10th ACM conference on Computer and Communications Security*, pp. 321–324, 2003.

[41] A. Orso and A. Zeller, (eds.) *Fifth International Workshop on Dynamic Analysis (WODA 2007)*, 2007.

[42] B. Arkin, S. Stender, and G. McGraw. "Software penetration testing." *IEEE Security & Privacy Magazine*, 3(1):84–87, January 2005.

[43] Y. Xie and A. Aiken. "Static detection of security vulnerabilities in scripting languages." *15th USENIX Security Symposium*, pp. 179–192, 2006.

[44] V. B. Livshits and M. S. Lam. "Finding security vulnerabilities in Java applications with static analysis." *Proceedings of the 14th USENIX Security Symposium*, pp. 271–286, 2005.

[45] E. Haugh and M. Bishop. "Testing C programs for buffer overflow vulnerabilities." *Proceedings of the 10th Network and Distributed System Security Symposium (NDSS)*, pp. 123–130, February 2003.

[46] J. H. Saltzer and M. D. Schroeder. "The protection of information in computer systems." *Proceedings of the IEEE*, 63(9):1278–1308, September 1975.

[47] I. Goldberg, et al. "A secure environment for untrusted helper applications: Confining the wily hacker. *Proceedings of the 6th USENIX Security Symposium*, pp. 1–13, July 1996.

[48] N. Provos. "Improving host security with system call policies." *Proceedings of the 12th USENIX Security Symposium*, August 2003.

[49] G. Necula. "Proof-carrying code." *24th ACM SIGPLAN-SIGACT, Symposium on Principles of Programming Languages*, pp. 106–119, January 1997.

[50] R. Thomas and J. Martin. "The Underground Economy: Priceless," *Login: The USENIX Magazine*, 31(6):7–16, 2006.

[51] G. Keizer. "Botnet Steals eBay Accounts. " *PC World*, http://www.pcworld.com/article/id,136729-c,onlinesecurity/article.html, September 4, 2007.

[52] M. Lesk. "The new front line. Estonia under cyberassault." *IEEE Security & Privacy Magazine*, 5(4):76–79, July–August 2007.

[53] S. Gaudin. *Storm Worm Botnet More Powerful Than Top Supercomputers. iTnews*, http://www.itnews.com.au/News/60752,storm-worm-botnetmore-powerful-than-top-supercomputers.aspx, September 7, 2007.

[54] M. Abu Rajab, et al. "My botnet is bigger than yours (maybe, better than yours): Why size estimates remain challenging." *Proceedings of the 1st Workshop on Hot Topics in Understanding Botnets (HotBots '07)*, April 10, 2007.

[55] S. Staniford, V. Paxson, and N. Weaver. "How to own the Internet in your spare time." *Proceedings of the 11th USENIX Security Symposium*, pp. 149–167, 2002.

[56] D. Moore, V. Paxson, S. Savage, C. Shannon, S. Staniford, and N. Weaver. "Inside the slammer worm." *IEEE Security & Privacy Magazine*, 1(4):33–39, July 2003.

[57] K. Poulsen. *Slammer Worm Crashed Ohio Nuke Plant Network*. Security Focus, http://www.securityfocus.com/news/6767, September 2003.

[58] C. Shannon and D. Moore. "The spread of the Witty worm." *IEEE Security & Privacy Magazine*, 2(4):46–50, July–August 2004.

[59] D. Moore, C. Shannon, D. Brown, G. M. Voelker, and S. Savage. "Inferring Internet denial-of-service activity." *ACM Transactions on Computer Systems*, May 2006.

[60] D. Denning. "Cyberwarriors: Activists and terrorists turn to cyberspace." *Harvard International Review*, 23(2), 2001.

[61] "Hacktivists attack websites in war protests." *Network Security*, (4):2, April 2003.

[62] D. Sevastopulo. "Chinese Military Hacked into Pentagon." *Financial Times*, http://www.ft.com/cms/s/0/9dba9ba2-5a3b-11dc-9bcd-0000779fd2ac.html, September 3, 2007.

[63] V. Prevelakis and D. Spinelis. "The Athens affair." *IEEE Spectrum*, 44(7):26–33, July 2007.

[64] S. Landau. "A gateway for hackers." *Washington Post*, p. A17, August 9, 2007.

[65] R. Singel. "Point, Click... Eavesdrop: How the FBI Wiretap Net Operates." *Wired Magazine*, http://www.wired.com/politics/security/news/2007/08/wiretap, August 29, 2007.

[66] S. Bellovin, et al. "Risking Communications Security: Potential Hazards of the Protect America Act." *IEEE Security & Privacy Magazine*, 6(1):24–33, January-February 2008.

[67] United States Government Accountability Office, *Critical Infrastructure Protection. Multiple Efforts to Secure Control Systems Are Under Way, But Challenges Remain*. Technical Report GAO-07-1036, Report to Congressional Requesters, 2007.

[68] J. Eisenhauer, et al. *Roadmap to Secure Control Systems in the Energy Sector*. Energetics Incorporated. Sponsored by the U.S. Department of Energy and the U.S. Department of Homeland Security, January 2006.

[69] R. J. Turk. *Cyber Incidents Involving Control Systems*. Technical Report INL/EXT-05-00671, Idaho National Laboratory, October 2005.

[70] A. Greenberg. "America's Hackable Backbone." *Forbes*, http://www.forbes.com/logistics/2007/08/22/scada-hackers-infrastructuretech-security-cx ag 0822hack.html, August 2007.

5

Mining Databases and Data Streams

Carlo Zaniolo and Hetal Thakkar

5.1 Introduction and Historical Perspective

The term *data mining* describes the processes and the techniques that discover and extract novel, nontrivial, and useful knowledge from large datasets [1].

The theory and practice of *data mining* (DM) have been attracting growing interest as a result of: (1) their success in many diverse application domains, (2) the progress in the enabling technology, and (3) the growing number, size, and accessibility of datasets available for fruitful mining. As early as the 1990s, the large databases created by commercial enterprises represented the first area of opportunities. It was soon realized that, although created to support specific operational needs, these databases contain troves of information about customers' purchasing patterns and preferences: once these are discovered via data mining, the knowledge so acquired can be invaluable for improving the profitability of the company. For instance, retailers can use this information for targeted marketing, airlines can use it to design fare structures that maximize their profits, and so on. These early DM applications were deployed by companies as part of their "decision-support" and "business-intelligence" activities that were based on *data warehouses* specifically built for these activities. Thus, data collected from a company's operational databases are integrated into a data warehouse to support business intelligence activities [2].

In the last decade, the DM field has been significantly broadened by the arrival of the Web, whereby the whole world became accessible as a global data warehouse, and by several new applications such as link analysis, security applications, and text mining. Furthermore, it became clear that much of the information that is continuously exchanged on the Internet should be mined as it flows on the wire rather than from a stored database. In fact, a more traditional store-now-and-mine-later approach is often not possible because of real-time response requirements or simply because the data stream is too massive. In these situations, the information must be processed and mined as the data arrives. While this new research area, namely data stream mining, shares many similarities with database mining, it also poses unique challenges and requirements, discussed in Section 5.4.

The progress made by DM technology is due to the confluent contributions of several research fields. Indeed, while failures tend to be orphans, DM has seen very many paternity claims emerge as a result of its success and interdisciplinary nature. We now address this topic briefly, and refer to [3] for a more comprehensive discussion of the subject. *Machine learning* is the artificial intelligence area and where work on data mining started and where research continues with very advanced methods for knowledge extraction. However, while learning on sparse data remains very difficult, great success has been achieved in applications working with large datasets. Then, scalability and computing efficiency become a paramount concern, prompting much research work to support and integrate data mining methods in database management systems (DBMS). The research vision of inductive DBMS [4, 5], where DM queries are supported as regular DBMS queries, has yet to be realized, but much progress has been made and is also discussed in Section 5.4.

Many other areas and disciplines, such as visualization and statistics, have contributed to the progress of DM. In particular, the field of statistics has been contributing to data analysis for more than a century, and is now challenged with the task of providing rigorous analytical foundations for data mining methods of proven empirical effectiveness [6]. Statistical data mining is discussed in [6], whereit is also observed that DM plays a complementary role with respect to traditional data analysis. While DM's exploratory data analysis aims at discovering a pattern or a model from a given dataset, data analysis often focused on "confirmatory data analysis," where an assumed model needs to be confirmed and its parameters derived.

DM systems and software packages recently produced by various groups and disciplines also deserve much of the credit for the popularity of DM. Along with stand-alone DM systems such as WEKA [7], there are those that are part of statistical-software packages such as STATISTICA, CLEMENTINE, and R [8–10], and those that are provided as extensions of commercial DBMS [11–13], besides the packages designed to support DM for specialized application domains.

5.1.1 Examples of DM Methods and Applications

Different DM methods are best suited for different applications. For instance, in border control, the risk factor of any crossing vehicles can be computed on the basis of various attributes, such as vehicle type, number of times the vehicle crossed the border in the last few months, driver's license type, driver's visa status, and so forth. Once the data have been integrated from various sources, *predictive* mining models, such as classification, are often used to determine the risk involved. These techniques have also been widely used in financial applications, where banks often perform loan payment prediction and risk assessment on the basis of attributes, such as income and credit history, describing the customer. Another well-known application is targeted marketing, where potential customers are segmented into groups of similar traits and characteristics. *Clustering* techniques represent the mining tool of choice in this second type of application. Samples that do not fall in the clusters are called *outliers*.

In crime analysis, outliers are often associated with individuals or situations that deserve to be further investigated. For instance, when monitoring computer networks for intrusion detection, outliers often denote anomalous activities by users, which in turn indicate a possible attack by hackers. Along with classification and clustering, *association* represents the third core mining method. The objective of this method is to discover patterns and predictive rules. For instance, by mining records of sale transactions we might be able to discover that a customer who buys certain books is also likely to buy others.

In the next section we discuss these three core methods—classification, association, and clustering—along with simple applications of such methods. More complex applications, including security applications, which often require a combination of several DM techniques, are discussed in Section 5.3.

For instance, to detect money laundering, we might employ text mining tools (to identify the entities involved in company documents, e-mails, text messages), link analysis tools (to identify links between different individuals, companies, documents, and accounts), outlier-detection tools (to detect unusual money transfers), and various visualization tools. Similarly, intrusion detection, which aims to prevent attacks to the integrity, confidentiality, and/or availability of computers, involves classification and detection of sequential patterns besides outlier analysis. Furthermore, all data mining applications, and in particular homeland security applications, must be preceded by extensive preparation steps to collect and clean the relevant data (e.g., record consolidation for name matching), discretization, and other preprocessing steps discussed in the next section.

The organization of this chapter reflects the historical evolution of the DM field and its data-driven applications: in the next section we discuss knowledge discovery from databases, then we discuss new applications, particularly those driven by the Web (which can be viewed as a global database), and Section 5.4 covers data stream mining.

5.2 Data Mining Methods

We now discuss the more established methods and techniques of *knowledge discovery in databases (KDD)*. KDD is the algorithmic analysis of large data repositories to identify novel and useful patterns in the data. The algorithms that are at the heart of the KDD process can be grouped into methods that perform similar DM tasks, such as classification, association, and clustering. These three core methods are described in more detail in this section.

5.2.1 Taxonomy of Data Mining Methods

Figure 5.1 provides a skeleton taxonomy of the main mining tasks discussed in the next section.

In addition to the hierarchy presented in Figure 5.1, distinction between supervised and unsupervised learning is also common in DM literature. Here again clustering and classification are at the opposite extremes: clustering requires little input from the users (e.g., the number of clusters to be constructed), while classification requires numerous examples that the classifier uses to learn and emulate the behavior of the system that has produced those examples.

5.2.2 Classification and Prediction

Classification is the task of predicting the value of a particular attribute in a tuple on the basis of the values of other attributes, using a model learned from training data. In other words, classifiers are trained on a large set of examples from which they must learn how to best assign a category to new tuples. Because of

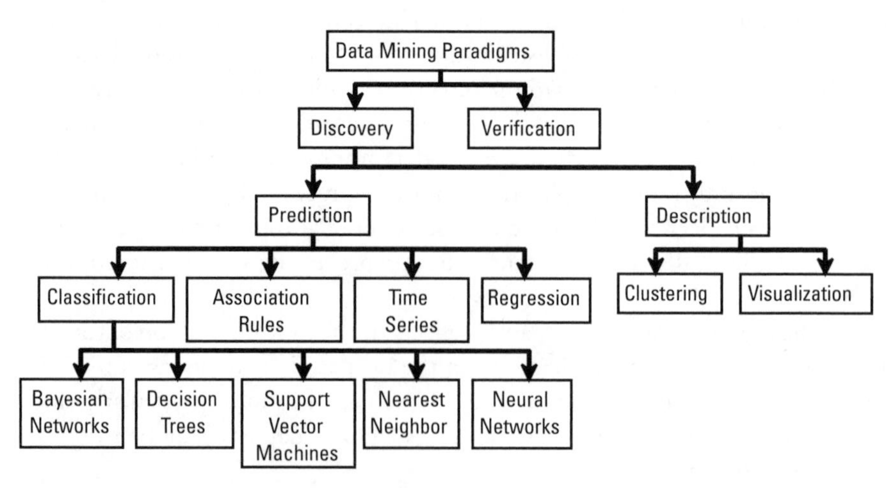

Figure 5.1 Taxonomy of data mining methods.

their ability to make predictions based on training examples, classifiers represent a fundamental DM method used in many and diverse applications. Examples include detecting spam e-mail messages, given the message header, content, and subject, or deciding whether a cargo entering the US is high risk (e.g., contains drugs) given the vehicle type, driver's license type, driver's age, and time of day. Classification is a supervised learning method insofar as it is driven by an initial set of examples describing the behavior that the classifier must learn to emulate. These examples are normally given in the form of a database relation such as that of Table 5.1.

In this table, we have a tuple with four attributes describing a vehicle crossing the border and a binary HighRisk attribute, whose value is either *Y* or *N*. Here *Y* (versus *N*) denotes if the vehicle may contain drugs or illegal weapons (versus not). This example shows data of potential interest to the automated commercial environment (ACE) program of the U.S. Department of Homeland Security [14]. ACE attempts to integrate data from different government agencies to enable accurate data analysis for border control [14]. Independent of the interpretation of the values in the table, the objective of classification is: (1) to learn the function $\mathcal{F}$ that maps the known vehicle attributes into the *Y*/*N* values (possible values) of HighRisk (unknown attribute), and (2) to evaluate the accu-

Table 5.1
Examples of Past Border Crossings

Vehicle Type	Driver Lic Type	Driver Age	Time of Day	High Risk
Passenger	Regular	> = 35	Night	N
Passenger	Regular	> = 35	Day	N
Workvan	Regular	> = 35	Night	Y
Truck	Disabled	> = 35	Night	Y
Truck	Commercial	<35	Night	Y
Truck	Commercial	<35	Day	N
Workvan	Commercial	<35	Day	Y
Passenger	Disabled	> = 35	Night	N
Passenger	Commercial	<35	Night	Y
Truck	Disabled	<35	Night	Y
Passenger	Disabled	<35	Day	Y
Workvan	Disabled	> = 35	Day	Y
Workvan	Regular	<35	Night	Y
Truck	Disabled	> = 35	Day	N

racy of $\mathcal{F}$. For this purpose the original set is divided into a training set of rows for building the classifier, and a testing set of rows for determining the accuracy of the classifier.[1] Thus, the known vehicle attributes of each testing row are fed to the learned classifier and the prediction is compared with the actual label. The number of correct decisions over the number of tuples tested defines the accuracy of the classifier. While high accuracy can be expected in many cases, this is normally well below 100% because of the following reasons: (1) the limitations of the classification algorithm used in learning complex $\mathcal{F}$ functions, and (2) the function $\mathcal{F}$ is nondeterministic—that a vehicle might be not involved in criminal activities although another with identical attribute values is. Several classifier algorithms have been proposed in the literature and are discussed next. The applicability and accuracy of the classification algorithms depend on the application domain, that is, not all algorithms apply to any application domain.

5.2.2.1 Naive Bayesian Classifiers

The training set for a mini-example, containing 14 tuples, is shown in Table 5.1 (note: real-life applications contain thousands or even millions of tuples and many more attributes). The *HighRisk* value is Y in 9 of these examples and N in the remaining 5. Thus, if we had to attempt a prediction with only this information, we would most likely go with Y, since the probability of Y, $pr(Y) = 9/14$, is larger than $pr(N) = 5/14$. However, the tuples also contain information about the vehicle attributes associated with the Risk values, thus we can compute conditional probabilities to improve the quality of our predictions. The conditional probabilities of the VehicleType being either passenger, workvan, or truck, for $HighRisk = Y$ is estimated (by counting the entries in the table) as:

$$pr\left(passenger\middle|Y\right)=2/9,\quad pr\left(workvan\middle|Y\right)=4/9,\quad pr\left(truck\middle|Y\right)=3/9$$

Similarly, for $HighRisk = N$ we find: $pr(passenger|N) = 3/5$, $pr(workvan|N) = 0$, and $pr(truck|N) = 2/5$. Similar count-based estimations of conditional probabilities can be easily obtained for the remaining vehicle attributes. Table 5.2 shows these probabilities with the unconditional probability in the last row.

Having collected this information, the classifier must predict which one of the Risk values, Y or N, is more likely given a vehicle tuple with four attributes: $\langle VehicleType, DriverLicType, DriverAge, TimeOfDay\rangle$. If X is the current value of this 4 tuple, then the right prediction is the value of C that maximizes the value of $pr(C|X)$. But by Bayes' theorem we obtain that:

$$pr\left(C\middle|X\right)=\frac{pr\left(X\middle|C\right)\times pr\left(C\right)}{pr\left(X\right)}$$

1. For instance, 70% of the rows could be used for training and 30% for testing.

Table 5.2
Probabilities for the Training Set of Table 5.1

Value	Ys	Ns
Passenger	2/9	3/5
Workvan	4/9	0
Truck	3/9	2/5
Regular	2/9	2/5
Disabled	4/9	2/5
Commercial	3/9	1/5
> = 35	3/9	4/5
<35	6/9	2/5
Day	3/9	3/5
Night	6/9	2/5
All	9/14	5/14

Since X is given, $pr(X)$ is a constant that can be ignored in finding the maximum C. Also, $pr(C)$ was previously estimated as $pr(Y) = 9/14$ and $pr(N) = 5/14$. However, estimating the value of $pr(X|C)$ to any degree of accuracy can be very difficult. If we make a naive simplifying assumption that the vehicle attributes are independent of each other, then we have:

$$pr\left(\langle X_1,\ldots,X_k\rangle|C\right) = pr\left(\langle X_1\rangle|C\right) \times \ldots \times pr\left(\langle X_K\rangle|C\right)\Big)$$

Then, we simply plug the correct values from Table 5.2 into this formula and compare the results to find the one with the highest probability. For instance, say that our input sample is $X = \langle truck, regular, >= 35, night\rangle$, then we have:

$$pr(X|Y) \times pr(Y) =$$
$$pr\left(truck|Y\right) \times pr\left(regular|Y\right) \times pr\left(>=35|Y\right) \times pr\left(night|Y\right) \times pr\left(all|Y\right) =$$
$$3/9 \times 2/9 \times 3/9 \times 6/9 \times 9/14 = 0.0105$$

and

$$pr(X|N) \times pr(N) =$$
$$pr\left(truck|N\right) \times pr\left(regular|N\right) \times pr\left(>=35|N\right) \times pr\left(night|N\right) \times pr\left(all|N\right) =$$
$$2/5 \times 2/5 \times 4/5 \times 2/5 \times 5/14 = 0.0182$$

Since 0.0182 is larger than 0.0105, our prediction is *N*.

Observe that these computations can be derived by restricting Table 5.2 to the rows containing the values *truck*, *regular*, > = 35, *night*, and *all* (unconditional) and then performing a vertical multiplication on the *Y* and *N* columns. Thus a naive Bayesian classifier (NBC) can be easily implemented on a spreadsheet system or on a relational DBMS using the basic aggregate functions in their standard language SQL.[2] Although so simple, NB classifiers normally produce accurate results, even in the situation where the naive assumption of independence between the different attributes is not satisfied. Simple generalizations of NBC are also at hand [15] for continuous attributes, whereby the actual driver age replaces the categorical > = 35 and <35 in the training set: then in Table 5.1, we can use entries that describe the statistical distribution of these values rather than the values themselves. Finally, it is possible to overcome the naive assumption of independence between attributes, and use Bayesian networks to describe their correlations. The more complex techniques required by this approach are outside the scope of this chapter.

5.2.2.2 Decision-Tree Classifiers

Although NB classifiers can provide efficient and accurate predictions, their ability to offer insights into the nature of the problem and the attributes involved is quite limited. On the other hand, decision-tree classifiers are effective at revealing the correlation between decisions and the values of the pertinent attributes. Furthermore, these correlations are expressed through simple rules.

For instance, Figure 5.2 shows a simple decision-tree classifier for the training set of Table 5.1. Each node of the tree represents an attribute and the branches leaving the nodes represent the values of the attribute. Much understanding of the nature of the decision process is gained through this tree. Thus, we conclude that when the VehicleType is *workvan*, then the decision is *Y*, but when the VehicleType is passenger then the decision also depends on DriverAge,

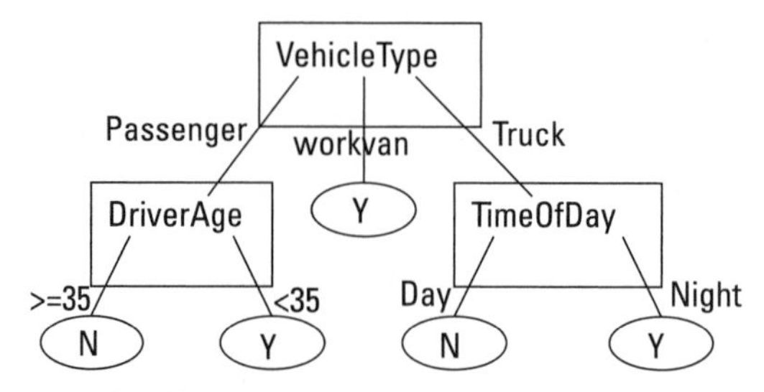

Figure 5.2 Desicion-tree classifier.

2. SQL is an acronym for Structured Query Language.

while TimeOfDay becomes important when it is a truck. A more formal statement of this understanding can come in the form of rules, one for each leaf node in the tree. For the example at hand, we have five rules as follows:

```
VehicleType = passenger,DriverAge >=35 →          N
VehicleType = passenger,DriverAge <35 →           Y
VehicleType = workvan →                           Y
VehicleType = truck,TimeOfDay = day →             N
VehicleType = truck,TimeOfDay = night →           Y
```

Clearly, more compact trees yield terser and more incise rules, and provide a better descriptive model for the data. Therefore, a number of techniques have been proposed for pruning the decision tree for limiting its growth, while assuring that the predictive accuracy of the resulting tree is not impaired.

The following two steps are repeated recursively to generate a decision tree: (1) select the "best" attribute, say, *B*, for classification, and then (2) partition the examples in the training set according to the value of attribute *B*. The selection of the best attribute is normally performed greedily to minimize the heterogeneity (a.k.a. impurity) of the generated partitions. Several criteria have been proposed to select the best partition, for instance, the Gini index[3] is probably the simplest among them. The Gini index of a partition *S* containing both *N* and *P* samples can be measured as:

$$Gini(S) = 1 - F_P^2 - F_N^2$$

where $F_P(F_N)$ denotes the number of *P*(*N*) samples divided by the total number of samples in *S*. Thus, when *S* contains the same number of *P* and *N* samples *Gini*(*S*) = 1/2. However, *Gini*(*S*) = 0 when *S* is pure (i.e., homogeneous) and contains only *P* samples or only *N* samples, with the objective of partitioning our training set into classes that are as homogeneous as possible. Thus, we select the attribute with the minimal Gini index, where the Gini index of an attribute is the weighted sum of the Gini index of each individual attribute value. Here weight is defined as the fraction of tuples that attain a given attribute value. For instance, we can group the rows in Table 5.1 by the *VehicleType* attribute and partition the table into three groups. The first group is the *workvan* group with four *Y*s

3. Named after the Italian statistician Corrado Gini who introduced it in his 1912 paper "Variabilità e mutabilità" ("Variability and Mutability").

and 0 *N*s: thus its Gini index is zero with weight 4/14. The second group is the *truck* group with three *Y*s and two *N*s: thus its Gini index is $(1-3^2/5^2 - 2^2/5^2)$ and its weight is 5/14. The third and final group is the *passenger* group with two *Y*s and three *N*s with Gini value $(1-2^2/5^2 - 3^2/5^2)$ and weight 5/14; thus the sum of the Gini values multiplied by their weight produces a Gini value of 9/35 for the *VehicleType* partition. The same computation repeated for the other three columns produces higher Gini indexes. Thus we put *VehicleType* at the root of our decision tree and proceed to partition the samples according to their *VehicleType* values and obtain three subtables, one for each value of *VehicleType*. Now we recursively repeat this attribute selection + partition procedure on each of these subtables. Thus, for the subtable where *VehicleType* = *passenger* we find that the best (lowest Gini index) partition is on *DriverAge*, while for *VehicleType* = *truck* the best partition is on *TimeOfDay*. However, the subtable generated from *VehicleType* = *workvan* is homogeneous since *HighRisk* = *Y* in all of these tuples: thus we simply label this branch as *Y*.

The situation where all tuples are homogeneous (or the level of heterogeneity is below a certain threshold) represents the first and most desirable stopping condition for the algorithm. Two less desirable outcomes are: (1) all attributes have already been used, but the qualifying tuples are not homogeneous, and (2) the last partition does not contain any qualifying tuples. Thus, any testing tuples that are processed at the case (1) node will have to be assigned a majority label, whereas a random label will have to be assigned for case (2).

Decision-tree classifiers are unique in their ability to achieve both high descriptive quality and high predictive accuracy. Much research work has been devoted to improve decision tree classifiers and produce robust/scalable algorithms. Some of these algorithms use information gain or x^2-based measures, as opposed to the Gini index, for attribute selection. Algorithms and strategies for decision-tree pruning have also been proposed. Some algorithms use multiway splitting rather than binary splitting, as described in the examples. Binary split algorithms can also be extended to deal directly with continuous data (i.e., without requiring them to be discretized) by simply finding the optimum split value *V* for a given attribute *A*. Thus, samples with an *A* value less than *V* are assigned to one branch and the remaining ones to the other branch. Detailed discussion on these advanced decision-tree methods can be found in [15, 16].

5.2.3 Association Rules

Whenever a market basket is pushed through the checkout counter of a grocery store, or a "checkout-and-pay" button is clicked during a Web transaction, a record describing the items sold is entered into the database. A business enterprise can then use such records to identify the frequent patterns in such transactions.

These patterns can then be described by the following rule stating that if a market basket contains both Diapers and Milk it also contains Beer.

$$r_1 : \text{Diapers, Milk} \rightarrow \text{Beer}$$

A marketing analyst would only find such rules useful if the following three conditions hold.

1. The rule is valid, at least to a certain level of statistical *confidence*.
2. The rule occurs frequently, i.e., has a certain level of *support*.
3. The rule is new and interesting, i.e., the rule was not previously known and can potentially be put to use to improve the profitability of company (e.g., by bundling these three items in a sale campaign or by placing them in nearby locations on the shelf).

Obviously condition 3 can only be decided by the analyst who is intimately familiar with the domain. Thus a DM system generates and presents to the analyst all rules that satisfy conditions 1 and 2: such rules can be found by counting the frequency of the itemsets in the database. Consider, for instance, the following database of Table 5.3, containing five transactions.

We are interested in rules that have a high level of *support* and *confidence*. The *support* of a rule is defined by the fraction of transactions that contain all the items in the rule. The example database contains five transactions and two of those contain all the items of rule r_1 (i.e., the itemset {Beer, Diapers, Milk}). So the support for r_1 is $S_1 = 2/5 = 0.4$.

The confidence for a rule is instead the ratio between the support of the rule and the support of its left side: the left side of r_1 is {Diapers, Milk}, which has support $S_2 = 3/5 = 0.6$. Therefore, the confidence for the rule is $S_1/S_2 = 0.66$. In many situations, only rules that have a high level of confidence and support are of interest to the analysts. For instance, if the analyst specifies a minimum confi-

Table 5.3
Examples of Market Basket Transactions

TID	Itemset
1	{Bread, Milk}
2	{Bread, Beer, Diapers, Eggs}
3	{Beer, Cola, Diapers, Milk}
4	{Beer, Bread, Diapers, Milk}
5	{Bread, Cola, Diapers, Milk}

dence ≥ 0.66 and a minimum support ≥ 0.6, then r_1 fails because of insufficient support. However, the following two rules meet this criteria:

$$r_2 : \text{Milk} \longrightarrow \text{Bread}$$
$$r_3 : \text{Bread} \longrightarrow \text{Milk}$$

In fact, both r_2 and r_3 have support equal to 3/5 = 0.6, and confidence equal to 3/4 = 0.75.

The symmetric role played by Bread and Milk in these rules illustrates the fact that the rules we construct only establish a strong correlation between items without establishing a causality implication. While a more in-depth analysis of the dataset can reveal more about the nature of the relationship between the items, the analyst is normally the best and final judge on the significance of the rule. For instance, a rule frequently discovered when mining the data of a large department store is that people who buy maintenance contracts are likely to buy large appliances. While this represents a statistically valid association, it is useless in predicting the behavior of customers. In fact, we know that the purchase of a maintenance contract represents the consequence, not the cause of purchasing a large appliance. Only the analyst can discard such trivial rules. Thus we must make sure that the algorithms do not overwhelm the analysts by returning too many candidate rules. Also, we are dealing with hundreds of thousands of transactions and finding all those rules can become computationally intractable as the database size increases. The problems of limiting the number of rules generated and the running time for generating them can both be effectively addressed by setting a high minimum support level. Under this assumption, there exist efficient algorithms to generate all frequent itemsets from very large databases of transactions, as we discuss next.

The a priori algorithm operates by generating frequent itemsets of increasing size: the algorithm starts with singleton sets and then from these it generates frequent pairs, then frequent triplets from frequent pairs, and so on. The algorithm exploits the downward closer property (*if an itemset is frequent then all its subsets must also be frequent*). Let us use the market basket example with minimum support of 0.60 to illustrate how this principle is applied.

Step 1: The algorithm first finds singleton items with frequency 3/5 = 0.60. Thus we get the following items (with frequency). {Beer} : 3, {Bread} : 4, {Diapers} : 4, {Milk} : 4.

Step 2: Next, the frequent pairs are obtained by combining the singleton frequent items: thus we obtain the following candidate pairs: {Beer, Bread}, {Beer, Diapers}, {Beer, Milk}, {Bread, Diapers}, {Bread, Milk}, {Diapers, Milk}.

Step 3: Then we filter these pairs to find the ones with a frequency greater than or equal to the minimum support threshold (frequency 3 in this case): {Beer, Diapers} : 3, {Bread, Diapers} : 3, {Bread, Milk} : 3, {Diapers, Milk} : 3.

Step 4: Next, the algorithm joins these pairs based on the shared items to find candidate triplets—note that these triplets may not qualify as candidates, if any of their subset is not frequent (the a priori principle). Thus, as the first two pairs are joined to form {Beer, Bread, Diapers}, we find that its subset {Beer, Bread} is not frequent: this triplet is not a candidate frequent itemset. Likewise, {Beer, Diapers, Milk} is missing the frequent subset {Beer, Milk}, so the only triplet that can possibly be frequent is {Bread, Diapers, Milk}.

Step 5: Then the algorithm counts the occurrences of {Bread, Diapers, Milk} for which we find three occurrences, which makes it frequent. With no pairs of triplets to combine into a quadruplet, the algorithm terminates.

Thus, the algorithm constructs all candidate sets with support better than 0.6. The minimum confidence requirement is also easy to implement since it is now trivial to compute their confidence. For instance, let us assume that the required minimum confidence is 0.9. We see that a rule Bread, Diapers ⟶ Milk qualifies since {Bread, Diapers, Milk} and {Bread, Diapers} have the same support and thus the rule has a confidence of 1. However, the rule Bread ⟶ Milk does not qualify since the support of {Bread, Milk} is 3/5 while that of {Bread} is 4/5; thus the confidence is only 0.75.

While a priori is rather simple and efficient, several algorithms, such as FP-growth [17], have been proposed to reduce the cost of finding frequent itemsets and to eliminate the need for performing multiple passes through the database.

5.2.3.1 Sequential Patterns

Often there is a temporal correlation between items purchased, whereby, say, customers who buy a suit and a shirt often return to purchase a belt and a tie. This behavior can be modeled as a sequence of three itemsets: S_1 = ⟨{*suit, shirt*}, {*belt*}, {*tie*}⟩. S_2 is said to be a subsequence of S_1 if it can be obtained from S_1 by eliminating some itemsets. Thus S_2 = ⟨{*suit*}, {*belt*}, {*tie*}⟩ and S_3 = {*suit, shirt*}, {*belt*}⟩ are both a subsequence of S_1. We can now determine the support for a temporal rule such as ⟨{*suit*}, {*belt*}⟩ ⟶ ⟨{*tie*}⟩ by simply counting the number of sequences in our database that have ⟨{*suit*}, {*belt*}, {*tie*}⟩ as their subsequence. For example, say that our database only contains sequences S_1, S_2, and S_3; its support of this rule is 2/3. The confidence in the rule is also 2/3. Observe that the a priori principle also holds here: if a sequence occurs *k* times in the database, then its subsequences must occur at least *k* times. Thus algorithms inspired by the a priori algorithm can be used to find sequences whose frequency and confidence exceed given thresholds.

5.2.4 Clustering Methods

Given a large set of objects described by their attributes, clustering methods attempt to partition them into groups such that the similarity is maximized between the objects in the same group. The similarity is computed based on the attributes of the objects. Clustering often represents the first step in a knowledge discovery process. For instance, in marketing applications clustering for customer segmentation makes it possible to learn the response of the population to a new service or a product by interviewing sample customers in each cluster. In intrusion detection, clustering is the first step to learn the behavior of regular users, making it possible for the analyst to concentrate on outliers and anomalous behavior.

Clustering methods have been used in a wider variety of applications and longer than any other DM method. In particular, biologists have long been seeking taxonomies for the living universe and early work in cluster analysis sought to create a mathematical discipline to generate such classification structures. Modern biologists instead use clustering to find groups of genes that have similar expressions. Librarians and archivists have long used clustering to better organize their collections. Moreover, as discussed in the next section, clustering represents a very valuable tool for organizing and searching the billions of pages available on the Web, and monitoring the innumerable messages exchanged as e-mail. Let's assume that we want to analyze these data to detect terrorist activities or other suspicious activities. These data are highly dynamic, which precludes the use of classification algorithms. Therefore, after appropriate cleaning, each document can be converted to a feature vector containing the terms and their occurrence in the document. Then the documents can be clustered based on these feature vectors such that the documents with similar feature vectors are assigned to the same cluster. Such clustering can result in interesting clusters that can be further investigated by human experts (e.g., a cluster where the top distinctive words are *biological virus* and *jihad*).

Many clustering algorithms have been proposed over the years and their detailed discussion is well beyond the scope of this chapter. Therefore, we instead group them into a crude taxonomy and briefly discuss sample algorithms from the main classes of the taxonomy.

- **Hierarchical Clustering Methods:** These methods assume an underlying conceptual structure that is strictly hierarchical. Thus agglomerative methods start with individual points and, at each step, merge the closest pair of clusters—provided that their distance is below a given threshold (otherwise we end up with one all-inclusive cluster containing all points). Divisive methods instead start from one all-inclusive cluster and at each step split clusters that contain dissimilar points.

- **Density-Based Clustering:** This method assigns points to clusters based on the neighborhood of the point, i.e., if many neighbors of a point belong to cluster C then the point is also assigned to cluster C, even if the center of cluster C is not close to the point. Therefore density-based methods can generate clusters of any shape and as such can describe trajectories and epidemic phenomena. A popular density-based algorithm called DBSCAN is described in Section 5.4.
- **Prototype-Based Methods:** These seek to cluster the objects near a set of prototype points that summarize the clusters—typically the mean or the median of the points in the cluster. A very popular method in this class is the *K-means* algorithm, which is discussed next.

The K-Means Algorithm:

1. Select K points as initial centroids.
2. *Repeat* the following two steps:
   (i) form K clusters by assigning each point to its closest centroid, and
   (ii) recompute the centroid of each cluster
3. *Until* the centroids do not change.

While clustering is considered an unsupervised learning method, most clustering algorithms require a fair degree of guidance from the analyst. Furthermore, the correct value of K must be specified by the user. A value of K that is too low will cause some clusters to be too sparse: on the other hand, a value of K that is too large will generate clusters that have centroids very close to each other and should be merged. In practice, therefore, the application of the algorithm could be an iterative process that, for example, starts with a low value of K and then increases it by splitting sparse clusters. Even after the correct value of K is determined, the selection of the initial K points remains a delicate operation. Indeed, the choice of different initial points can result in different final centroids. One solution to this problem consists of randomly selecting a small sample of the dataset and clustering it using hierarchical clustering methods.

A centroid is computed as the average (i.e., the center of mass) of its points. This point is overly sensitive to noise and outliers, and rarely corresponds to an actual object. An alternative, and often preferable, solution consists of using *medoids* (i.e., the mean points of the clusters). Another decision that the analyst must make is which matrix to use for similarity between the objects. Common matrices include Euclidian distance, Manhattan distance, and so forth. Other notions of similarity are often preferable for particular applications. For instance, the cosine distance is often used for documents and the Jaccard and Tanimoto coefficients are used as similarity measures for binary data.

5.2.5 Other Mining Techniques

The core techniques described so far are supplemented by an assortment of other techniques and methods, which we only summarize due to space limitations. Among the many techniques proposed for predictions, we find the support vector machine (SVM), a computationally expensive classification method that achieves excellent accuracy and avoids the "curse of dimensionality" problem. The goal of SVM is to separate sets of examples belonging to different classes by hyperplanes to maximize the distance between the hyperplane and the examples in each class. Nearest-neighborhood classifiers instead attempt to achieve compact representations for the training set so that any new sample can be classified on the basis of the most similar *k* examples in the training set [15]. Thus, rather than providing a higher-level descriptive model, nearest-neighborhood classifiers justify their decision on the basis of similar instances from the training samples. Moreover, both the descriptive model and intuitive connections with training samples can be lost when using *neural networks* [15]. This is a popular method that requires a long training phase but can deliver good predictive performance in applications such as handwritten character recognition and speech generation.

Regression is a popular statistical method based on well-developed mathematical theories, which is not restricted to linear functions, but can also be used for nonlinear, multivariate functions, and event probabilities (logistic regression). Even so it cannot match the performance and generality of classifiers in highly nonlinear problems. Take, for instance, the analysis of *time series* and *event sequences*, such as stock quotations on a ticker tape. To perform trend analysis in this environment, the analyst can use regression-based methods (after suitable smoothing of the data). However, in most advanced applications the analyst is interested in detecting patterns, such as the "double bottom" pattern, that provide a reliable predictor of future market gains. Special techniques are used to recognize these patterns and similar ones defined by relaxing the matching requirements both in the amplitude and time (time warping). Special query language constructs, indexing, and query optimization techniques (Section 5.4) are used for this purpose [18].

In many respects, *outlier analysis* complements these methods by identifying data points that do not fit into the descriptive mining model. Therefore, outlier mining is often a "search and destroy" mission since outliers are either viewed as noisy data (e.g., incorrect values due to faulty lines) or true but inconvenient data that can hamper the derivation of mining models. However, applications such as fraud analysis or intrusion detection focus on the recognition and in-depth study of outliers since these are frequently an indication of criminal activities. An assortment of other techniques ranging from OLAP data cubes to logistic regression [19] can be used for outlier detection.

5.2.6 The KDD Process

Figure 5.3 summarizes the overall KDD mining process consisting of several steps, including:

- Data collection, integration, and cleaning;
- Data exploration and preparation;
- Selection and application of DM tasks and algorithms;
- Evaluation and application of DM results.

Because of its exploratory nature the DM process is often iterative: if at any point in the process the analyst becomes dissatisfied with the results obtained so far, he or she returns to previous steps to revise them or restart from scratch. We can now elaborate on these steps.

5.2.6.1 Data Collection, Integration, and Cleaning

Before any mining can begin, data must be collected and integrated from multiple sites. Furthermore, they must be "cleaned" to remove noise and eliminate discrepancies (e.g., by consolidating different names used for the same entity). These operations constitute time-consuming and labor-intensive tasks that lead to construction of a *data warehouse*. A data warehouse is an integrated database into which a company consolidates the information collected from its operational databases for the purpose of performing business-intelligence and decision-support functions. These functions often include a library of data mining

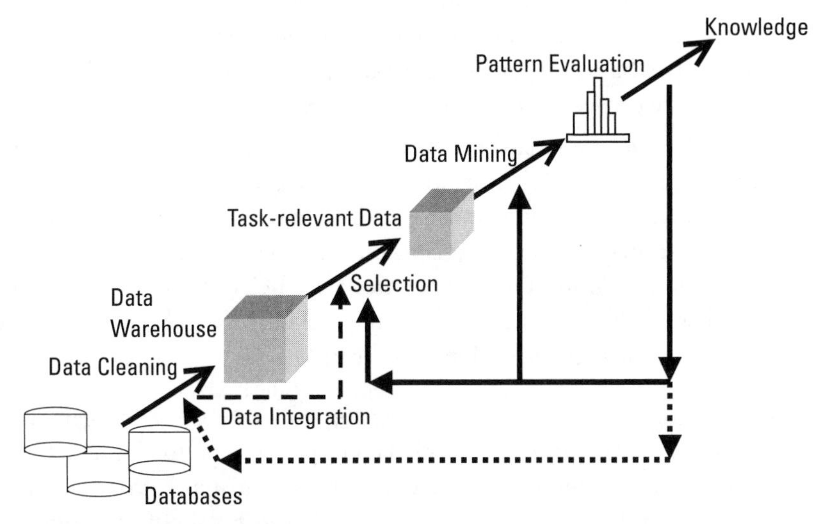

Figure 5.3 The KDD process: main steps.

methods—along with simpler tools for online analytical processing (OLAP), visualization, and statistical analysis. We next discuss these functions and various data-preparation tasks that often precede and pave the way to the actual deployment of DM methods.

- *OLAP Analysis:* OLAP tools have met great success in business applications, inasmuch as they provide users with easy-to-understand multidimensional views of their data, and they are now supported by most commercial DBMSs. Thus, by a simple SQL query, the manager analyzing the sales can retrieve the summary of the recent sales for each region, city, and each store within the city—and thus quickly identify outliers requiring further analysis. This analysis can be performed in the dimensions of location, hierarchically structured by region, city, and store; time, hierarchically structured by year, season, month, week, and day; and product types. Thus a simple OLAP query can return a hypercube-based representation of all the sales of interest hierarchically partitioned along the dimensions of interest. Similar SQL queries can also produce roll-up, drill-down, slice, and dice views of this data [2]. An application of OLAP analysis to associate criminal events is discussed in the next section.
- *Visualization:* People's ability to absorb large amounts of information and recognize patterns therein is limited by the available views of the data. Histograms, pie charts, scatter plots, density plots, vector plots, and spatiotemporal charts are very effective in visualizing low-dimensional data and the results produced by statistical and OLAP functions. On the other hand, visualization methods suffer from the "dimensionality curse" since they cannot handle high-dimensional data well—although techniques based on parallel coordinates, star coordinates, and Chernoff faces have proved effective in special cases.
- *Data Statistics:* Standard notions from statistics can be used to describe continuous data by their central tendency and dispersion. Measures of central tendency include mean, median, and midrange, while measures of data dispersion include quantiles, interquantile range, and variance. These statistics and related analytics are efficiently supported on large databases by OLAP queries of commercial DBMS, and can be visualized via histograms, quantile plots, and box plots.

DM methods, and in particular clustering, can also be used to gain a better understanding of the structure of the data in preparation for the actual mining tasks. The natural outcome of the data exploration step is that the analyst achieves a better understanding of which DM algorithms should be applied to the data. However, before such an algorithm can be applied, one or several of the following preprocessing steps are frequently needed:

- *Eliminating Missing Values:* Special symbols called *nulls* are normally used to represent information in the collected dataset. In certain cases, the DM algorithm can treat nulls as any other value and produce satisfactory results. In many cases, however, this approach results in errors or decreased accuracy. Thus, null values must be eliminated, by either removing the table rows containing these null values altogether, or by replacing each null with a probable value derived as the mean of the column (non-null) values or by other probability-based estimation.
- *Discretization of Numerical Values:* Table attributes can either be categorical (i.e., allowing only a small set of values) or numerical (i.e., representing continuous variables, such as access time or driver's age). Many mining algorithms support continuous attributes or perform the discretization as part of the algorithm itself. But in some situations, such as driver's age in border control, or employee salary in intrusion detection, discretization is best performed on the basis of the analyst's understanding of semantic significance of the various intervals in the domain.
- *Dimensionality Reduction:* In some large applications this step can be used to either reduce the number of rows or columns in the dataset. Sampling and aggregation over a dimension (see OLAP cubes) can be used to reduce the number of rows. While discretization does not reduce the number of rows involved, it can make their representation more compact; compression techniques based on wavelet transforms and principal component analysis methods have also been used for the same purpose. Feature selection methods can instead be used to reduce the number of columns by identifying and eliminating the columns that are not critical for the mining task.

5.2.6.2 Evaluation of DM Results

The quality and usefulness of the results produced must be evaluated for the application at hand. For predictive tasks, the quality of the model is evaluated in terms of the accuracy of the predictions. The usefulness of the results is normally evaluated in terms of: (1) the novelty of the patterns discovered, and (2) their generality. For (1) it is clear that while confirmation of preexisting knowledge produces greater confidence in the accuracy of the DM approach, it does not deliver immediate business value to the company. The business value of new knowledge is often measured by its ability to inspire actions that increase the profitability of the business (e.g., by detecting a previously unknown pattern of frauds or vulnerability in the system). Finally, the benefits grow with the realm of applicability of the new rule or pattern that is learned: a widely applicable business rule is better than a very specific one.

5.3 Web Mining and New Applications

The last decade has seen a significant growth in DM technology and applications; in particular, the Web has tremendously expanded the information available for DM and created new application opportunities. Two new research areas have thus emerged: one is Web mining— that is, the global mining of information retrievable from Web sites. The other is data stream mining—that is, the online mining of information exchanged over the communication lines of the Internet. In the meantime, much progress has been achieved in DM methods, applications, and systems, and success stories from fields as diverse as security, biology, and marketing have inflated perceptions of the power and reach of DM technology. As a result, major information management software vendors are moving aggressively to provide systems that integrate a rich assortment of DM methods and can support the whole mining process discussed in the previous section. A parallel development has been the creation of large datasets for mining that, eventually, can be shared between cooperating companies and agencies, or even published for research purposes. This trend has met with serious concerns about the privacy of individuals being compromised as a result, and the new research area of *privacy-preserving DM* has emerged to address this concern. In this section, we briefly discuss Web mining, security, and privacy.

5.3.1 Web Mining

The Web can be regarded as the largest database available and thus offers endless opportunities for knowledge discovery. However, the discovery process is more difficult here than in traditional databases because of the lack of uniformity between Web sites, and the prevalence of unstructured and semistructured information.

5.3.1.1 Web Usage Mining

Web usage mining is the study of the data generated by the Web surfers' sessions. For instance, users' click streams recorded in the Web logs of the server can be mined for frequent association patterns, which can then be used to improve the design of the Web site and to improve the recommendations and ads presented to the users.

5.3.1.2 Web Content Mining

In many respects, the growth of the web is intimately connected with that of search engines that use crawlers to systematically download pages and documents from the Web and information retrieval (IR) methods to index and rank the Web pages on the basis of their contents, as to promptly return the URL of the relevant pages in response to keyword-based searches. While content mining of multimedia documents represents a topic of growing interest, the mining of text documents remains the cornerstone of search engines, which often combine classical IR techniques with Web-oriented enhancements (e.g., those based on link analysis).

In the keyword-based approach of IR, the document categorization is performed using the terms/words that occur in the documents. Then *stop words* are removed: these are words such as "article" that are very common in the languages and thus have little discriminative value. Additionally, word consolidation is often applied whereby synonyms and words represented by the same stems are merged.

Therefore, a document is now represented by a vector of terms, where a frequency is associated with each term. These frequencies are first normalized relative to the total number of words in the document; it is further scaled down by using a log scale. Finally, to increase the discriminating power of the term-frequency vector, the term frequency is also scaled down in proportion to the number of documents in the collection containing this particular term.

Now, when a query containing a vector of search terms is submitted by a user, the system computes the normalized inner product of the query vector with the term-frequency vector (cosine measure) for each document in the collection and ranks the documents based on the cosine measure.

While the techniques described above apply to traditional text documents and Web pages alike,[4] the latter also contain *tags* and *links*, which are critical pieces of information not found in the former.

Tags define the structure of a Web page and can also be used to pinpoint the semantics of its elements: it can thus provide precise descriptors to Web applications and search engines. While opportunities created by tag mining have yet to be fully exploited, this situation can be reasonably expected to change once the semantic Web progresses toward actual realization [20]. On the other hand, link mining has already made a dramatic impact through Google and social networks.

5.3.1.3 Link Analysis

Link analysis is the activity of mining the Web to determine its structure [21]. The model is a graph where nodes represent pages and arcs represent links. More often than not, links to a page are expressions of interest and act as approval for the page; therefore, the number of links pointing to a page is normally a measure of its quality and usefulness. Thus, for a particular topic, an *authoritative page* is one which is pointed to by many other pages while a *hub* is a page that provides many links to such authorities. In scientific literature, for instance, survey papers play the role of hubs by referring to authorities who introduced or best described particular topics and techniques.

4. In fact, similar distances and techniques are often used to cluster documents in libraries.

A major development in link analysis came with the introduction of the PageRank algorithm, and the extraordinary success of Google that used it to prioritize the relevant pages. Indeed, given the enormity of the Web, keyword searches tend to return large number of URLs, whereas users tend to explore only a limited number of such URLs—typically those displayed in the first page (or first few pages) of their Web browser. Thus, it is essential that the top URLs returned by the search engine are those that are approved by the largest number of most reliable users. Given that approval of a page is denoted by linking to it, in the PageRank algorithm the rank of a page is determined by the number of pages that point to it and their ranks. This recursive definition leads to equations that are akin to those of Markov models where links to a Web page can be seen as transition probabilities. A number of efficient numeric and symbolic techniques are at hand to solve these equations and determine the PageRank of each page discovered by the Web crawler.

5.3.1.4 User Modeling

Link analysis is also used to discover accounts, companies, and individuals in a network of terrorists or other criminal organizations. Link analysis can find hidden links between the accounts and find a cluster of such accounts that are interrelated. These links include deposits to/withdrawals from the account, wire transfers, and shared account owners. Once such groups are identified, additional information about the group plans and goals can be obtained. Similarly, link analysis can also discover noncriminal *social networks* by identifying highly connected subgraphs in the Web. We note that individuals/companies participating in social or terror networks are normally connected in many different ways in addition to links in their Web pages. For instance, these entities may work together, share e-mails or phone calls or text messages, and participate in same recreational activities. Much of this information might not be available in electronic form, or could be protected by privacy. Even with these restrictions the discovery of such networks can be invaluable for homeland security, investigation of criminal activities, and viral marketing.

Viral marketing is based on the idea that recommendations (about movies, and products) spread through word of mouth, e-mail, and other informal contacts (i.e., in ways similar to those in which viral infections spread through the population). Thus, a very cost-effective strategy for marketing is to focus on a small subset of well-connected people in the network and turn them into enthusiastic early adopters who will then propagate the message to the social network. For terrorist cell detection, user modeling takes the form of assigning roles to the entities involved in the cell. This identifies financial sources of the cell and influential entities in the cell. This labeling is based on many different attributes such as node degree, closeness, connectedness, and the degree of separation.

In traditional targeted marketing applications, potential customers are first partitioned into segments through clustering algorithms that group them according to their age, sex, income, education, ethnicity, and interests. Then the responses from samples collected from these segments are used to decide on a marketing strategy. However, in many cases, this approach cannot be used on the Web since most users prefer to order goods and services without revealing personal information. A technique called *collaborative filtering* has been developed to address this problem. In this approach, the system makes recommendations during live customer transaction. For instance, if a customer is buying a book, the system identifies customers, called neighbors, who have interests similar to the customer involved in the transaction (e.g., similar purchase patterns or product ratings) and then recommends other related items frequently bought (or highly rated) by neighbors. Many shopping sites such as Amazon.com and movie rental sites such as Netflix employ this technique with great success. Collaborative filtering can also be used to analyze criminal records and predict the behavior of felons based on the behavior of "neighbor" felons. Thus collaborative filtering has attracted much research and commercial interest [21].

5.3.2 Security Applications

Data mining is now regarded as a key technology in security applications that use DM techniques of increasing complexity and sophistication. For instance, simple OLAP techniques have been used to associate outlier crimes committed by the same perpetrator [22]. The significance of the outliers can be explained by the following example [22]:

1. Of 100 robberies, in 5 the weapon-used attribute had value "Japanese sword," while a gun was used in the other 95. Obviously, the Japanese swords are of more discriminative significance than gun.
2. For the method-of-escape attribute we have 20 different values: "by car," "by foot," and so forth. Although both "Japanese sword" and "by car" both occurred in 5 crimes, they should not be treated equally.
3. The combination of "Japanese sword" and "by car" is more significant than "Japanese sword" and/or "by car" alone.

Thus, the basic assumption is that criminals keep their modus operandi and reoccurrences of uncommon traits in successive crimes suggest that they are probably due to the same perpetrator. An OLAP-based approach can be used to count the occurrences of crimes for different values of attributes such as weapon used and method of escape and for all possible combinations of these attributes. OLAP aggregates also support natural generalizations over attribute values, whereby, for example, city blocks can be generalized into larger quarters, and person-heights might be represented by coarser granularities. As a preprocessing step, clustering was used to select the most significant attributes for analysis and partition their values into discrete intervals. Finally, an outlier score function was used [23] to measure the extremeness level of outliers, and an association method based on these scores was employed to associate criminal incidents. The method was applied to the robbery incident dataset in Richmond, Virginia, in 1998, with promising results [23].

Effective border control represents a top priority for homeland security [14]. Simple classifiers such as those discussed in the previous section represent an important DM technique that can be used in such applications. More sophisticated DM techniques or a combination of techniques can be used in more advanced applications. For instance, telecommunication companies have been early adopters of DM technology and are now sophisticated users of this technology in diverse applications, including identification of network faults, overload detection, customer profiling, marketing, discovery of *subscription* fraud, and *superimposition* fraud. *Subscription* fraud is very common in the business world and is perpetrated when new subscribers use deception in applying for calling cards, telephone accounts, and other services for which they are not qualified, and/or have no intention to ever pay for. Mining techniques such as classification and clustering have been used with success in these and other risk-management situations (e.g., approving loan applications).

Cellular cloning fraud is a classical example of *superimposition* fraud. This fraud is perpetrated by programming into one or more clones the identity of a legitimate cellular phone (whose account is then charged with the calls made from the second one). More traditional techniques based on rules such as "no two calls can be made within a short time from two distant places" have now been successfully enhanced with DM methods based on building a typical behavior profile for each customer and detecting deviation from this typical behavior using techniques such as neural nets and classifiers. An *anomaly detection* approach has also proven effective in *intrusion detection* applications, as discussed next.

The fast growth of the Internet has been accompanied by a surge of activities to attack the security of the network and intrude into its nodes via increasingly sophisticated tricks and approaches. As a result, intrusion detection tools have become indispensable for network administrators. Early systems based on signatures that describe known intrusion scenarios have now evolved into data mining algorithms that first build a model of "normal" behavior and then detect anomalies using techniques such as classifiers, clustering, and outlier detection [2]. Due to the transient and dynamic nature of intrusion these tasks must be performed in a data stream mining environment, and will thus be discussed in more details in the next section.

A third and more elaborate kind of fraud is *collusive agent fraud* that is perpetrated when several agents collaborate in fraudulent or illegal activities. An example of this is the laundering of dirty money resulting from drug trafficking. Of more recent interest is the analysis of funding mechanisms for terrorism—e.g., the situation where charity money is used for weapons and terroristic training. The approach used aims at identifying suspicious and otherwise unusual electronic transactions while minimizing the number of "false positives" returned by the system. Traditional methods based on simple rules written by experts often lack precision and cannot cope with new money laundering schemes continuously devised by the criminals. Relational data mining methods for the discovery of rare patterns have been used to overcome these problems [24]. Link analysis can also play a major role in detecting large networks of colluding agents. For instance, the following quote is taken from Wikipedia [25].[5]

> Data mining has been cited as the method by which the U.S. Army unit Able Danger supposedly had identified the September 11, 2001 attacks leader, Mohamed Atta, and three other 9/11 hijackers as possible members of an al Qaeda cell operating in the U.S. more than a year before the attack.

Able Danger (a classified military intelligence program, created in 1999, against transnational terrorism—specifically al Qaeda) relied on link analysis techniques to associate publicly available information with classified information in an attempt to make associations between individual members of terrorist groups. Whether factual or overstated, Able Danger's reported success in investigating social (or criminal) networks by combining secret information with public-domain information has added to the fear about the government breaching the privacy of individuals by mining their massive databases.[6] An even more pressing concern is that malicious users and organizations could employ similar mining techniques and public-domain databases for nefarious purposes. Propelled by these concerns, privacy-preserving DM has emerged as a very active topic of current research.

---

5. According to statements by Lt. Col. Anthony Shaffer and those of four others, Able Danger had identified the September 11, 2001, attack leader Mohamed Atta, and three of the 9/11 plot's other 19 hijackers, as possible members of an al Qaeda cell linked to the 1993 World Trade Center bombing. This theory was heavily promoted by Representative Curt Weldon, vice chairman of the House Armed Services and House Homeland Security committees. In December 2006, an investigation by the U.S. Senate Intelligence Committee concluded that those assertions were unfounded.... However, witness testimony from these hearings is not publicly available.
6. Indeed in the United States, Able Danger is only one (and hardly the largest) of the many classified government programs that use data mining for homeland security applications [28].

Propelled by these concerns, privacy-preserving DM has emerged as a vibrant area of new technological developments which, besides K-anonymity and other techniques, include data perturbation and secured multiparty computations, discussed next. Besides the obvious privacy issues that DM applications for Homeland Security face in the data acquisition phase, the sharing of data between security agencies is also stifled by statutes and organizational cultures. Therefore, before releasing their data to other agencies, security organizations might decide to sanitize them with the data perturbation techniques used for privacy-preserving data mining [26], since these tend to provide stronger privacy guarantees than K-anonymity. Moving to deeper shades of paranoia, we find intelligence organizations that will never share or reveal their "resources," but often have to join forces to perform specific tasks. Methods of secure multiparty computation address this problem through cryptographic computations. They assure perfect privacy when data owners want to compute a function over the entire data without revealing their sensitive information. For instance, an efficient protocol to securely compute a decision tree from two private databases is discussed in [27].

5.3.3 Privacy-Preserving Data Mining

In this Information Age, a great deal of scientific, medical, and economic information is collected on every aspect of human activity. The benefits of this information can be maximized by sharing it among agencies and researchers. Unfortunately, this can also lead to misuses and the privacy-breach of the citizens.A vivid example of this problem was provided by Latanya Sweeney who, in 2002 [29], started from a large public domain database of medical diagnoses and procedures performed on people. The database had been sanitized to remove information, such as Social Security number, and address, which could be used to identify the actual individuals. However, the person's birthdate, zip code, and sex were still reported in the database. Sweeney bought the voter registration list for Cambridge, Massachussets (56,000 records), listing the birthdate, zip, and sex for each voter, and was thus able to derive the medical record of William Weld, who was state governor at the time. The problem here is clearly identifiable in the fact that birthdate, zip, and sex are a quasi-key: it uniquely identifies 87% of the population in the United States. The concept of *K-anonymity* was thus introduced to avoid the quasi-key problem. *K*-anonymity requires that, at least, $K >> 1$ records share the same values for any combination of attribute values. In order to make a database publishable, while avoiding the quasi-key problem, researchers are now seeking to provide techniques to introduce modifications in the database to achieve *K*-anonymity, while still preserving the statistics of the datasets to enable accurate mining. However, *K*-anonymity is only a necessary condition that must be satisfied when seeking privacy-preserving data mining. The problem of finding sufficient conditions is much harder inasmuch as it has been shown that the mere publication of mining results, such as association rules or decision trees, can compromise privacy in certain situations. This problem provides a fertile ground for much current research.

5.4 Mining Data Streams

The tremendous growth of the Internet has created a situation where a huge amount of information is being exchanged continuously in the form of data streams. A store-now-and-process-later approach is often impractical in this environment, either because the data rate is too massive, or because of applications' real-time (or quasi-real-time) requirements. A new generation of information management systems, called data stream management systems (DSMSs) are thus being designed to support advanced applications on continuous data streams. For instance, in a fraud detection system for Web purchases, clicks have to be marked as spam or nonspam as soon as they arrive so that the rest of the system does not get affected by the fraudulent clicks. We next discuss intrusion detection, and network monitoring which represent key application areas for this technology.

5.4.1 Intrusion Detection and Network Monitoring

The threat to homeland security posed by intrusions became obvious in 2007 when cyber war against Estonia broke out after the "Bronze Soldier" statue was removed from the Russian World War II memorial in the center of Tallinn. This cyber assault disabled most of the Web services of government ministries, political parties, newspapers, and banks in this Baltic nation, whereby NATO scrambled to the rescue by sending its top terrorism experts [30]. The crisis was triggered by hackers who, using malware,[7] penetrated masses of computers and launched synchronized bursts of requests to overwhelm computer servers and bring them to a halt. This distributed denial of service attack illustrates that these "virtual" cyber attacks can jeopardize the infrastructure of whole countries. Therefore, intrusion detection systems (IDSs) are now widely used to protect computer networks from cyber attacks by monitoring audit data streams collected from networks and systems for clues of malicious behavior. The misuse detection techniques of more traditional IDSs detect intrusions by directly matching them against known patterns of attacks (called signatures or rules). These methods have a low rate of false alarms, but they cannot protect the systems from new patterns of attack. Moreover, the process of learning and encoding signatures accurately is slow and laborious and prevents a fast response to attacks. This problem can been addressed effectively by using classifiers and other predictive methods that learn from instances of IP logs labeled as "normal" or "intrusive" [31].

7. A generic name for evil software—http://en.wikipedia.org/wiki/Malware.

More recently approaches based on anomaly detection have found interesting uses in IDS. These approaches focus on modeling the regular patterns of normal user activities; then user activities that deviate from normal behavior patterns are regarded as possible intrusion activities (by malware disguising as the user). While statistical techniques can detect anomalous user behavior with only limited accuracy, the density-based clustering algorithm DBSCAN was proposed in [32] to cluster the salient parameter of past behavior for the user, as to enable sudden changes in such behavior to be detected as outliers in the data. However, gradual shifts in user behavior should be allowed without raising false alarms. To incorporate the new data without requiring recomputation from scratch, incremental clustering algorithms are used in [32], making the approach more suitable for real-time computations on data streams.

5.4.1.1 Data Stream Management Systems

Since data-intensive applications often involve both streaming data and stored data (e.g., to validate a requested credit card transaction against historical account data), the approach taken by most general purpose DSMSs is that of generalizing database languages and their technology to support continuous queries on data streams. Research in this vibrant area has led to many new techniques for response time/memory optimization, synopses, and quality of service [33–35]. DSMSs have now moved beyond the research phase [36–39] and are used in many important application areas, including publish/subscribe, traffic monitoring, sensor networks, and algorithmic trading. Unfortunately, DSMSs cannot yet support mining functions and queries and research advances in two main areas are needed to correct this situation by:

- Enriching database query language with the power and extensibility to express generic data stream mining functions;
- Taming the computational complexity of mining queries by developing faster data stream mining algorithms and better synoptic primitives (e.g., windows and sampling) for complex mining tasks such as frequent itemset mining for association rules.

(The goal of supporting mining queries in database management systems proved very difficult to achieve because of technical challenges—discussed next—since they help clarify similar problems faced by DSMS on the first point above.) Then we address the second item and discuss "fast and light" mining algorithms for mining data streams.

5.4.2 Data Mining Systems

In the mid-1990s, DBMS vendors were able to integrate support for OLAP and data warehousing functions into their systems via limited extensions to SQL.

This extraordinary technical and commercial success was promptly followed by abject failure when the same vendors tried to integrate support for data mining methods into their SQL systems. Indeed, performing data mining tasks using DBMS constructs and functions proved to be an exceedingly difficult problem even for DBMSs supercharged with OR-extensions [40]. Efforts to solve this difficult problem developed along two main paths that were described as the high-road approach and the low-road approach in [4].

In their visionary paper, Imielinski and Mannila [4] called for a quantum leap in the functionality and usability of DBMSs to assure that mining queries can be formulated with the same ease of use as current queries in relational DBMSs. The notion of inductive DBMSs was thus born, which inspired much research from the fields of knowledge discovery and databases [5], including a number of proposals for new data mining query languages such as MSQL [41], DMQL [42], and Mine Rule [43]. These proposals made interesting research contributions, but suffered limitations in terms of generality and performance.

DBMS vendors entering the DM market instead opted for a rather low-road approach based on cache-mining. In this approach, the data of interest are first moved from the database into a file or main memory, where the cached data are then mined using the special DM functions from the vendor-provided library. Various enhancements, including graphical user interfaces and integration tools with the DBMS, have also been proposed. For instance, *IBM DB2 intelligent miner* [12] provides a predefined set of mining functions, such as decision-tree classification, neural network–based classification, clustering, sequential pattern mining algorithm, linear regression, and other mining functions, along with selected preprocessing functions. These functions are supported by a combination of Java stored procedures and virtual mining views that enhance the usability of the system, however, other limitations remain. In particular, the system incurs data transfer overhead and does not allow modification or introduction of new mining algorithms (at least not without significant effort). *Oracle Data Miner* supports similar functionality and closer integration with the DBMS, via PL/SQL [11]. Furthermore, it supports more complex data mining functions, such as support vector machines. We next discuss in more detail the Microsoft SQL Server that, among database vendors, claims to achieve the closest integration and best interoperability of the mining functions with the DBMS [13].

5.4.2.1 OLE DB for DM

Microsoft SQL Server supports OLE DB for DM,[8] which views mining models as first-class objects that can be created, viewed, and modified, just like other

8. Stands for object linking and embedding data base for data mining.

database tables. OLE DB for DM maintains schema rowsets, which are special system tables that store meta-data for all defined mining objects. These schema rowsets allow easy management and querying of mining objects. For instance, a "create model" construct, which is similar to the "create table" construct of SQL, can be used to create, say, classifiers. Then, "insert into model" constructs can be used to insert new examples into the model, thus training the classifier just created. Finally, predicting a class of testing tuples requires a special join, called *prediction join*, which takes a trained data mining model and a database table with testing data to classify the tuples in the table. Therefore, SQL is further extended to allow a "prediction join" operator between a mining model and a database table. With these special constructs, OLE DB for DM achieves a closer integration of mining with SQL and the DBMS.

To compensate for the fact that their proprietary DM extensions lack in flexibility and user extensibility, DBMS vendors also allow inspecting and importing/exporting descriptive models through the XML-based representation called PMML (Predictive Model Markup Language). PMML is a markup language proposed to represent statistical and data mining information [44]. Even so, DBMS-based mining systems have only gained limited popularity with respect to the many existing stand-alone DM packages, and analytics software packages such as STATISTICA [10], MATHEMATICA [45], SAS [46], and R [9] (open source) that are now supporting a rich set of DM methods. Among the several stand-alone systems at hand, we next describe the WEKA system [7], from the university of Waikato, which is very popular among data mining researchers.

5.4.2.2 WEKA

The Waikato Environment for Knowledge Analysis is a general-purpose tool for machine learning and data mining, implemented in Java. It supports many algorithms for different data mining needs such as data cleaning, visualization, filtering, classification, clustering, and association rule mining. Furthermore, these algorithms are supported *generically*, that is, independently from the schema used by the tables containing the data. This is achieved via a configuration file, which describes the data to DM functions. The configuration file lists the attributes of the table, their types, and other meta-information. Therefore, the main advantages of WEKA are as follows:

- All algorithms follow standard Java interface for extensibility.
- A large set of mining algorithms for classification, clustering, and frequent itemsets mining are also available.
- A comprehensive set of data preprocessing and visualization tools are provided.

5.4.3 DSMSs and Online Mining

Much research interest has recently emerged on the problem of mining data streams to support a variety of applications including click stream analysis, surveilliance, and fraud detection. These applications require the ability to process huge volumes of continuously arriving data streams and to deliver the results in a timely fashion even in the presence of bursty data. Thus we are faced with twin challenges of providing mining algorithms that are fast and light enough to process bursty streams in a timely fashion, and data stream mining systems with functionalities parallel to those provided by, say, WEKA or Microsoft OLE DB on stored data. To achieve this goal, the power of DSMSs and their query languages must be extended with the ability of expressing and supporting continuous mining queries efficiently, while using the primitives and techniques currently provided by DSMSs, which include buffering, sampling, load shedding, scheduling, windows, punctuation, approximate query answering, and many others.[9]

5.4.4 Stream Mill: An Inductive DSMS

Research on DSMSs has led to the development of many DSMS prototypes [47–49] and DSMSs that are now being used in the commercial world [50]. Most of these systems support some dialect of SQL as their continuous query language and thus inherit the limitations of SQL in terms of their ability to support data mining applications. In fact, these limitations are also restricting DSMSs, such as Aurora/Borealis [39, 48], which provide attractive "boxes and arrows" graphical interface improving the usability, but not the expressive power of the system.

Moreover, the expressive power of SQL and other query languages on data stream applications is further diminished by the fact that only nonblocking queries [33] can be supported on data streams.[10] Furthermore, as shown in [33], nonblocking queries are exactly monotonic queries, but if we disallow the use of nonmonotonic operators (such as EXCEPT), in SQL, we also lose the ability of expressing some of its monotonic queries. Thus, SQL is not complete with respect to nonblocking queries [33]. Fortunately, the following positive result was also proved in [33]: SQL becomes Turing-complete, and also complete with respect to nonblocking queries, once it is extended with user-defined aggregates (UDAs). The Stream Mill system, developed at UCLA [37], builds on this important result by supporting the Expressive Stream Language (ESL) [37], which extends

9. On the other hand, data mining systems on databases make less use of DSMS primitives such as, say, transaction support and query optimization [39].
10. Blocking queries are those where the result can only be returned after all input tuples have been processed, whereas nonblocking queries are those that can return results incrementally as the data arrive.

SQL with the ability of defining UDAs. We now briefly discuss the constructs used in ESL to efficiently support window-based synopses on UDAs.

5.4.4.1 User-Defined Aggregates (UDAs)

Example 5.1 defines a continuous UDA equivalent to a continuous version of the standard AVG aggregate in SQL. The second line in Example 5.1 declares a local table state, where the sum and the count of the values processed so far are kept. Furthermore, while in this particular example state contains only one tuple, it is in fact a table that can be queried and updated using SQL statements and can contain any number of tuples. The INITIALIZE clause inserts the value taken from the input stream and sets the count to 1. The ITERATE statement updates the tuple in state by adding the new input value to the sum and 1 to the count and returns the ratio between the sum and the count as a result.

Example 5.1

Defining the Standard Aggregate Average

```
AGGREGATE avg(Next Real) : Real
{
  TABLE state(tsum Real, cnt Int);
  INITIALIZE :
  {
    INSERT INTO state VALUES (Next, 1);
  }
  ITERATE :
  {
    UPDATE state SET tsum=tsum+Next, cnt=cnt+1;
    INSERT INTO RETURN SELECT tsum/cnt FROM state;
  }
}
```

In addition to the INITIALIZE and ITERATE state ESL also allows the use of a TERMINATE state to specify computation to be performed once the end of the input is detected. Thus, to implement the standard AVG aggregate of SQL the line INSERT INTO RETURN SELECT tsum/cnt FROM is moved from the ITERATE state to the TERMINATE. This change produces a blocking aggregate that can be used ondatabase tables, but not on data streams [33]. Instead UDAs without TERMINATE, such as that of Example 5.1, can be freely applied over data streams, since they are nonblocking. As shown in [51], data mining algorithms can be expressed through UDAs (typically containing 40 lines of code rather than the 6 lines above): in fact, as shown in Section 5.2, all DM algorithms can be naturally viewed as complex aggregates that compute complex statistics over a given dataset. In ESL, the user is not restricted to write aggregates using SQL: external functions coded in a procedural language can also be used, for example, to speed up execution or take advantage of external libraries [37].

5.4.4.2 Windows

Because of memory limitation, only synopses of past data can be kept, such as samples and histograms of most recent arrivals. (ESL provides powerful constructs for specifying synopses, such as windows, slides, tumbles on arbitrary UDAs, not just built-in ones [37].) These synoptic constructs are critical in data stream applications. For instance, in a traffic monitoring application, users might be interested in retrieving every minute of the average traffic speed at each sensor over the last 10 minutes. Such queries require a computing average over a moving window of 10 minutes with a slide of 1 minute [37]. To optimize the execution over these windows, ESL allows windowed UDAs to be customized with a special EXPIRE state that is invoked when a tuple expires out of a window. For instance, a windowed version of AVG aggregate is defined in Example 5.2.

Example 5.2

Windowed Average

```
WINDOW AGGREGATE avg(Next Real) : Real
{
  TABLE state(tsum Real, cnt Int);
  INITIALIZE :
  {
    INSERT INTO state VALUES (Next, 1);
  }
  ITERATE :
  {
    UPDATE state SET tsum=tsum+Next, cnt=cnt+1;
    INSERT INTO RETURN SELECT tsum/cnt FROM state;
  }
  EXPIRE :
  {
    UPDATE state SET tsum=tsum-oldest().Next, cnt=cnt-1;
  }
}
```

Thus, in the EXPIRE state we perform customized delta maintenance for AVG by decrementing the sum with the expiring value (oldest value in the window) and the count with 1, respectively. This delta-maintenance technique results in efficient implementation of many UDAs, as discussed in [37]. Moreover, by combining the basic blocking version and the window version of a given UDA, the ESL compiler derives efficient implementations for logical (time-based) windows, physical (count-based) windows, and other complex synopses based on slides and tumbles [37]. Arbitrary windowed UDAs can then be invoked in ESL using the SQL:2003 OLAP aggregate syntax.

5.4.4.3 Genericity

Unlike the simple aggregates such as AVG, complex mining aggregates are often applied to tables and streams comprised of an arbitrary number of columns. Therefore, like WEKA, mining algorithms implemented in the system should be generically applicable over different data streams. Given a data tuple, WEKA converts it to a homogeneous record that can have an arbitrary number of columns—this technique in general is called *verticalization*, since a table tuple is verticalized into a record. Thus, all the mining algorithms can operate over this generic record. This record is essentially an array of real values, where categorical values are stored as the sequence index of the attained value. For example, if an attribute can get one of three values, such as passenger, workvan, or truck, then internally WEKA stores 1, 2, or 3, for passenger, workvan, or truck, respectively. Date and time columns are converted to real, whereas integer and real are stored as is. Much in the same way as WEKA, data mining algorithms implemented as ESL aggregates should also be generically applicable over various application domains.

5.4.5 Data Stream Mining Algorithms

Although the mining methods used on data streams are conceptually similar to those on databases, the actual algorithms must be significantly modified or replaced with new ones to minimize their computational costs and the number of passes required over the data. The state of the art here has progressed differently for different mining tasks. For instance, while the prediction task is effectively supported using classifier ensembles, the problem of computing frequent itemsets for association rules on data streams is significantly more challenging and has been the focus of many research efforts [52–55]. Windows containing only the most recent data are critical for mining data streams since they provide better performance and adaptability in the presence of *concept shift*. Concept shift (concept drift) denotes an abrupt (gradual) change of the statistical and logical properties of the data, whereby the old classifier stops being an accurate predictor and a new classifier is required. Ensembles of classifiers can provide effective solutions to the concept shift/drift problem and are discussed next.

5.4.5.1 Classifier Ensembles

Continuous adaptation is possible for Bayesian classifiers that use count-based statistics, which can be updated upon the arrival or expiration of each tuple via ESL aggregates with the expire clause, as discussed previously (delta maintenance). However, for most mining algorithms, including decision-tree classifiers, delta maintenance is not possible. For those cases, we instead prefer to build a new model every N tuples. Thus we partition the incoming stream into windows (these are often called tumbling windows to distinguish them from continuously sliding windows used, say, for Bayesian classifiers): we build a new classifier for each window and use an ensemble containing K classifiers to achieve more accurate predictions via advanced techniques, such as *bagging* and *boosting* [56, 57]. Ensemble methods have long been used to improve the predictive accuracy of classifiers built from different samples of the same dataset. However, for data streams, classifiers are continuously built from successive, nonoverlapping windows. The records in the new window are used as training data to build a new classifier. Furthermore, the same training data are also used to assess the accuracy of the old classifiers. Since older classifiers are continuously replaced with the ones built on a new window, a simple adaptation to concept drift is automatically achieved. Whenever a rapid concept shift occurs, this causes a sudden fall in the accuracy of several classifiers, whereby several (or all) of the old classifiers are discharged and we restart by rebuilding a largely (or totally) new ensemble [57].

Using bagging, the decision on how to classify each new arriving tuple is made by combining the votes of each classifier, either by straight majority or by weights based on their accuracy. Therefore, with bagging, the errors of classifiers can impact the weight assigned to their votes, thus impacting the decision phase. However, boosting affects the training phase, since each newly arriving training tuple is assigned a weight that increases with the classification errors incurred by the previous classifiers. In other words, the new classifier is trained to compensate for the errors of its predecessors. Boosting ensembles can learn fast and produce accurate answers using weak learners, such as shallow decision-tree classifiers that are constructed in fast and light computations [56]. However, boosting methods are known to be very sensitive to noise.

Stream Mill allows ensemble methods, based on either bagging or boosting, to be applied over arbitrary classifiers implemented as UDAs.

For instance, Figure 5.4 shows the generic support for weighted bagging in Stream Mill. The training stream is first fed to a UDA, named "classifier building," which learns the next classifier model to be stored with other models. The training stream is also sent to a "classification" UDA that predicts the class for each tuple using each of the existing classifiers in the ensemble. These predictions are then used to assign weights to the classifiers (based on accuracy) for the next window of testing tuples. The newly arriving testing tuples are first classified using

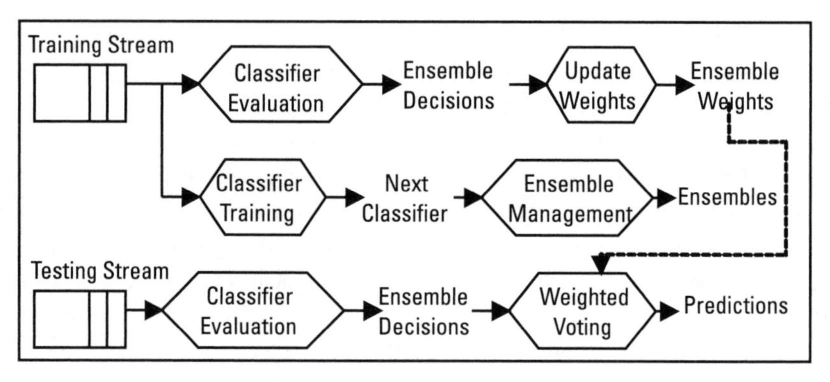

Figure 5.4 Generalized weighted bagging.

each of the classifiers from the ensemble. Then, a weighted voting scheme is employed to determine the final classification of the tuples. The general flow of data tuples, both training and testing, does not depend on the particular classification algorithm. In fact, only the boxes labeled "classifier building" and "classification task" are dependent on the particular classifier. Thus, any classification algorithm that provides implementation for these two "boxes" can be used as a base classifier for the weighted bagging approach.

5.4.5.2 Other Mining Tasks

Other mining tasks in addition to classification are also supported by the Stream Mill system. For instance, online frequent itemsets mining algorithms are also supported in the Stream Mill system. Research over online frequent itemsets mining has led to many algorithms, for instance, Moment [53] and CanTree [34]. However, these algorithms have performance problems when dealing with large windows. Therefore, Stream Mill uses the IMLW algorithm to check and discover efficiently frequent itemsets over large sliding windows [55]. Mining of large sliding windows leads to more reliable results as compared with small sliding windows or large tumbling windows, since both of the latter cases suffer from lack of statistical support. Furthermore, IMLW can handle large itemsets and can also maintain the frequency of any interesting patterns, not just the ones that are frequent ones. Stream Mill fully supports this algorithm by allowing user-specified confidence and support thresholds. Furthermore, the user can also specify a set of itemsets that should always be maintained.

For clustering, Stream Mill also supports a stream-oriented version of *K*-means and DBSCAN [58, 59], both tumbling and continuous window settings. Unlike *K*-means that tend to produce spherical clusters around centroids, DBSCAN is a density-based method that can discover clusters of arbitrary shape; as such, it is quite useful in detecting the spreading of viral diseases and other trajectories. Both algorithms can be expressed quite naturally using ESL UDAs.

Sequence queries represent a very useful tool for identifying sequential patterns in time series and data streams. As demand for sequence pattern queries has grown in both database and data stream applications, SQL standard extensions are being proposed by collaborating DBMS and DSMS companies [60]. These extensions are based on allowing Kleene-closure expressions in the SQL query. These constructs, and their supporting query optimization techniques, were first introduced in SQL-TS [61], which is currently supported in Stream Mill [62].

5.5 Conclusions

The explosive growth of data that are available in the modern Information Age has created a fertile application ground for data mining (DM) technology—that is methods and techniques for knowledge discovery from databases. The DM technology is powerful and it has been successfully used in a broad spectrum of advanced applications from business to science to security. However, DM remains a complex interdisciplinary field that combines techniques from different areas and requires significant application domain expertise for a successful deployment in real-world applications. In this chapter, we have provided an up-to-date introduction to DM designed for researchers and technical experts coming from other fields. Thus, we have first discussed the more established DM methods and core techniques, including classification, association, and clustering. Then, turning our attention to the latest trends and applications, we discussed Web mining and data stream mining.

References

[1] G. Piatetsky-Shapiro and W. J. Frawley (eds.), *Knowledge Discovery in Databases.* AAAI/MIT Press, 1991.

[2] J. Han and M. Kamber. *Data Mining, Concepts and Techniques.* Morgan Kaufmann, 2001.

[3] J. M. ˙Zytkow and W. Klosen. "Interdisciplinary contributions to knowledge discovery," in W. Klosen and J. M. Żytkow (eds.), *Handbook of Data Mining and Knowledge Discovery,* pp. 22–32. Oxford University Press, 2002.

[4] T. Imielinski and H. Mannila. "A database perspective on knowledge discovery." Communications of the *ACM*, 39(11):58–64, 1996.

[5] C. Zaniolo. "Mining databases and data streams with query languages and rules. Keynote paper." *KDID 2005: Knowledge Discovery in Inductive Databases, Fourth International Workshop.* Volume 3933 of *Lecture Notes in Computer Science*, pp. 24–37. Springer, 2006.

[6] C. R. Rao, E. J. Wegman, and J. L. Solka. *Handbook of Statistics, Volume 24: Data Mining and Data Visualization (Handbook of Statistics).* North-Holland Publishing Co., 2005.

[7] WEKA 3: Data Mining with Open Source Machine Learning Software in Java. http://www.cs.waikato.ac.nz. August 2005.

[8] Clementine. http://www.spss.com/clementine. August 2005.

[9] The R Project. http://www.r-project.org/. December 2007.

[10] Statistica. http://www.statsoft.com/products/products.htm. December 2007.

[11] ORACLE. Oracle Data Miner Release 10gr2. http://www.oracle.com/technology/products/bi/odm, March 2007.

[12] IBM. DB2 Intelligent Miner. http://www.306.ibm.com/software/data/iminer. August 2005.

[13] Z. Tang, J. Maclennan, and P. Kim. "Building data mining solutions with OLE DB for DM and XML analysis." *SIGMOD*, 34(2):80–85, 2005.

[14] R. Skinner. Automated Commercial Environment. http://www.dhs.gov/xoig/assets/mgmtrpts/oig_06-56_aug06.pdf. December 2007.

[15] P.-N. Tan, M. Steinbach, and V. Kumar. *Introduction to Data Mining (First Edition)*. Addison-Wesley, 2005.

[16] R. Quinlan. C4.5 Decision Tree Classifier. http://www.rulequest.com/personal/c4.5r8.tar.gz. June 2005.

[17] J. Han, J. Pei, and Y. Yin. "Mining frequent patterns without candidate generation." *SIGMOD*, 2000.

[18] D. Shasha and Y. Zhu. *High Performance Discovery in Time Series: Techniques and Case Studies (Monographs in Computer Science)*. Springer Verlag, 2004.

[19] J. Gao and P.-N. Tan. "Converting output scores from outlier detection algorithms into probability estimates." *ICDM '06: Proceedings of the Sixth International Conference on Data Mining*, pages 212–221. IEEE Computer Society, 2006.

[20] L. Kagal, et al. "Using semantic web technologies for policy management on the web." *AAAI*, 2006.

[21] S. Chakrabarti. *Mining the Web: Discovering Knowledge from Hypertext Data*. Morgan-Kaufmann, 2002.

[22] S. Lin and D. E. Brown. "An outlier-based data association method for linking criminal incidents." *Proceedings of Third SIAM International Conference on Data Mining*. San Francisco, CA, May 1–3, 2003.

[23] S. Lin and D. E. Brown. "Criminal incident data association using the OLAP technology." H. Chen et al., (eds.), *Intelligence and Security Informatics*, pp. 13–26. Springer, 2003.

[24] O. Maimon and L. Rokach, (eds.), Chapter 57 of *The Data Mining and Knowledge Discovery Handbook*. Springer, 2005.

[25] Wikipedia on Data Mining. http://en.wikipedia.org/wiki/data mining. August 2006.

[26] R. Agrawal and R. Srikant. "Privacy preserving data mining." *ACM SIGMOD International Conference on Management of Data, SIGMOD 2000*, Dallas, TX, 2000, pp. 439–450.

[27] Y. Lindell and B. Pinkas. "Privacy preserving data mining." *Journal of Cryptology*, Vol. 15, No. 3, 2002, pp. 177–206.

[28] J. Seifert. "Data mining and homeland security: An overview." Technical report, Congressional Research Service: Resources, Science, and Industry Division, 2007.

[29] L. Sweeney. "Achieving K-anonymity privacy protection using generalization and suppression." *International Journal of Uncertainty, Fuzziness, and Knowledge-Based Systems*, 2002.

[30] M. Sweney. "Is Eastern Europe's Cyberwar the Shape of Things to Come?" http://blogs.guardian.co.uk/news/2007/05/eastern europes.html. May 17, 2007.

[31] W. Lee and S. J. Stolfo. "A framework for constructing features and models for intrusion detection systems." *ACM Transactions on Information and System Security*, 3(4):227–261, 2000.

[32] S.-H. Oh and W.-S. Lee. "Anomaly intrusion detection based on dynamic cluster updating," Z.-H. Zhou, H. Li, and Q. Yang (eds.), *Advances in Knowledge Discovery and Data Mining, Eleventh Pacific-Asia Conference: PAKDD 2007*, pp. 737–744. Springer, 2007.

[33] Y.-N. Law, H. Wang, and C. Zaniolo. "Data models and query language for data streams." *VLDB*, pp. 492–503, 2004.

[34] B. Babcock, et al. "Models and issues in data stream systems." *PODS*, 2002.

[35] N. Tatbul, et al. "Load shedding in a data stream manager." *VLDB*, 2003.

[36] A. Arasu, S. Babu, and J. Widom. "Cql: A language for continuous queries over streams and relations." *DBPL*, pp. 1–19, 2003.

[37] Y. Bai, H. Thakkar, C. Luo, H. Wang, and C. Zaniolo. "A data stream language and system designed for power and extensibility." *CIKM*, pp. 337–346, 2006.

[38] C. Cranor, et al. "Gigascope: A stream database for network applications." *SIGMOD*, pp. 647–651. ACM Press, 2003.

[39] D. Abadi et al. "Aurora: A new model and architecture for data stream management." *VLDB*, 12(2):120–139, 2003.

[40] S. Sarawagi, S. Thomas, and R. Agrawal. "Integrating association rule mining with relational database systems: Alternatives and implications." *SIGMOD*, 1998.

[41] T. Imielinski and A. Virmani. "MSQL: A query language for database mining." *Data Mining and Knowledge Discovery*, 3:373–408, 1999.

[42] J. Han, et al. "DMQL: A data mining query language for relational databases." *Workshop on Research Issues on Data Mining and Knowledge Discovery (DMKD)*, pp. 27–33. Montreal, Canada, June 1996.

[43] R. Meo, G. Psaila, and S. Ceri. "A new SQL-like operator for mining association rules." *VLDB*, pp. 122–133. Bombay, India, 1996.

[44] Data Mining Group (DMG). Predictive Model Markup Language (PMML). http://sourceforge.net/projects/pmml. August 2005.

[45] Mathematica. http://www.wolfram.com/products/mathematica/index.html. December 2007.

[46] SAS: Statistical Analysis Software. http://www.sas.com/technologies/analytics/statistics/stat/. August 2005.

[47] A. Arasu and J. Widom. "Resource sharing in continuous sliding-window aggregates." *VLDB*, pp. 336–347, 2004.

[48] D. Abadi et al. "The design of the Borealis stream processing engine." CIDR, 12(2):120–139, 2005.

[49] S. Chandrasekaran et al. "Telegraphcq: Continuous dataflow processing for an uncertain world." *CIDR*, 2003.

[50] Stream Base. http://www.streambase.com/. September 2007.

[51] H. Wang and C. Zaniolo. "ATLaS: A native extension of SQL for data mining." *SIAM International Conference on Data Mining (SDM)*, San Francisco, CA, May 5, 2003.

[52] W. Cheung and O. R. Zaiane. "Incremental mining of frequent patterns without candidate generation or support." *DEAS*, 2003.

[53] Y. Chi, H. Wang, P. S. Yu, and R. R. Muntz. "Moment: Maintaining closed frequent itemsets over a stream sliding window." *Proceedings of the 2004 IEEE International Conference on Data Mining (ICDM '04)*, November 2004.

[54] C.K.-S. Leung, Q. I. Khan, and T. Hoque. "CanTree: A tree structure for efficient incremental mining of frequent patterns." *ICDM*, 2005.

[55] B. Mozafari, H. Thakkar, and C. Zaniolo. "Verifying and mining frequent patterns from large windows over data streams." *International Conference on Data Engineering (ICDE)*, 2008.

[56] F. Chu and C. Zaniolo. "Fast and light boosting for adaptive mining of data streams." *PAKDD*, Vol. 3056, 2004.

[57] H. Wang, et al. "Mining concept-drifting data streams using ensemble classifiers." *SIGKDD*, 2003.

[58] M. Ester, et al. "A density-based algorithm for discovering clusters in large spatial databases with noise." *Second International Conference on Knowledge Discovery and Data Mining*, pp. 226–231, 1996.

[59] C. Luo, H. Thakkar, H. Wang, and C. Zaniolo. "A native extension of SQL for mining data streams." *SIGMOD*, pp. 873–875, 2005.

[60] F. Zemke, et al. "Pattern matching in sequences of rows." Technical report, Oracle and IBM, 2007.

[61] R. Sadri, C. Zaniolo, A. Zarkesh, and J. Adibi. "Optimization of sequence queries in database systems." *PODS*. Santa Barbara, CA, May 2001.

[62] Y. Bai, et al. *Stream Data Management*. Kluwer, 2004.

6

Private Information Retrieval: Single-Database Techniques and Applications[1]

Rafail Ostrovsky and William E. Skeith III

6.1 Introduction

A single-database private information retrieval (PIR) scheme is a game between two players: a user and a database. The database holds some public data (for concreteness, an n-bit string). The user wishes to retrieve some item from the database (such as the ith bit) without revealing to the database which item was queried (i.e., i remains hidden). We stress that in this model the database data is public (such as stock quotes) but centrally located; the user, without a local copy, must send a request to retrieve some part of the central data.[2] A naive solution is to have the user download the entire database, which of course preserves privacy. However, the total communication complexity in this solution, measured as the number of bits transmitted between the user and the database, is n. Private information retrieval protocols allow the user to retrieve data from a public database with communication strictly smaller than n (i.e., with smaller communication than just downloading the entire database). The application to homeland security is that PIR allows a classified machine to retrieve data from a public or unclas-

1. This material is based on a plenary talk presented at PKC-2007.
2. PIR should not be confused with a private-key *searching on encrypted data* problem, where the user uploads his own encrypted data to a remote database and wants to privately search over that encrypted data without revealing any information to the database. For this model, see the discussion in [1, 2] and references therein.

sified source (or a network with a lower classification level) without revealing to that source what information is being sought. This allows retrieval of data from lower classification security levels and has applications to national security.[3]

6.2 Single-Database Private Information Retrieval

PIR was introduced by Chor, Goldreich, Kushilevitz, and Sudan [3] in 1995 in a setting in which there are many copies of the same database and none of these copies are allowed to communicate with each other. In the same paper, Chor et. al. [3] showed that single-database PIR does not exist (in the information-theoretic sense). Nevertheless, two years later (assuming a certain secure public-key encryption) Kushilevitz and Ostrovsky [4] presented a method for constructing single-database PIR. The communication complexity of their solution is $O\left(2^{\sqrt{\log n \log\log N}}\right)$, which for any $\epsilon > 0$ is less then $O(n^{\epsilon})$. Their result relies on the algebraic properties of the Goldwasser-Micali public-key encryption scheme [5]. In 1999, Cachin, Micali, and Stadler [6] demonstrated the first single-database PIR with polylogarithmic communication under the so-called ϕ-hiding number-theoretic assumption. Chang [7], and Lipmaa [8] showed $O(\log^2 n)$ communication complexity PIR protocols (with a multiplicative security parameter factor) using a construction similar to the original [4] but replacing the Goldwasser-Micali homomorphic encryption with the Damgård and Jurik variant of the Pailler homomorphic encryption [9]. Gentry and Ramzan [10] also showed the current best bound for communication complexity of $O(\log^2 n)$ with an additional benefit that, if one considers retrieving more than one bit, and in particular many consecutive bits (which we call blocks), then the ratio of block size to communication is only a small constant. The scheme of Lipmaa [8] has the property that, when acting on blocks, the ratio of block size to communication actually approaches 1, yet the parameters must be quite large before this scheme becomes an advantage over that of [10]. In general, the issue of *amortizing* the cost of a PIR protocol for many queries has received a lot of attention. We discuss it separately in the next section.

All the works mentioned above exploit some sort of algebraic properties, often coming from homomorphic public-key encryptions. In the work of [11], Kushilevitz and Ostrovsky have shown how to construct single-database PIR without the use of any algebraic assumptions, and instead rely on the existence of one-way trapdoor permutations. However, basing the protocol on more minimal assumptions comes with a performance cost: they show how to achieve $\left(n - O\left(\frac{n}{k} - k^2\right)\right)$ communication complexity and, additionally, the protocol requires more than one round of interaction.

3. For more discussion on this topic, see [12] and www.stealthsoftwareinc.com/technologies.html.

In this survey, we give the main techniques and ideas behind all these constructions (and in fact show a generic construction from any homomorphic encryption scheme with certain properties) and attempt to do so in a unified manner.

6.2.1 Amortizing Database Work in PIR

Instead of asking to retrieve blocks, one can ask what happens if one wants to retrieve k out of n bits of the database (not necessarily consecutive). Indeed, this was considered by Ishai, Kushilevitz, Ostrovsky, and Sahai [13]. In this setting, in addition to communication complexity (of retrieving k out of n bits) there is another important consideration: the total amount of *computation* needed to be performed by the database to compute all k PIR answers. Observe that for a single PIR query, the amount of computation required by the database must be linear, since the database's function for answering a query must involve every bit of the database: if it is independent of any bit, then the database can safely deduce that this bit is not among those being retrieved, violating the user's privacy. Now, what is the total computation required to retrieve k different bits? A naive solution is to just run one of the PIR solutions k times. It is easy to see that using *hashing* one can do better: the user, with indices $i_1,\ldots, i_k$, picks at random a hash function h that sends all n entries of the database to k buckets and where the selection of h is made independently from $i_1,\ldots, i_k$. The user sends h to the database. Note that the expected size of each bucket is about n/k. The database partitions its database into buckets according to h (that is gets from the user), and treats every bucket as a new "tiny" database. For an appropriate choice of a hash family, this ensures that with probability $1–2^{-\Omega(\sigma)}$, the number of items hashed to any particular bucket is at most $\sigma \log k$. Now the user can apply the standard PIR protocol $\sigma \log k$ times to each bucket. Except for the $2^{-\Omega(\sigma)}$ error probability, the user will be able to get all k items. Note that the cost is much smaller than the naive solution. In particular, counting the length of all PIR invocations, the total size of all databases on which we run standard PIR is $\sigma \log k \cdot n$, instead of the naive kn. This idea is developed further, and in fact the error probability is removed, and better performance is derived via explicit *batch codes* [13] instead of hashing.

Note, however, that this approach requires that it is the *same* user that is interested in all k queries. What happens if the users are different? In this case, assuming the existence of *anonymous communication*, nearly optimal PIR in all parameters can be achieved in the multiuser case [14].

6.2.2 Connections: Single-Database PIR and OT

Single-database PIR has a close connection to the notion of oblivious transfer (OT), introduced by Rabin [15]. A different variant of oblivious transfer, called

1-out-of-2 OT, was introduced by Even, Goldreich, and Lempel [16] and, more generally, 1-out-of-n OT was considered in Brassard, Crepeau, and Robert [17]. Informally, 1-out-of-n OT is a protocol for two players: a *sender* who initially has n secrets $x_1, \ldots, x_n$ and a *receiver* who initially holds an index $1 \leq i \leq n$. At the end of the protocol the receiver knows x_i but has no information about the other secrets, while the sender has no information about the index i. Note that OT is different from PIR in that there is no communication complexity requirement (beyond being polynomially bounded) but, on the other hand, "secrecy" is required for *both* players, while for PIR it is required only for the user. All oblivious transfer definitions are shown to be equivalent [18]. As mentioned, communication-efficient implementation of 1-out-of-n OT can be viewed as a single-server PIR protocol with an additional guarantee that only one (out of n) secret is learned by the user and the remaining $n-1$ remain hidden. In [4], it is noted that their protocol can also be made into a 1-out-of-n OT protocol,[4] showing the first 1-out-of-n OT with sublinear communication complexity. Naor and Pinkas [20] have subsequently shown how to turn any PIR protocol into 1-out-of-n protocol with one invocation of a single-database PIR protocol and logarithmic number of invocations of 1-out-of-2 OT. Di Crescenzo, Malkin, and Ostrovsky [21] showed that any single-database PIR protocol implies OT. In fact, their result holds even if the PIR protocol allows the communication from the database to the user to be as big as $n-1$. Thus, [21] combined with [20] tells us that any single-database PIR implies 1-out-of-n OT. In [11], it is shown how to build 1-out-of-n OT based on any one-way trapdoor permutation with communication complexity strictly less than n.

6.2.3 Connections: PIR and Collision-Resistant Hashing

Ishai, Kushilevitz, and Ostrovsky [22] showed that any one-round single-database PIR protocol is also a collision-resistant hash function. Simply pick an index i for the PIR query at random, and generate a PIR query. Such a PIR query is the description of the hash function. The database contents serve as the input to the hash function and the evaluation of the PIR query on the database is the output of the hash function. It is easy to see that the PIR function is both length decreasing and collision resistant. It is length decreasing by the nontriviality of PIR protocol, since it must return an answer with length, which is less than the size of the database. It is collision resistant since if the adversary can find two different databases that produce the same PIR answer, then these two databases

4. 1-out-of-n OT in the setting of *multiple copies of the database* where none of the copies is allowed to talk to each other was treated in [19] and renamed *symmetric private information retrieval* (SPIR), though for single-database PIR, the definition of SPIR is identical to the more established notion of 1-out-of-n OT.

must differ in at least one position, say, j. Finding such a position tells us that $j \neq i$, hence it reveals information about i. This violates the PIR requirement that no information about i should be revealed.

6.2.4 Connections: PIR and Function-Hiding PKE

A classic view of a public-key encryption/decryption paradigm is that of an identity map: it takes a plaintext message m and creates a ciphertext, which can be decrypted back to m. However, in many applications, instead of an identity map, there is a need for a public-key encryption to perform some *secret computation* during encryption. That is, the key-generation algorithm takes as an additional input a function specification $f(\cdot) \in \mathcal{F}$ from some class $\mathcal{F}$ of functions and produces a public key. The resulting public-key is not much bigger than the description of a typical $f' \in \mathcal{F}$, yet the public-key should not reveal which f from $\mathcal{F}$ have been used during the key-generation phase. The encryption/decryption maps m to $f(m)$. The definition becomes nontrivial (in the sense that one cannot push all the work of computing $f(\cdot)$ to the decryption phase) when for all $f \in \mathcal{F}$ it holds that $|f(m)| < |m|$, and we insist that the ciphertext size must be smaller than the size of m.

Any single-round PIR can be used to achieve this notion for the class of encryption functions that encrypts a single bit out of the message, hiding *which* bit they encrypt: simply publish in your public-key both the PIR query and an additional public-key encryption (with small ciphertext expansion, compared to the plaintext, such as [9, 23]). When encrypting the message, first compute the PIR answer, and then encrypt the resulting answer with the public-key encryption. (Some specific PIR constructions do not need this additional layer of encryption.)

What makes the function-hiding PKE notion interesting is that there are many examples of functions beyond PIR-based projection maps. For example, as was shown by Ostrovsky and Skeith [12], one can construct an encryption scheme that takes multiple documents and encrypts only a subset of these documents—only those that contain a set of hidden keywords, where the public-key encryption function does not reveal which keywords are used as selectors of the subset.[5]

6.2.5 Connections: PIR and Complexity Theory

Dziembowski and Maurer have shown the danger of mixing computational and information-theoretic assumptions in the bounded-storage model. The key tool to demonstrate an attack was a computationally private PIR protocol [24]. The

5. Reference [12] discusses additional applications for homeland security privacy-preserving solutions. For applications see: http://www.stealthsoftwareinc.com/technologies.html.

compressibility of NP languages was shown by Harnick and Naor to be intimately connected to computational PIR [25]. In particular, what they show is that if a certain NP language is compressible, then one can construct a single-database PIR protocol (and a collision-resistant hash function) that can be built (in a nonblack-box way) based on any one-way function. Naor and Nissim [26] have shown how to use computational PIR (and oblivious RAMs [2]) to construct communication-efficient secure function evaluation protocols.

There is an interesting connection between zero-knowledge arguments and Single-Database PIR. In particular, Tauman-Kalai and Raz have shown (for a certain restricted class) an extremely efficient zero-knowledge argument (with preprocessing) assuming single-database PIR protocols [27].

Another framework of constructing efficient PIR protocols is with the help of additional servers, such that even if some of the servers leak information to the database, the overall privacy is maintained [28]. The technique of [28] is also used to achieve PIR *combiners* [29], where given several PIR implementations, if some are faulty, they can still be combined into one nonfaulty PIR.

6.2.6 Public-Key Encryption That Supports PIR Read and Write

Consider the following problem: Alice wishes to maintain her e-mail using a storage provider, Bob (such as Yahoo! or Hotmail e-mail account). She publishes a public key for a semantically secure public-key encryption scheme, and asks all people to send their e-mails encrypted under her public-key to the intermediary Bob. Bob (i.e., the storage provider) should allow Alice to collect, retrieve, search, and delete e-mails at her leisure. In known implementations of such services, either the content of the e-mails is known to the storage provider Bob (and then the privacy of both Alice and the senders is lost) or the senders can encrypt their messages to Alice, in which case privacy is maintained, but sophisticated services (such as search by keyword, and deletion) cannot be easily performed by Bob. Boneh et al. [30] (solving the open problem of [31]) have shown how to create a public-key that allows arbitrary senders to send Bob encrypted e-mail messages that support PIR queries over these messages and the ability to modify (i.e., to do PIR writing) Bob's database, both with small communication complexity (approximately $O(\sqrt{n})$). It may be interesting to note, however, that manipulating the algebraic structures of currently available homomorphic encryption schemes cannot achieve PIR writing with communication better than $\Omega(\sqrt{n})$, as shown in the recent work of Ostrovsky and Skeith [32].

6.2.7 Organization of the Rest of the Chapter

In the rest of the chapter we give an overview of the basic techniques of single-database PIR. It is by no means a complete account of all of the literature, but we hope that it rather serves as an introduction, and a clear exposition of the techniques that have proved themselves most useful. We begin with what we feel are the most natural and intuitive settings, which are based upon homomorphic encryption, and we attempt to give a fairly unified and clear account of this variety of PIR protocols. We then move to PIR based on the Φ-hiding assumption, and to a construction based upon one-way trapdoor permutations. Throughout, our focus is primarily on the intuition behind these schemes; for complete technical details, one can of course follow the references. We have attempted to make this as self-contained as possible, however, there are some prerequisites for certain sections. Section 6.4 requires undergraduate-level algebra and some elementary number theories, as does Section 6.6. For an introductory algebra reference, see Herstein's text [33], or for a more advanced reference, see [34]. Section 6.5 also has a very minimal amount of algebraic overhead, but much of this section could be understood with only high school–level algebra. If the reader has questions regarding algebraic notational conventions, please consult [33, 34].

6.3 Background and Preliminaries

Here we'll provide a few basic definitions and outline some basic techniques that will be useful in constructing PIR protocols. We'll begin with modern definitions of security for a cryptosystem since they are quite intuitive and also contain the essential ideas for many other types of protocols (e.g., PIR).

6.3.1 Encryption Schemes

We define a cryptosystem to be a triple of algorithms, $(\mathcal{K},\mathcal{E},\mathcal{D})$, which perform the functions of key generation, encryption, and decryption, respectively. In many respects, we can think of these just as abstract functions, which is the view we'll use for the moment. However, as we'll see shortly, there are a number of other issues with this view (it is necessary that they be efficiently computable functions, and as it turns out they will often be required to involve some amount of randomness). The algorithm $\mathcal{K}$ takes a security parameter $s \in \mathbb{Z}^+$ and outputs public and private information, A_{public} and $A_{private}$. And of course, for every message m in the plaintext set, we have that $\mathcal{D}(\mathcal{E}(m))= m$.

In the last half century, there have been a number of new ideas about what the "right" definition of security is for such a system. Ideally, the definition of security would be that the ciphertext contains no information *whatsoever* about the plaintext. However, as shown by Shannon [35], this can only be achieved by rather impractical means (one must secretly exchange a private-key that is equal in length to the total length of the subsequent secret communication). The definition below tries to capture the essence of perfect security with respect to *what can be computed*, which has much more practical realizations. That is, we would like a cryptosystem to have the property that the ciphertext contains no information about the plaintext with respect to what is *efficiently computable* (i.e., anything that can be efficiently computed from a ciphertext could also be computed *without it*). Note one consequence of this definition for public-key systems: a public-key system satisfying this condition *must* be probabilistic. For if an adversary knows that a message comes from a small set, or if perhaps the message space itself is small, then by computing encryptions of all possible messages, the plaintext in its entirety could be learned from a ciphertext. More generally, if the message space is arbitrary, then for some specific message m_0 the answer to the question "does this ciphertext correspond to the message m_0?" could be efficiently computed, which would already violate our definition. This is in part why the following standard definition involves probability. To put a metric on this new view of security, we'll use the following definition.

Definition 6.1 *A function* $g : \mathbb{N} \longrightarrow \mathbb{R}^+$ *is said to be negligible if for any* $c \in \mathbb{N}$ *there exists* $N_c \in \mathbb{Z}$ *such that* $n \geq N_c \Rightarrow g(n) \leq \frac{1}{n^c}$, *i.e., if g is asymptotically smaller than the reciprocal of any polynomial.*

Definition 6.2 *Let* $(\mathcal{K},\mathcal{E},\mathcal{D})$ *be a cryptosystem as defined above. Consider a game between a challenger C and an adversary A consisting of the following steps:*

1. *C runs the algorithm* $\mathcal{K}(s)$ *on input* $s \in \mathbb{Z}^+$ *and sends* A_{public} *to A. (Note:* A_{public} *may be empty.)*
2. *A repeatedly requests* $\mathcal{E}(m)$, *an encryption of a message m, and C responds correctly. (Note: if the system is public-key, this step is not necessary as A can compute these encryptions without the help of C.)*
3. *A chooses messages* m_0, m_1 *of equal length from the plaintext set and sends both back to C.*
4. *C picks* $b \in \{0,1\}$ *at random, and sends* $c = \mathcal{E}(m_b)$ *to A.*
5. *A outputs a guess* $b' \in \{0,1\}$.

We say that A wins the game if $b' = b$, *and loses otherwise. We define the adversary's advantage in this game to be*

$$\mathrm{Adv}_A(s) = \left| \Pr(b' = b) - \frac{1}{2} \right|$$

where this probability is over all internal randomness in the algorithms. If the function $\mathrm{Adv}_A(s)$ *is negligible in the security parameter s, we say that the cryptosystem* $(\mathcal{K},\mathcal{E},\mathcal{D})$ *is* semantically secure, *or also* IND-CPA secure.

We have introduced probability into our definition, but at the moment it may not be clear what the distribution is. This is where a strict functional model of cryptosystem algorithms breaks down. In general, the key generation and the encryption algorithm involve randomness. If one wishes to maintain the strict functional model, one must extend the function to include a parameter for randomness. In this case, the distribution we analyze would be that which is induced by sampling the random parameter uniformly from its domain, and then evaluating the extended function. If one wishes to avoid the introduction of new parameters for randomness, one can view the function's output (on fixed input) as a random sample from a probability distribution, which is defined by the function and input parameters as above. For our purposes, the functional model will often suffice for the construction of protocols and, hence, we will not always explicitly discuss random parameters as they often confuse notation. For a more detailed, introductory reference, see [36].

It may not be immediately obvious, but indeed, Definition 6.2 says that no function of the plaintext can be efficiently computed from a corresponding ciphertext. For simplicity, consider Boolean functions (i.e., predicates) since any other function could easily be modified to be a predicate. If there exists such a predicate that one can compute better than just guessing, say, with probability ε better than any distribution that is independent of the predicate, then an adversary could win the game in Definition 6.2 as follows. Choose m_0, m_1 so that m_1 satisfies the predicate, but m_0 does not, and then put the ciphertext from the challenger into the algorithm that computes the predicate, and return the result, unchanged. Then the advantage for computing the predicate translates to an advantage in the game from Definition 6.2. So by the converse, if one can not gain an advantage in this game, then no function of the plaintext can be computed with any nonnegligible advantage. Rigorous proof of this is simple, using only the most elementary items from probability. As mentioned above, this definition encapsulates the computational analog of perfect security: with perfect security, the adversary's guess will be completely independent from the ciphertext (no matter what class the algorithm A comes from) and hence $\mathrm{Adv}_A(s)$ would be identically 0, whereas here, we have that as s linearly increases, the adversary's guess becomes exponentially close to being independent of the ciphertext. We, of course, cannot guarantee that the advantage is 0 for every algorithm, since in general the ciphertext contains information about the plaintext, which may be extracted with some extremely small probability. For example, in the case of a public-key system, the ciphertext contains *complete* information about the plaintext, and furthermore, given a plaintext, ciphertext, and randomness, it is efficiently computable to check whether or not they correspond (by using the public-encryption function). So, guesses can be efficiently verified, and hence with some small probability, an algorithm can extract information from a ciphertext—we just demand that this probability is unreasonably small for any adversary. This justifies our definition as the natural extension of perfect security. There are, however, more stringent definitions of security, but they are all essentially the same as the above definition, only with a few added tools given to the adversary.

Note that any *public-key* cryptosystem satisfying Definition 6.2 must clearly use randomness in encryption: otherwise any adversary could trivially win the game with probability 1.

Many cryptosystems have a construction that is algebraic in nature, since many relate their difficulty to standard, number-theoretic problems that are believed to be intractable. As a result, perhaps, it is often the case that various constructions of cryptosystems naturally preserve some sort of algebraic structure of the plaintext set in the ciphertext set. The ability to algebraically manipulate plaintext when only given corresponding ciphertext values is extremely useful, especially with respect to PIR. We'll call such cryptosystems *homomorphic*. More formally, we have the following definition for the case of groups.

Definition 6.3 *Let* $(G_1,\cdot)$, $(G_2,*)$ *be groups. Let* $(\mathcal{K},\mathcal{E},\mathcal{D})$ *be a cryptosystem as defined above, with plaintext set* G_1 *and ciphertext set* G_2. *The encryption scheme is said to be* group homomorphic *if the encryption map* $\mathcal{E}: G_1 \rightarrow G_2$ *has the following property*:

$$\forall a,b \in G_1,\ \mathcal{D}(\mathcal{E}(a \cdot b)) = \mathcal{D}(\mathcal{E}(a) * \mathcal{E}(b))$$

Note that since encryption is, in general, probabilistic, we have to phrase the homomorphic property using $\mathcal{D}$, instead of simply saying that $\mathcal{E}$ is a homomorphism. Equivalently, if $\mathcal{E}$ is onto G_2, one could say that the underlying function of $\mathcal{D}$ is a homomorphism of groups (in the usual sense), with each coset of $ker(\mathcal{D})$ corresponding to the set of all possible encryptions of an element of G_1. Also, as standard notation when working with homomorphic encryption as just defined, we use id_{G_1}, id_{G_2} to be the identity elements of G_1, G_2, respectively.

6.3.2 Private Information Retrieval

Recall that the problem of PIR applies to a setting in which a user U wants to retrieve a particular item from a database X, but wishes to do so without revealing this item to the owner of X. One simple solution is to just request every item in X, but as it turns out, there are much more efficient solutions (in terms of communication). The trivial solution we've just described requires the database owner to communicate n bits of information, and if one demands privacy in the sense of Shannon [35], the trivial solution (with communication $\Omega(n)$) is in fact optimal [3]. However, just as we've seen with encryption, if we relax our definition to what is efficiently computable, we can achieve the computational analog of perfect security with much greater efficiency. This is the motivation for the following definition of a private information retrieval protocol:

A *PIR protocol* consists of the following PPT algorithms.

- $K(s)$, which, on input of a security parameter k, outputs public and private system parameters.
- $Q(i, n)$, which generates a query description $q_i \in \{0,1\}^k$ to retrieve item i from a database of size n.
- $R(q_i, X)$, which responds to a query with some description $r_i \in \{0,1\}^l$.
- $D(r_i)$, which decodes a query response.

Definition 6.4 *(Correctness) We say that a PIR protocol (K, Q, R, D) is* correct *if for any set of system parameters generated by K, the composition of the algorithms yields the correct database item, i.e., for a database* $\{x_j\}_{j=1}^{n}$, *we have that* $D(R(Q(i, n), X)) = x_i$.

Definition 6.5 *(Security) For a PIR protocol (K, Q, R, D), we define security in terms of a game between a challenger C and an adversary A (that holds a database* $\{x_j\}_{j=1}^{n}$*), which consists of the following steps:*

1. *K is run by C on input* $s \in \mathbb{Z}^+$ *and the public parameters are distributed.*
2. *A may repeatedly request the output of* $Q(i,n)$ *for* $i \in [n]$ *of its choice, and C responds with the correct output.*
3. *A sends a pair* (i_0, i_1) *with* $i_0 \neq i_1 \in [n]$ *of its choice to C.*
4. *C randomly selects a bit* $b \in \{0,1\}$ *and sends the output of* $Q(i_b,n)$ *to A.*
5. *C responds with a guess* $b' \in \{0,1\}$.

We define the adversary's advantage in this game to be

$$\mathrm{Adv}_A(s) = \left|\Pr(b' = b) - \frac{1}{2}\right|$$

where this probability is taken over all internal randomness used in the algorithms. If the function $\mathrm{Adv}_A(s)$ *is negligible in s, then we say that the PIR protocol (K, Q, R, D) is secure.*

6.3.3 Balancing the Communication Between Sender and Receiver

Virtually every single-database private information retrieval protocol is somewhat comparable to every other in that they all:

- Adhere to a strict definition of privacy.
- Necessarily have $\Omega(n)$ computational complexity (where n is the size of the database).[6]

6. In order to preserve privacy, the database's computation must involve every database element.

As such, it is the case that the primary metric of value or quality for a PIR protocol is the total amount of *communication* required for its execution. Therefore, it may be useful to examine a somewhat general technique for minimizing communication complexity in certain types of protocols, which we'll be able to apply to single database PIR. Suppose that a protocol $\mathcal{P}$ is executed between a user **U** and a database **DB**, in which **U** should privately learn some function $f(X)$ where $X \in \{0,1\}^n$ is the collection of data held by **DB**. By "privately," we mean that **DB** should not gain information regarding certain details of f. Let $g(n)$ represent the communication from **U** to **DB** and $h(n)$ be the communication from **DB** to **U** involved in the execution of $\mathcal{P}$. So, $g,h:\mathbb{Z}^+ \rightarrow \mathbb{Z}^+$. As a simplifying assumption to illustrate the idea, suppose that:

1. The function of the database $f(X)$ that **U** wishes to compute via the protocol depends only on a single bit of X.
2. g,h can be represented, or at least estimated, by polynomial (or rational) functions in n.

If all of these conditions are satisfied, then we often have a convenient way to take the protocol $\mathcal{P}$, and derive a protocol $\mathcal{P}'$ with lower communication, which will just execute $\mathcal{P}$ as a subroutine. The idea is as follows: since the function of X we are computing is highly local (it depends only on a single bit of X), we can define $\mathcal{P}'$ to be a protocol that breaks down the database X into y smaller pieces (of size n/y) and executes $\mathcal{P}$ on each smaller piece. Then, the desired output is obtained in one of the y executions of $\mathcal{P}$. Such a protocol exhibits total communication $T_n(y) = g(n/y) + yh(n/y)$. It may be the case that this will increase the communication of **U** or **DB**, but will reduce the total communication involved. If indeed all functions are differentiable, as we've assumed, then we can use standard calculus techniques to minimize this function (for any positive n) with respect to y. For example, suppose that the user's communication is linear, and the database's communication is constant. Let $g(n) = rn+s$ and $h(n) = c$, so that $T_n(y) = yc + s + \frac{rn}{y}$. Solving the equation $\frac{d}{dy}T_n(y) = 0$ on $(0,\infty)$ gives

$$y = \frac{\sqrt{crn}}{c}$$

This value of y is easily verified to be a local minimum, and we see that by executing the protocol $\mathcal{O}(\sqrt{n})$ times on pieces of size $\mathcal{O}(\sqrt{n})$ we can minimize the total communication.

More generally, similar techniques can of course be applied when the function f depends on more than one bit of X, as long as there is a uniform way (independent of f) to break down the database X into pieces that contain the relevant bits. These techniques can be applied to more general situations still, in which the function depends on many database locations; however, in this case one needs a method of reconstructing the output from the multiple protocol returns (in our simple example, the method is just selecting the appropriate value from all the returns). Also, for this technique to be of value in such a situation, it is generally necessary to have a uniform way to describe the problem on smaller database pieces.

6.4 Semantically Secure Homomorphic Encryption Schemes: Examples

It may be useful to have a few concrete examples of homomorphic cryptosystems in mind before we examine basic PIR protocols in detail. In this section, we illustrate a few semantically secure, homomorphic systems in light to moderate detail.

6.4.1 Encryption Based on Quadratic Residues

The first proposed cryptosystem satisfying Definition 6.2 was that of Goldwasser and Micali [5], based on the indistinguishability of quadratic residues mod n, where n is a large composite number. This system is group homomorphic with plaintext set $\mathbb{Z}_2$, and ciphertext set $H < \mathbb{Z}_n$, where H has index 2 in $\mathbb{Z}_n^*$. We briefly describe this system next.

6.4.1.1 Quadratic Residuosity Assumption

Define $Q_n : \mathbb{Z}_n^* \longrightarrow \{0,1\}$ as follows: let $y \in \mathbb{Z}_n^*$. Then $Q_n(y) = 0$ if $\exists w \in \mathbb{Z}_n^*$ such that $w^2 = y$, and $Q_n(y) = 1$ otherwise. The y for which $Q_n(y) = 0$ are called quadratic residues mod n(QR), or else we say that y is a quadratic nonresidue mod n (QNR). Denote the subgroup of quadratic residues by

$$R = \left\{x^2 | x \in \mathbb{Z}_n^*\right\}$$

The problem of distinguishing quadratic residues from nonresidues mod n is thought to be the most difficult when $n = pq$, where p,q are primes of equal length. In this case $\mathbb{Z}_n^* \simeq \mathbb{Z}_p^* \times \mathbb{Z}_q^*$. Recall also that for prime p, $Q_p(y)$ is determined by the Legendre symbol $\left(\frac{y}{p}\right)$: y is a QR mod p if $\left(\frac{y}{p}\right) = 1$ and it is a QNR if $\left(\frac{y}{p}\right) = -1$. Finally, remember that the Jacobi (Legendre) symbol is multiplicative: $\left(\frac{y}{p}\right)\left(\frac{y}{q}\right) = \left(\frac{y}{n}\right)$. Putting these facts together, you can see that the Jacobi symbol tells us about $Q_n(y)$ in some cases. Since $\mathbb{Z}_n^* \simeq \mathbb{Z}_p^* \times \mathbb{Z}_q^*$, the only way for $y \in \mathbb{Z}_n^*$ to be a QR mod n is for y mod p to be a QR in $\mathbb{Z}_p^*$ and for y mod q to be a QR in $\mathbb{Z}_q^*$. So, if the Jacobi symbol of $y \in \mathbb{Z}_n$ is -1, you know that either $\left(\frac{y}{p}\right) = -1$ or $\left(\frac{y}{q}\right) = -1$ since $\left(\frac{y}{p}\right)\left(\frac{y}{q}\right) = \left(\frac{y}{n}\right)$, and hence it must be that $Q_n(y) = 1$ since y is not a residue for both p and q. Since the Jacobi symbol is easy to compute from n alone, without the factorization, this seems like a problem for the hardness of the quadratic residuosity assumption, but it is easy to fix, just restricting attention to the subgroup of elements with a Jacobi symbol equal to 1:

$$H = \left\{ y \in \mathbb{Z}_n^* \middle| \left(\frac{y}{n}\right) = 1 \right\}$$

This is perfect, since

$$R < H \quad \text{and} \quad [H : R] = 2$$

So now, we have a very nice, symmetric problem, and it leaves no apparent way to distinguish the quadratic residues from nonresidues without knowing the factorization of n. We can then state the quadratic residuosity assumption as follows, using the terminology established earlier.

Let $x \in H$ be chosen uniformly at random. Then, for all adversaries $A \in PPT$,

$$\left| \Pr[Q_n(x) = A(x)] - \frac{1}{2} \right|$$

is a negligible function in the length of n (i.e., in $\log(n)$).

Building a cryptosystem from this is now simple: define the key-generation algorithm in the obvious way to select an appropriate n, and then for encryption, for $b \in \mathbb{Z}_2$, compute

$$\mathcal{E}(b) = x^2 y^b$$

where $x \in \mathbb{Z}_n^*$ is uniformly random, and $y \in H$ is a fixed quadratic nonresidue. Then of course $\mathcal{E}(0)$ is uniformly random in R as squaring is a 4-to-1 map, and $\mathcal{E}(1)$ is uniformly random in $H \backslash R$. Finally, decryption is a simple process given the factorization of n. So, if we want to make this system public-key, we can just publish (n, y) as the public parameters.

The semantic security of this system is clearly equivalent to the quadratic residuosity assumption. This system can of course be extended to a homomorphic system with plaintext set $\mathbb{Z}_2^k$, for any $k \in \mathbb{Z}^+$, just by stringing together encryptions of single bits to form a ciphertext. This, too, is semantically secure. One can show that if an adversary can distinguish between two messages, then it can distinguish between two bits. Just note that for probability distributions, if D_1 is computationally indistinguishable from D_2, and D_2 is computationally indistinguishable from D_3, then D_1 is also indistinguishable from D_3. Hence, we can construct an intermediate string of distributions between two messages that only differ by one bit at a time, and if the beginning can be distinguished from the end, then at some point two adjacent distributions must be distinguishable, meaning that the adversary can distinguish a single bit, which now directly violates the quadratic residuosity assumption.

Oftentimes, an encryption that is homomorphic over $\mathbb{Z}_2$ is useful, however, for more general purposes, this is clearly not ideal for space efficiency as the ciphertext-to-plaintext size ratio is equal to the security parameter. The following subsection illustrates a few examples of more efficient systems, which still satisfy our strong definitions of security.

6.4.2 The ElGamal Cryptosystem

The ElGamal cryptosystem [37] is based on the hardness of the Diffie-Hellman problem, which seems closely related to the well-known discrete logarithm problem. Below, we describe the Diffie-Hellman problem, build a cryptosystem based upon it (as in [37]), and finally prove its security.

6.4.2.1 The Diffie-Hellman Problem

For a (multiplicative) cyclic group G of order p, the Diffie-Hellman problem is as follows: let g be a generator of G, and let $x, y \in \mathbb{Z}_p^*$. Then the computational Diffie-Hellman problem is to compute g^{xy} given only (g, g^x, g^y). The decisional version of this problem is to distinguish the distribution (g, g^x, g^y, g^{xy}) from (g, g^x, g^y, g^z), where $x, y, z \in \mathbb{Z}_p^*$ are selected uniformly at random.

Next, we describe the cryptosystem of [37]. The correctness of the system is quite straightforward, so we focus our efforts on proving security.

6.4.2.2 Algorithm Descriptions

The plaintext set of this system is the group G, and the ciphertext set is $G \times G$.

- $\mathcal{K}(s)$—This algorithm selects a group G of order p (an s-bit number) where the Diffie-Hellman problem is assumed to be hard, randomly selects a generator g of G, randomly selects $x \in \mathbb{Z}_p^*$, and sets $h = g^x$. The algorithm outputs $A_{public} = (G, g, h)$ and saves x as the private key.
- $\mathcal{E}(m)$—For $m \in G$, set

$$\mathcal{E}(m) = (g^r, mh^r)$$

  where $r \in \mathbb{Z}_p^*$ is uniformly random.
- $\mathcal{D}(c)$—Let $(g^r, mh^r) = (u, v)$ be a ciphertext. Then to decrypt using the private key x, just compute

$$\frac{v}{u^x}$$

  which clearly evaluates to the message m.

6.4.2.3 Security

Now how do we prove security? Clearly recovering a plaintext given a ciphertext is equivalent to the computational Diffie-Hellman problem, as we need to compute the mask $h^r = g^{xr}$, given (g, g^x, g^r). But what about semantic security (Definition 6.2)? This can be proved equivalent to the decisional Diffie-Hellman problem as follows. Suppose that there exists an adversary A that can gain a nonnegligible advantage ϵ in the semantic security game for the ElGamal system. Then one could use A as a subroutine to gain an advantage in distinguishing Diffie-Hellman tuples from random tuples. Proceed as follows, where C is an algorithm that provides us with instances of the decisional Diffie-Hellman problem.

1. Begin interacting with C to obtain a group G and a tuple (g, g^x, g^y, g'), where g' is equal to g^{xy} with probability 1/2, and is uniformly random in G with probability 1/2.
2. Begin interacting with A, sending A the tuple (G, g, g^x) as system parameters for the ElGamal system.
3. A will return two messages $m_0, m_1 \in G$.
4. Select $b \in \{1,0\}$ at random and create a "potential" encryption of m_b by setting ciphertext $c = (g^y, m_b g')$. Send c to A.
5. A will return a guess b'.
6. If $b' = b$, guess that the tuple we received from C is a Diffie-Hellman tuple, and otherwise, guess that it is not.

If A guesses correctly with probability $(1/2+\epsilon)$, then our guess to C is correct with probability $(1/2+\epsilon/2)$. This can be seen as follows. Our behavior when interacting with A is completely indistinguishable from an actual challenger in the semantic security game, until step 4. In step 4, we give A a "potential" encryption $c=(g^y, m_b g')$ of one of the messages, m_b. Now if $g' = g^{xy}$, then indeed c is a valid ElGamal encryption of m_b, using randomness y, and this case we have emulated the challenger in Definition 6.2 perfectly, and hence A will guess $b' = b$ with probability $1/2+\epsilon$. However, if g' were uniformly random in G, then the value $m_b g'$ is also uniformly random in G, and independent of b. Hence, no matter what A does in this situation, we have $\Pr(b' = b) = 1/2$. So, the probability that our guess is correct for the decisional Diffie-Hellman problem is $\frac{1}{2} + \frac{\epsilon}{2}$, and we have gained a nonnegligible advantage.

6.4.3 The Paillier Cryptosystem

We now present a modern public-key, group homomorphic cryptosystem satisfying the definitions above under a new assumption that can be viewed as an extension/generalization of the hardness of quadratic residues [5], and is also related to the RSA problem [38].

This system is based on an intractability assumption called the "composite residuosity assumption," which is something of a generalization of the hardness of distinguishing quadratic residues, and also can be reduced to the RSA problem [38]. This assumption (which we will abbreviate as CRA) is about distinguishing higher-order residue classes. The Paillier system and its extensions are additively homomorphic, and have a very low ciphertext-to-plaintext ratio, as well as several other desirable properties.

For this system, the proofs of security are actually somewhat easier than those of correctness (and are also quite similar to the proofs of the previous sections). Therefore we focus only on describing the basic algorithms for this system. For a more detailed summary and proofs of correctness, the reader may consult [12].

6.4.3.1 Preliminaries

Let $n=pq$ be an RSA number, with $p<q$. We make the additional minor assumption that $p \nmid q-1$ i.e., that $(n, \varphi(n))=1$. The plaintext for the Paillier system is represented as elements of $\mathbb{Z}_n$ and the ciphertext is represented as elements of $\mathbb{Z}^*_{n^2}$. Note the following:

$$\mathbb{Z}^*_{n^2} \simeq \mathbb{Z}_n \times \mathbb{Z}^*_n$$

This can be proved using nothing more than elementary facts from number theory and group theory. Given this structure of $\mathbb{Z}^*_{n^2}$, it is not hard to see that the factor of the direct product that is isomorphic to $\mathbb{Z}^*_n$ is in fact the *unique* subgroup of order $(p-1)(q-1)$. Let $H< \mathbb{Z}^*_{n^2}$ denote this subgroup of order $(p-1)(q-1)$. Now define G to be the quotient

$$G = \mathbb{Z}^*_{n^2}/H$$

By our above remarks, we have the structure of G to be cyclic of order n: $G \simeq \mathbb{Z}_n$. We are now ready to state the composite residuosity class problem.

6.4.3.2 The Composite Residuosity Class Problem

Let $g \in \mathbb{Z}^*_{n^2}$ such that $\langle gH \rangle = G$ and let w be an arbitrary element in $\mathbb{Z}^*_{n^2}$. Then, since gH generates $G = \mathbb{Z}^*_{n^2}/H$, we have $w = g^i h$ for some $i \in \{0, 1, 2, \ldots, n-1\}$ and $h \in H$. Given g and w, the composite residuosity class problem is simply to find i.

Note that there is also a decisional version of this problem: given w, g as above, and $x \in \{0, \ldots, n-1\}$, determine if $w = g^x h$ for some $h \in H$. This decision version of the problem is clearly equivalent to distinguishing nth residues mod n^2 (which is the special case of $x = 0$) since H is exactly the subgroup of nth residues.

Note also that these problems have several random self-reducibility properties. Any instance of the problem can be converted to a uniformly random instance of the problem with respect to w (just by multiplying by $g^a b^n$, with $a \in \mathbb{Z}_n, b \in \mathbb{Z}_n^*$, and subtracting a from the answer). Also, the problem is self-reducible with respect to the generator g. In fact, one can show that any instance with generator g can be transformed into an instance with generator g'. So, the choice of g has no effect on the hardness of this problem. If there are any easy instances, then all instances are easy.

Now that we have formalized the hardness assumption, we are prepared to build a cryptosystem. The following subsection contains descriptions of basic algorithms for such a cryptosystem.

6.4.3.3 Algorithm Descriptions

As mentioned before, there are several variants and extensions of this cryptosystem. We state below a variant of the system that we believe to be the clearest and most simple. Let $(\mathcal{K},\mathcal{E},\mathcal{D})$ be the key-generation, encryption, and decryption algorithms, respectively. They are implemented as follows:

- $\mathcal{K}(s)$—This algorithm randomly selects an s-bit RSA number $n=pq$, with $p<q$ and the additional property that $p \nmid q-1$ (which is satisfied with overwhelming probability when p,q are randomly chosen). It outputs n as the public parameters and saves the factorization as the private key.
- $\mathcal{E}(m)$—For a plaintext message $m<n$, choose $r \in \mathbb{Z}_n^*$ at random and set the ciphertext, c, as follows:

  $$c = (1+n)^m r^n \in \mathbb{Z}_{n^2}^*$$

  Recovering m from c is precisely an instance of CRCP, since r^n is a random element in the subgroup H, and the coset $(1+n)H$ generates all of G.

  Note 1: Due to the random self-reducibility of CRCP, $(1+n)$ is just as good of a choice of g as any other. Note 2: although it may seem more natural to choose $r \in \mathbb{Z}_{n^2}^*$, letting $r \in \mathbb{Z}_n^*$ is just as good. (See [12] for details.)
- $\mathcal{D}(c)$—Let ciphertext $c = (1+n)^m r^n \bmod n^2$. To recover the message m, first look at this equation mod n rather than n^2:

  $$c = (1+n)^m r^n \bmod n$$

  which becomes

$$c = r^n \bmod n$$

Now this equation is something familiar: finding r from c is an instance of the RSA problem (since we are given n, which is relatively prime to $\varphi(n)$ and an exponentiation of r mod n). And since the factorization $n = pq$ is known to us, we can just use RSA decryption as a subroutine to recover r. Now that we have r, it is a simple process to obtain m.

To begin, compute r^n mod n^2 and divide c by this value:

$$\frac{c}{r^n} = (1+n)^m \bmod n^2$$

Now use the binomial theorem:

$$(1+n)^m = \sum_{i=0}^{m} \binom{m}{i} n^i$$

Reducing mod n^2 gives us

$$(1+n)^m = \sum_{i=0}^{1} \binom{m}{i} n^i = 1 + mn \bmod n^2$$

So finally, we have

$$m = \frac{\frac{c}{r^n} - 1}{n}$$

6.4.3.4 A Few Words About Extensions to the System

Ivan Damgård and Mads Jurik [9] made a very natural extension to the Paillier system that uses larger groups for its plaintext and ciphertext. This extension works for any $s \in \mathbb{Z}^+$. In the extended system, the plaintext is represented by an element in $\mathbb{Z}_{n^s}$, and the ciphertext is an element of $\mathbb{Z}^*_{n^{s+1}}$. There are two very appealing properties of this system: first, the ratio of plaintext length to ciphertext length approaches 1 as s tends to ∞. Second, just as in the original Paillier scheme, the public and private information can be simply n and its factorization, respectively. You need not share s ahead of time. In fact, the sender of a message can choose s to his or her liking based on the length of the message to be sent. Then, except with negligible probability, the receiver can deduce s from the length of the ciphertext. So, the public (and private) parameters remain extremely simple.

6.5 PIR Based on Group-Homomorphic Encryption

The original work on computational PIR by Kushilevitz and Ostrovsky [4] presents a private information retrieval protocol based upon homomorphic encryption. Such techniques are often very natural ways to construct a variety of privacy-preserving protocols. It is often the case with such protocols based upon homomorphic encryption that, although the protocol is designed with a specific cryptosystem, there is a more fundamental, underlying design that could be instantiated with many different cryptosystems in place of the original, and furthermore this choice of cryptosystem can have a very nontrivial impact on performance. For example, the work of [4] used the homomorphic cryptosystem of Goldwasser and Micali [5] to create a PIR protocol, and, in the following years, many other similar protocols were developed based upon other cryptosystems (e.g., the work of Chang [7], which is based upon the cryptosystem of Paillier [23], and also the work of Lipmaa [8]). However, the method of [4] is actually quite generic, although it was not originally stated in such a way. In this section, we present an abstract construction based upon any group-homomorphic encryption scheme that has [4, 7] as special cases, as well as capturing the work of [8]. Hopefully, this section provides the reader with general intuition regarding private information retrieval, as well as a pleasant way to understand the basics of a moderate amount of the literature in computational PIR.

Let $(\mathcal{K},\mathcal{E},\mathcal{D})$ be a cryptosystem, the symbols representing the key-generation, encryption, and decryption algorithms, respectively. To construct a PIR protocol, we only need a secure cryptosystem (see Definition 6.2) that is homomorphic (see Definition 6.3) over an abelian group, G. If the cryptosystem $(\mathcal{K},\mathcal{E},\mathcal{D})$ has plaintext set G, and ciphertext set G', where G,G' are groups, then we have that

$$\mathcal{D}(\mathcal{E}(a) \star \mathcal{E}(b)) = a * b$$

where $a,b \in G$, and $*,\star$ represent the group operations of G,G', respectively, that is, the cryptosystem allows for oblivious distributed computation of the group operation of G (again, recall that we reduce the equivalence to hold modulo decryption since the encryption algorithm $\mathcal{E}$ must be probabilistic in order to satisfy our requirements for security).

For such a cryptosystem to be of any conceivable use, we of course have that $|G|>1$. Hence, there is at least one element $g \in G$ of order greater than 1. Suppose that $\mathrm{ord}(g)=m$. If the discrete log problem[7] in G is easy (as will often be the case, e.g., when G is an additive group of integers), then we can represent our database as $X=\{x_i\}_{i=1}^{n}$, where each $x_i \in \{0,\ldots, m-1\}$. Otherwise, we just restrict the values of our database to be binary, which is the traditional setting for PIR (i.e., $x_i \in \{0,1\}$ for all $i \in [n]$). As it turns out, a homomorphic encryption protocol alone is enough to create a PIR protocol.

7. We refer to the problem of inverting $\mathbb{Z}$-module action on an abelian group G as the "discrete log problem in G."

6.5.1 Basic Protocols from Homomorphic Encryption

In what follows, we provide a sequence of examples of PIR from homomorphic encryption, each becoming slightly more refined and efficient. Let us suppose that the i^* position of the database is desired by a user $\mathcal{U}$. Keeping the notation established above, let G, G' be groups that correspond to the plaintext and ciphertext of our homomorphic cryptosystem, respectively, and let $g \in G$ be a nonidentity element. As a first attempt at a PIR protocol, a user $\mathcal{U}$ could send queries of the form $Q = \{q_i\}_{i=1}^{n}$, where each $q_i \in G'$ is such that

$$\mathcal{D}(q_i) = \begin{cases} g & \text{if } i = i^* \\ \text{id}_G & \text{otherwise} \end{cases}$$

The database can respond with

$$R = \sum_{i=1}^{n} x_i \cdot q_i$$

using additive notation for the operation of G' (and of G from this point forward) and using the symbol $\cdot$ to represent the $\mathbb{Z}$-module action. Now $\mathcal{U}$ can recover the desired database bit as follows, computing

$$\mathcal{D}(R) = \mathcal{D}\left(\sum_{i=1}^{n} x_i \cdot q_i\right) = \sum_{i=1}^{n} x_i \cdot \mathcal{D}(q_i) = x_{i*} \cdot \mathcal{D}(q_{i*}) = x_{i*} \cdot g$$

and hence $\mathcal{U}$ determines $x_{i^*} = 1$ if and only if $\mathcal{D}(R) = g$, or, in the case where the discrete log is easy in G, $\mathcal{U}$, would compute x_{i^*} as the log of $\mathcal{D}(R)$ to the base g (just by division, in the case of an additive group of integers). This protocol is clearly correct, but it is also easily seen to be private. The only information received by $\mathcal{DB}$ during the protocol is an array of ciphertexts, which by our assumptions on the cryptosystem comes from (computationally) indistinguishable distributions, and hence contains no information that can be efficiently extracted. For a formal proof, one can apply a standard hybrid argument.

However, although our protocol is both correct and private, it unfortunately requires the communication of information proportional in size to the entire database in order to retrieve a single database element. This could have just as easily been done by sending the entire database to the user, which would also maintain the user's privacy. Setting $k = \log |G'|$ as a security parameter, the user must communicate $\mathcal{O}(nk)$ bits. Fortunately, this can be modified into a more communication-efficient protocol without much effort. To begin, one can organize the database as a square, $X = \{x_{ij}\}_{i,j=1}^{\sqrt{n}}$, and if the (i^*, j^*)position of the database is desired, the user can send a query of the form $Q = \{q_i\}_{i=1}^{\sqrt{n}}$ defined just as before (we ignore the j^* index for reasons that become clear shortly). Then, the database can compute $R_j = \sum_{i=1}^{\sqrt{n}} x_{ij} \cdot q_i$ for each j and send $\{R_j\}_{j=1}^{\sqrt{n}}$ back to the user as the query response. Now as we've seen, from $\mathcal{D}(R_j)$, $\mathcal{U}$ can recover $\{x_{i^*j}\}_{j=1}^{\sqrt{n}}$ just as before. In particular, $\mathcal{U}$ can compute $x_{i^*j^*}$ (even though much more information is actually received). Note that the total communication involved in the protocol has now become nontrivially small: it is now proportional to $\sqrt{n}$ for each party as opposed to the $\mathcal{O}(n)$ communication required by our original proposal and the trivial solution of communicating the entire database to $\mathcal{U}$. However, we can make further improvements still.

6.5.2 Optimizing Via an Integer Map

Suppose that

$$\phi : G' \hookrightarrow \mathbb{Z}$$

is an injective map such that for all $y \in G'$, each component of $\phi(y)$ is less than ord(g) (i.e., one can think of the map as $\phi : G' \hookrightarrow \mathbb{Z}^l_{\text{ord}(g)}$). Any such map will do—we only require that both ϕ and ϕ^{-1} are efficiently, publicly computable. Note that in general we always have $l > 1$ since $\text{ord}(g) \leq |G|$ and $|G| < |G'|$, the latter inequality following from the fact that the encryption scheme is always probabilistic ($\mathcal{D}$ is never injective, but of course is always surjective). Again, note that we do not ask for any algebraic conditions from the map ϕ; it can be any easily computed injective set map. (For example, we could just break down a binary representation of elements of G' into sufficiently small blocks of bits to obtain the map ϕ.) Now, we can refine our query, and send $\left[Q = \{q_i\}_{i=1}^{\sqrt{n}}, \{p_j\}_{j=1}^{\sqrt{n}}\right]$ where q_{i^*} and p_{j^*} are set to encryptions of g, but all others encrypt id_G. Then, the database will initially proceed as before, computing

$$R_j = \sum_{i=1}^{\sqrt{n}} x_{ij} \cdot q_i$$

but then further computing

$$\overline{R}_t = \sum_{j=1}^{n} \phi(R_j)_t \cdot p_j$$

where $\phi(R_j)_t$ represents the t-th component of $\phi(R_j)$. This is sent as the query response to $\mathcal{U}$. To recover the desired data, $\mathcal{U}$ computes for every $t \in [l]$

$$\mathcal{D}(\overline{R}_t) = \phi(R_{j*})_t \cdot g$$

from which $\phi(R_{j^*})$ can be computed. Then, since ϕ^{-1} is efficiently computable, $\mathcal{U}$ can recover $R_{j^*} = \sum_{i=1}^{\sqrt{n}} x_{ij^*} \cdot q_i$, and as we have seen $x_{i^*j^*}$ is easily recovered from $\mathcal{D}(R_{j^*})$.

So now what amount of communication is required by the parties? The database sends $\mathcal{O}(l \log(|G'|)) = \mathcal{O}(lk)$ bits of information; meanwhile the user $\mathcal{U}$ sends $\mathcal{O}(2k\sqrt{n})$ bits of information, which is generally a large improvement on the database side. We can naturally extend this idea to higher-dimensional analogs. Representing the database as a d-dimensional cube (we have just seen the construction for d=1, 2), we accomplish the following communication complexity: $\mathcal{O}(kd\sqrt[d]{n})$ for the user's query, and $\mathcal{O}(l^{d-1}k)$ communication for the database's response.

The preceding construction is essentially that of [4, 7]. Both are simply special cases of what has been described earlier.

The work of [4] is based upon the cryptosystem of [5], which is homomorphic over the group $\mathbb{Z}_2$, having ciphertext group $\mathbb{Z}_N$ for a large composite N. In this case, it is simply the binary representation of a group element that plays the role of the map $\phi: G \hookrightarrow \mathbb{Z}^l$. Hence, we have l=k, the security parameter, and $\phi: \mathbb{Z}_N \hookrightarrow \mathbb{Z}^k$ takes an element $h \in \mathbb{Z}_N$ and maps it to a sequence of k integers, each in {0, 1} corresponding to a binary representation of h.

The work of [7] is also a special case of this construction. Here, the protocol is based upon the cryptosystem of [23], and we have $G = \mathbb{Z}_N$ and $G' = \mathbb{Z}^*_{N^2}$, hence we can greatly reduce the parameter l in comparison to the work of [4]. In this case, it is easy to see that we only need l=2, which is in fact minimal, as we have discussed before. The author of [7] uses the map $\phi: \mathbb{Z}^*_{N^2} \hookrightarrow \mathbb{Z}_N$ defined by the division algorithm, dividing by N to obtain a quotient and remainder of appropriate size. Roughly speaking (and using C-programming notation), he uses the map $x \mapsto (x/N, x\%N)$. However, as we have seen before, this is not necessary: any map could have been used (appropriately partitioning bits). Note also that since the discrete log in G is not hard (as we have defined it), we do not need to restrict our database to storing bits. Database elements could be any numbers in $\mathbb{Z}_N$.

These quite generic methods also capture the work of [8], as long as the appropriate cryptosystem is in place.

6.5.3 Further Improvements: Length-Flexible Cryptosystems

Consider a "length-flexible" cryptosystem, for example, that of Damgård and Jurik [9]. Such a cryptosystem has the property that given a message of arbitrary length, and given a *fixed* public key, one can choose a cryptosystem from a family of systems based on that key so that the message *fits in one ciphertext block* regardless of the key and the message length. Using this, we can further reduce the database's communication in our PIR protocol, using essentially the very same generic technique described above. We demonstrate the following.

Theorem 6.1 *For all $d \in \mathbb{Z}^+$ there exists a PIR protocol based on the homomorphic cryptosystem of Damgård and Jurik with user communication of $\mathcal{O}\left(kd^2 \sqrt[d]{n}\right)$ and database communication of $\mathcal{O}(kd)$ where k is a security parameter and n is the database size.*

What the Damgård and Jurik system affords us is the following: instead of having only one plaintext and ciphertext group G, G', we now have a countable family at our disposal:

$$\{G_i, G_i'\}_{i=1}^{\infty}$$

all of which correspond to a *single* public key. These groups are realized by $G_i \simeq \mathbb{Z}_{N^i}, G_i' \simeq \mathbb{Z}^*_{N^{i+1}}$, and hence we have natural inclusion maps of $G_i' \hookrightarrow G_{i+1}$. These, along with the observation that G_i is cyclic for all i, are essentially the only important facts regarding this system that we utilize. So, G_i is always cyclic, and we have a natural (although not algebraic) map

$$\psi_i : G_i' \to G_{i+1}$$

This is all we need to modify our generic method. We just replace the map ϕ with the maps ψ_i, and accordingly, we modify our query so that the vector for the ith dimension encrypts id_{G_i} in all positions except for the index of interest, which encrypts a generator of G_i. With only these minor substitutions to the abstract construction, the protocol follows exactly as before. This yields a protocol with communication complexity for the user $\mathcal{U}$ of

$$\sum_{i=1}^{d} ik\sqrt[d]{n} = \mathcal{O}\left(kd^2\sqrt[d]{n}\right)$$

and for the database, we require only

$$\mathcal{O}(kd)$$

as opposed to the previous exponential dependence on the dimension d of the cube used! Optimizing the parameters, setting $d = \frac{\log(n)}{2}$, we have $\mathcal{O}(k \log^2(n))$ communication for the user and $\mathcal{O}(k \log(n))$ for the database. So, as one can see, even a completely generic method can be quite useful, producing a near-optimal, polylogarithmic protocol.

6.6 Private Information Retrieval Based on the Φ-Hiding Assumption

Cachin, Micali, and Stadler [6] developed a new cryptographic assumption called the Φ-hiding assumption, and successfully used it to build a PIR protocol with logarithmic communication. Roughly, this assumption states that given two primes p_0, p_1 and a composite $m=pq$ such that either $p_0|\phi(m)$ or $p_1|\phi(m)$, it is hard to distinguish between the two primes. (Here, $\phi(m)$ is the Euler ϕ function, so that $\phi(m) = \phi(p-1)(q-1)$.) The assumption also of course states that given a small prime p, it is computationally feasible to find a composite m such that $p|\phi(m)$. Such an m is said to *ϕ-hide p*. A query for the ith bit of the database essentially contains input to a prime sequence generator, a composite m that ϕ-hides p_i (the ith prime in the sequence) and a random $r \in \mathbb{Z}_m^*$. The database algorithm returns a value $R \in \mathbb{Z}_m^*$ such that, with very high probability, R has p_ith roots if and only if the database bit at location i is 1.

6.6.1 Preliminaries

To understand the protocol, let us start with some very basic algebraic observations. Let G be a finite abelian group, and let $k \in \mathbb{Z}^+$. Consider the following map:

$$\varphi_k : G \to G \text{ defined by } x \mapsto x^k$$

Since G is abelian, it is clear that φ_k is a homomorphism for all $k \in \mathbb{Z}^+$. What is $\varphi_k(G)$? Clearly it is precisely the set of all elements in G that possess a kth root in G, that is,

$$\text{Im}(\varphi_k) = \left\{ x \in G \,\middle|\, \exists y \in G \ni x = y^k \right\}$$

We denote this set by H_k=(Im)(φ_k). Clearly it is a subgroup since it is the homomorphic image of a group. The size of this subgroup of course depends on k and G. If, for example, $(k, |G|)=1$, then it is easy to see that $\varphi_k(G)=(G)$, since if Ker$(\varphi_k)\neq\{e\}$, then there are nonidentity elements of order dividing k, which is clearly impossible. In the case that $(k, |G|)>1$, how big is Ker(φ_k)? It is at least as big as the largest prime divisor of $(k, |G|)>1$, by Cauchy's theorem if you like. For example, if k is a prime such that $k||G|$, then the map φ_k is at least a k-to-1 map.

Finally, let's take a look at the subgroups $H_k=\varphi_k(G)=\text{Im}(\varphi_k)$. We just need the following observation:

$$\forall k \in \mathbb{Z} \;\; H_K \lhd\!\lhd G$$

Here the symbol $H_k \lhd\lhd G$ signifies that H_k is a *characteristic* subgroup of G, which is to say that the subgroups H_k are fixed by *every* automorphism of G (compare with *normal* subgroups, which are those fixed by every *inner* automorphism of G). Note that for any finite G, if $H \lhd\lhd G$ and $\varphi \in \text{Aut}(G)$ then $\varphi(x) \in H \Leftrightarrow x \in H$ for all $x \in G$.

Let us summarize the few facts that will be of importance to us, and also narrow our view to correspond more directly to what we will need. Suppose that $p \in \mathbb{Z}$ is a prime, and define the maps φ_p as before. Then,

1. $\varphi_p \in \text{Aut}(G) \Leftrightarrow p \nmid |G|$.
2. φ_p is at least a p-to-1 map if $p \mid |G|$.
3. $\forall p_1\ H_p \lhd\lhd G$ (although this is trivial in the case that $p \nmid |G|$ and hence $H_p = G$). So for any $\varphi \in \text{Aut}(G)$ and $x \in G$, we have $\varphi(x) \in H_p \Leftrightarrow x \in H_p$.

6.6.2 A Brief Description of the Protocol

We now have enough information for a basic understanding of how and why the PIR protocol of [6] works. First, we will begin with the "how." Continuing with our preceding notation, suppose that $X = \{x_i\}_{i=1}^n$ is our database, with each $x_i \in \{0,1\}$, and again, suppose that the index of interest to $\mathcal{U}$ is i^*. The protocol executes the following steps, involving a database $\mathcal{DB}$ and a user $\mathcal{U}$.

1. $\mathcal{U}$ sends a random seed for a publicly known prime sequence generator to $\mathcal{DB}$, the primes being of intermediate size.[8]
2. $\mathcal{U}$ computes p_{i^*}, the i^*th prime in the sequence based on the random seed.
3. $\mathcal{U}$ finds a composite number m that ϕ-hides p_{i^*} and sends m to $\mathcal{DB}$. In particular, we have that $p_{i^*} \mid \phi(m)$. Recall that $\phi(m) = |\mathbb{Z}_m^*|$.
4. $\mathcal{DB}$ selects $r \in \mathbb{Z}_m^*$ at random, and computes $R \in \mathbb{Z}_m^*$ as follows:

$$R = \varphi_{p_n^{x_n}} \circ \varphi_{p_{n-1}^{x_{n-1}}} \circ \ldots \circ \varphi_{p_1^{x_1}}(r)$$

$$= r^{\prod_{i=1}^n p_i^{x_i}} \bmod m$$

5. $\mathcal{U}$ receives R from $\mathcal{DB}$ as the response, and determines that $x_{i^*} = 1$ if and only if $R \in H_{p_{i^*}}$.

8. Revealing a large prime dividing $\phi(m), (p > \sqrt[4]{m})$ enables one to factor m, so the primes must be chosen to be small.

These steps are essentially the entire protocol at a high level. However, it may not be immediately obvious that the statement $R \in H_{p_{i^*}}$ has much to do with the statement $x_{i^*} = 1$. But using the 3 facts we established early on, it isn't too hard to see that these are in fact equivalent with very high probability.

From our first fact, we know that $\varphi_{p_i} \in \text{AUT}(G)$ whenever $i \neq i^*$ with overwhelming probability, since the only way for this to not be the case is if $p_i \mid \phi(m)$. However, due to the fact that there are at most only a logarithmic number of prime divisors of $\phi(m)$ out of many choices, this event is extremely unlikely.[9] So, all of the φ_{p_i} are automorphisms, except for $\varphi_{p_{i^*}}$.

From our next fact, we know that with very high probability $r \notin H_{p_{i^*}}$, where $r \in \mathbb{Z}_m^*$ was the element randomly chosen by $\mathcal{DB}$. Since the map is at least p_{i^*} to 1, the entire group is at least p_{i^*} times the size of $H_{p_{i^*}}$. So, if we were to pick an element at random from $\mathbb{Z}_m^*$, there is at best a $\frac{1}{p_{i^*}}$ chance that it is in $H_{p_{i^*}}$. So, in the length of our primes p_i there is an exponentially small probability that a random r will be in $H_{p_{i^*}}$.

Finally, we noted that the subgroups H_{p_i} are characteristic subgroups, and hence are fixed by every automorphism of $\mathbb{Z}_m^*$. In particular, $H_{p_{i^*}} \lhd\lhd \mathbb{Z}_m^*$ So, all of the automorphisms $\{\varphi_{p_i}\}_{i \neq i^*}$ preserve this group: things outside remain outside, and things inside remain inside, and of course $\varphi_{p_{i^*}}$ moves every element into $H_{p_{i^*}}$, that is,

$$\varphi_{p_{i*}}(x) \in H_{p_{i*}} \forall x$$

and if $i \neq i^*$, then

$$\varphi_{p_i}(x) \in H_{p_{i*}} \Leftrightarrow x \in H_{p_{i*}}$$

We can trace the path that r takes to become R and see what happens: We have that the element r begins outside of the subgroup $H_{p_{i^*}}$ and then r is moved by many maps, all of which come from the set

$$\{\varphi_1, \ldots, \varphi_{i*-1}\} \cup \{\text{Id}\}$$

depending on whether or not $x_i = 1$. But what is important is that all of these maps are automorphisms, which therefore fix $H_{p_{i^*}}$. So, no matter what the configuration of the first $i^* - 1$ elements of the database, r has not moved into $H_{p_{i^*}}$ at

9. According to the prime number theorem, there are approximately $\frac{N}{2 \log N}$ primes of bit length equal to that of N. Our chances of picking m such that another p_i inadvertently divides $\phi(m)$ are approximately $\frac{\text{polylog}(m)}{m}$, which is negligibly small as the length of m in bits (i.e., $\log m$, the security parameter) increases.

this point. Next, we conditionally apply the map $\varphi_{p_{i^*}}$ depending on whether or not $x_{i^*} = 1$, which conditionally moves our element into $H_{p_{i^*}}$. This is followed by the application of more automorphisms, which as we have seen have no effect on whether or not the response R is in $H_{p_{i^*}}$. So, since $H_{p_{i^*}}$ is fixed by every automorphism, the only chance that r has to move from outside $H_{p_{i^*}}$ to inside $H_{p_{i^*}}$ is if the map $\varphi_{p_{i^*}}$ is applied, which happens if and only if $x_{i^*} = 1$. Hence, we have that (with overwhelming probability) $R \in H_{p_{i^*}}$ if and only if $x_{i^*} = 1$.

The privacy of this protocol can be proved directly from the Φ-hiding assumption, although it may be more pleasant to think of this in terms of the indistinguishability of the subgroup H_{p_i} to a party not knowing the factorization of m. Now, let us take a look back and examine the communication to see why this was useful. The challenge of creating PIR protocols is usually to minimize the amount of communication. A PIR protocol with linear communication is quite trivial to construct: just transfer the entire database. This is, of course, not very useful. The PIR protocol we have described above, however, has nearly optimal communication. The database's response is a single element $R \in \mathbb{Z}_m^*$, which has size proportional to the security parameter alone (which must be at least logarithmic in n), and the user's query has the size of the security parameter, and the random input to a prime sequence generator, which could also be as small as the logarithmic in n. So, we have constructed a PIR protocol with only logarithmic communication, which is, of course, optimal: if $\mathcal{DB}$ wants to avoid sending information proportional to the size of the database, then $\mathcal{U}$ must somehow communicate information about what index is desired, which requires at least a logarithmic amount of communication. However, with the recommended parameters for security, the total communication is approximately $\mathcal{O}(\log^8 n)$.

6.6.3 Generalizations: Smooth Subgroups

Gentry and Ramzan [10] have generalized some of the fundamental ideas behind these methods, creating protocols based on *smooth subgroups*, which are those that have many small primes dividing their order. Somewhat similar to CMS [6], a list of primes is chosen corresponding to the positions of the database, and a query for position i essentially consists of a description of a group G such that $|G|$ is divisible by p_i. However, the work of [10] is designed to retrieve blocks of data at a single time (CMS [6] must be repeatedly executed to accomplish this functionality). Rather than repeatedly exponentiating by all of the primes, the database is represented as an integer e such that when reduced mod p_i, the value is the ith block of the database (such an integer always exists, of course, by the chinese remainder theorem). Now to recover the data (which is just e mod p_i), a discrete log computation can be made in the (small) subgroup of order p_i.

6.7 Private Information Retrieval from Any Trapdoor Permutation

In 2000, Kushilevitz and Ostrovsky [11] demonstrated that the existence of one-way trapdoor permutations suffices to create a *nontrivial* PIR, where nontrivial simply means that the total communication between the parties is strictly smaller than the size of the database. Although the protocol requires multiple rounds of interaction, the basic construction remains fairly simple in the case of an honest but curious server. In case of a malicious server the construction is more complicated and the reader is referred to the original paper for details. Here, we only illustrate the basic idea of the honest-but-curious case.

6.7.1 Preliminaries

For this construction, the existence of one-way trapdoor permutations(f, f^{-1}) is assumed, as well as Goldreich-Levin hardcore bits.

Another tool (used in the honest-but-curious case) is the universal one-way hashing of Naor and Yung [39]. For the dishonest case, universal one-way hash functions are replaced with an *interactive hashing* protocol [40], and on top of that some additional machinery is needed. However, for the honest-but-curious case the proof is far more simple. Recall that universal one-way hash functions satisfy a slightly weaker type of collision resistance. Basically, if one first picks any input x from the domain, and then independently a hash function h from a universal one-way family, it is computationally infeasible to find $x' \neq x \in h^{-1}(h(x))$.

The PIR protocol we'll discuss here uses universal one-way hash functions, which are 2 to 1 (i.e., for all y in the codomain, $|h^{-1}(y)| = 2$) and each function will map $\{0,1\}^k \rightarrow \{0,1\}^{k-1}$ for some integer k.

6.7.2 Outline of the Protocol

At a very high level, the protocol revolves around the following idea: the server takes an n-bit database and partitions it into consecutive blocks of length k (k is the input length to a trapdoor permutation f). It collapses every block of the database by one bit, and sends this (slightly) reduced-size database back to the user. The user then selects and sends to the server some information that allows for the determination of the one missing bit of information for the block in which he/she is interested. Now, using communication-balancing techniques similar to what we've described in the introduction, we can hold on to the constant advantage (below n) given to us by the server collapsing the one bit of every database block. The trick, of course, is to avoid revealing information about which block the user is interested in when recovering this last bit. The solution is quite simple. As mentioned, the database collapses a bit of each database block before sending this information to the user. There are many obvious ways to do this, for example, just sending all but one bit of each block. However, in these situations the database knows exactly the two possibilities that arise from the collapsed data sent to the user, as well as knowing the actual value in the database. This would seemingly make it quite difficult for the user to determine which of the two possibilities exists in the database without the database gaining information. So instead, a method is devised in which the database collapses a bit of each block *without knowing the other possibility.* This enables the user to determine which possibility exists for a given block without revealing in what block he or she is interested.

6.7.3 Protocol Details

As we alluded to in the outline, we need to provide a way for the database to collapse a bit of each block, but without knowing the other possibility. This is accomplished precisely via a family $\mathcal{F}$ of universal one-way hash functions, and in fact, the original construction of such a family by Naor and Yung [39] is used. The important point is that the only assumption needed to build this family of universal one-way hash functions was the existence of one-way permutations, and furthermore, because they were constructed via one-way permutations, a party holding the trapdoor *can find collisions.* To summarize, here are the important properties we need from the family $\mathcal{F}$.

1. Each function of $\mathcal{F}$ is efficiently computable.
2. Each function has the property of being 2 to 1.
3. Given only x, $f(x)$ for $f \in \mathcal{F}$, it is computationally infeasible to find $x' \neq x \in f^{-1}(f(x))$ without trapdoor information.
4. With trapdoor information, it is feasible to find collisions in every function $f \in \mathcal{F}$.

The protocol proceeds as follows.

The database is divided into blocks of size K, one of which the user is interested in. Furthermore, the database is organized into pairs of blocks, denote them by $z_{i,L}$ and $z_{i,R}$ (L, R standing for "left" and "right"). A query consists of two descriptions of universal one-way hash functions, f_L, f_R, to which the user has the trapdoors. Upon receipt of the query, the database computes the values of $f_L(z_{i,L})$ and $f_R(z_{i,R})$ for each block of the database, and returns these values to the user. The user, who has trapdoors, can compute both possible preimages (z, z') that may correspond to the block of the database of interest. It only remains to have the database communicate which one, while maintaining privacy. This is accomplished via hardcore predicates. Without loss of generality, suppose the user wishes to retrieve the left block, say, $z_{s,L}$. Then, the user selects two hardcore predicates, r_L, r_R according to the conditions that $r_L(z_{s,L}) \neq r_L(z'_{s,L})$, yet $r_R(z_{s,R}) = r_R(z'_{s,R})$. These predicates are sent to the database, which responds with $r_L(z_{i,L}) \oplus r_R(z_{i,R})$ for every pair of blocks. Now, regardless of the possibilities of the right block, the hardcore predicates will be the same, hence the user can solve for the left hardcore predicate, and hence the left block, as we assumed the predicates evaluated on the two choices to be distinct. This completes a basic description of the protocol.

The descriptions of f_L, f_R, r_L, r_R are all $\mathcal{O}(K)$, which is the only communication from the user to the database. The communication from the database to the user is easily seen to be $n - \frac{n}{2K}$ bits in the initial round, and one more bit in the final response. Hence, the protocol does achieve smaller than n communication, for $n > O(k^2)$. Next we argue that the protocol is also secure. The only information sent to the database that contains any information about what block the user is interested in is that of the hardcore predicates, r_L, r_R. The value of the hardcore predicates *on the two possible preimages of a hash value* is exactly what gives us the information regarding the user's selection. We only need to show that given such predicates, they do not reveal information about the selected block. Informally, this is the right approach, as the definition of hardcore predicate states that the outcomes are hard to predict better than random when only given the output of a function. Indeed, as the fairly straightforward hybrid argument shows, this is the case.

6.8 Conclusions

We provided a general survey of single-database PIR and its many connections to other cryptographic primitives. We also discussed several implementations of single-database PIR, including a generic construction from homomorphic encryption. As well studied as single-database PIR seems to be, many open problems remain. For example, reducing the communication of a PIR protocol based on general trapdoor permutations, and exploring the connections of PIR to other primitives both in cryptography and communication complexity.

References

[1] R. Curtmola, et al. "Searchable symmetric encryption: improved definitions and efficient constructions." *ACM Conference on Computer and Communications Security, CCS 2006*, pp. 79–88, 2006.

[2] O. Goldreich and R Ostrovsky. "Software protection and simulation on oblivious RAMs." *Journal of the ACM,* 43(3): 431–473, 1996.

[3] B. Chor, et al. "Private information retrieval." *Proceedings of the 36th Annual IEEE Symposium on Foundations of Computer Science*, pp. 41–51, 1995. Journal version: *Journal of the ACM*, 45:965–981, 1998.

[4] E. Kushilevitz and R. Ostrovsky. "Replication is not needed: Single database, computationally-private information retrieval." *Proceedings of the 38th Annual IEEE Symposium on Foundations of Computer Science*, pp. 364–373, 1997.

[5] S. Goldwasser and S. Micali. "Probabilistic encryption." *Journal of Computer and System Sciences,* 28(1): 270–299, 1984.

[6] C. Cachin, S. Micali, and M. Stadler. "Computationally private information retrieval with polylogarithmic communication," in J. Stern, editor, *Advances in Cryptology—EUROCRYPT '99*, vol. 1592 of *Lecture Notes in Computer Science*, pp. 402–414. Springer, 1999.

[7] Y. C. Chang. "Single database private information retrieval with logarithmic communication." *ACISP*, 2004.

[8] H. Lipmaa. "An oblivious transfer protocol with log-squared communication." *ISC*, pp. 314–328, 2005.

[9] I. Damgård and M. Jurik. "A generalisation, a simplification and some applications of Paillier's probabilistic public-key system." *Public Key Cryptography (PKC 2001).*

[10] C. Gentry and Z. Ramzan. "Single database private information retrieval with constant communication rate." *ICALP 2005, LNCS 3580*, pp. 803–815, 2005.

[11] E. Kushilevitz and R. Ostrovsky. "One-way trapdoor permutations are sufficient for non-trivial single-server private information retrieval." *EUROCRYPT*, pp. 104–121, 2000.

[12] R. Ostrovsky and W. Skeith. "Private searching on streaming data." *Advances in Cryptology—CRYPTO 2005.* Full version in *Journal of Cryptology*, 4: 2007.

[13] Y. Ishai, E. Kushilevitz, R. Ostrovsky, and A. Sahai. "Batch codes and their applications." *STOC*, pp. 262–271, 2004.

[14] Y. Ishai, E. Kushilevitz, R. Ostrovsky, and A. Sahai. "Cryptography from anonymity." *FOCS*, pp. 239–248, 2006.

[15] M. O. Rabin. "How to exchange secrets by oblivious transfer." Technical Memo TR-81. Aiken Computation Laboratory. Harvard University, 1981.

[16] S. Even, O. Goldreich, and A. Lempel. "A randomized protocol for signing contracts." *Communications of the ACM*, 28: 637–447, 1985.

[17] G. Brassard, C. Crepeau, and J.-M. Robert. "All-or-nothing disclosure of secrets." *Advances in Cryptology: Proceedings of Crypto '86*, pp. 234–238. 1987.

[18] C. Crépeau. "Equivalence between two flavors of oblivious transfers." *Proceedings of CRYPTO '87*, pp. 350–354, 1988.

[19] Y. Gertner, Y. Ishai, E. Kushilevitz, and T. Malkin. "Protecting data privacy in private information retrieval schemes." *Proceedings of the 30th Annual ACM Symposium on the Theory of Computing*, pp. 151–160, 1998.

[20] G. Di Crescenzo, T. Malkin, and R. Ostrovsky. "Single-database private information retrieval implies oblivious transfer." *Advances in Cryptolog—EUROCRYPT 2000*, pp. 122–138, 2000.

[21] M. Naor and B. Pinkas. "Oblivious transfer and polynomial evaluation." In *Proceedings of the 31th Annual ACM Symposium on the Theory of Computing*, pp. 245–254, 1999.

[22] Y. Ishai, E. Kushilevitz, and R. Ostrovsky. "Sufficient conditions for collision-resistant hashing." *TCC*, pp. 445–456, 2005.

[23] P. Paillier. "Public key cryptosystems based on composite degree residue classes." *Advances in Cryptology—EUROCRYPT 99*, LNCS 1592: 223–238. Springer Verlag, 1999.

[24] S. Dziembowski and U. Maurer. "On generating the initial key in the bounded-storage model." *EUROCRYPT*, pp. 126–137, 2004.

[25] D. Harnik, and M. Naor. "On the compressibility of NP instances and cryptographic applications." *FOCS*, pp. 719–728, 2006.

[26] M. Naor and K. Nissim. "Communication complexity and secure function evaluation." *Electronic Colloquium on Computational Complexity (ECCC)*, 8(062): 2001.

[27] Y. Tauman Kalai and R. Raz. "Succinct non-interactive zero-knowledge proofs with preprocessing for LOGSNP." *FOCS*, pp. 355–366, 2006.

[28] G. DiCrescenzo, Y. Ishai, and R. Ostrovsky. "Universal service-providers for database private information retrieval." *Proceedings of the 17th Annual ACM Symposium on Principles of Distributed Computing*, pp. 91–100, 1998. Full version in *Journal of Cryptology*, 14(1): 37–74, 2001.

[29] R. Meier, and B. Przydatek. "On robust combiners for private information retrieval and other primitives." *CRYPTO*, pp. 555–569, 2006.

[30] D. Boneh, et al. "Public key encryption that allows PIR queries." *CRYPTO*, pp. 50–67, 2007.

[31] D. Boneh, et al. "Public key encryption with keyword search." *EUROCRYPT*, pp. 506–522, 2004.

[32] R. Ostrovsky and W. Skeith. "Algebraic lower bounds for computing on encrypted data." In *ECCC, Electronic Colloquium on Computational Complexity*, 2007.

[33] I. N. Herstein. *Abstract Algebra*. Prentice-Hall, 1986, 1990, 1996.

[34] T. W. Hungerford. *Algebra*. Springer-Verlag, Berlin, 1984.

[35] C. Shannon. "Communication theory of secrecy systems." *Bell System Technical Journal*, 28(4): 656–715, 1949.

[36] O. Goldreich. Foundations of Cryptography. Manuscript, 2001. Available at http://www.wisdom.weizmann.ac.il/oded/books.html.

[37] T. ElGamal. "A public-key cryptosystem and a signature scheme based on discrete logarithms." *IEEE Transactions on Information Theory*, IT-31(4): 469–472, 1985, or *CRYPTO 84*, pp. 10–18.

[38] R. L. Rivest, A. Shamir, and L. Adleman. "A method for obtaining digital signatures and public key cryptosystems." *Commun. ACM*, 21: 120–126, 1978.

[39] M. Naor and M. Yung. "Universal one-way hash functions and their cryptographic applications." *Proceedings of the 21st Annual ACM Symposium on Theory of Computing.* May 15–17, 1989. Seattle, WA, USA.

[40] R. Ostrovsky, R. Venkatesan, and M. Yung. "Fair games against an all-powerful adversary." Presented at *DIMACS Complexity and Cryptography workshop*, October 1990. Princeton. Prelim. version in *Proceedings of the Sequences II workshop,* 1991. Springer-Verlag, pp. 418–429. Final version in *AMS DIMACS Series in Discrete Mathematics and Theoretical Computer Science*, Vol. 13, *Distributed Computing and Cryptography*, J.-Y. Cai, editor, pp. 155–169. AMS, 1993.

7

Tapping Vehicle Sensors for Homeland Security

Mario Gerla and Uichin Lee

7.1 Introduction

Vehicular ad hoc networks (VANETs) are acquiring commercial relevance because of recent advances in intervehicular communications and decreasing costs of related equipment. This is stimulating a brand new family of visionary services for vehicles, from entertainment applications to tourist/advertising information, from driver safety to opportunistic transient connectivity to the fixed Internet infrastructure [1–4]. In particular, vehicular sensor networks (VSNs) are emerging as a new tool for effectively monitoring the physical world, especially in urban areas where a high concentration of vehicles equipped with onboard sensors is expected [5]. Vehicles are typically not affected by strict energy constraints and canbe easily equipped with powerful processing units, wireless transmitters, and sensing devices of some complexity, cost, and weight (GPS, chemical spill detectors, still/video cameras, vibration sensors, acoustic detectors). VSNs represent a significantly novel and challenging deployment scenario, considerably different from more traditional wireless sensor network environments, thus requiring innovative, specific solutions. In fact, different from wireless sensor nodes, vehicles usually exhibit constrained mobility patterns due to street layouts, junctions, and speed limitations. In addition, they usually have no strict limits on processing power and storage capabilities. Most important, they can host sensors that may generate huge amounts of data, such as multimedia video streams, thus making impractical the instantaneous data reporting solutions of conventional wireless sensor networks.

VSNs offer a tremendous opportunity for different large-scale applications, from traffic routing and relief to environmental monitoring and distributed surveillance. In particular, there is an increasing interest in proactive urban monitoring services where equipped vehicles continuously sense events from urban streets, maintain sensed data in their local storage, autonomously process them (e.g., recognize license plates), and possibly route messages to vehicles in their vicinity to achieve a common goal (e.g., to permit police agents to track the movements of specified cars). For instance, proactive urban monitoring could usefully apply to post-facto crime scene investigation. Reflecting on tragedies such as 9/11 and the London bombings, VSNs could have actually helped emergency recovery and forensic investigation/criminal apprehension. In the London bombings police agents were able to track some of the suspects in the subway using closed-circuit TV cameras, but they had a hard time finding helpful evidence from the double-decker bus; this has motivated the installation of more cameras in fixed locations along London streets. VSNs could be an excellent complement to the deployment of fixed cameras/sensors. The completely distributed and opportunistic cooperation among sensor-equipped vehicles has the "deterrent" effect of making it harder for potential attackers to disable surveillance.

Another less sensational but relevant example is the need to track the movements of a car, used for a bank robbery, in order to identify thieves, say. It is highly probable that some sensor-equipped vehicles have spotted the unusual behavior of the thieves' car in the hours before the robbery, and might be able to identify the threat by *opportunistic* correlation of their data with other vehicles in the neighborhood. It would be much more difficult for the police to extract that information from the massive number of multimedia streams recorded by fixed cameras. As for privacy, let us briefly note that people are willing to sacrifice privacy and to accept a reasonable level of surveillance when the data can be collected and processed only by recognized authorities (with a court order) for forensic purposes and/or for counteracting terrorism and common crimes.

As shown by these examples, the reconstruction of a crime and, more generally, the forensic investigation of an event monitored by VSNs require the collection, storage, and retrieval of massive amounts of sensed data. This is a major departure from conventional sensor network operations where data is dispatched to *sinks* under predefined conditions such as alarm thresholds. Obviously, it is impossible to deliver all the streaming data collected by video sensors to a police authority sink because of sheer volume. Moreover, input filtering is not possible because nobody knows which data will be of use for future investigations. The problem becomes one of searching for sensed data in a massive, mobile, opportunistically collected, and completely decentralized storage. The challenge is to find a completely decentralized VSN solution, with low impact on other services, good scalability (up to thousands of nodes), and tolerance of disruption caused by mobility and attacks.

To exploit this interesting scenario and to meet its challenge, we are presenting MobEyes, an urban monitoring middleware recently developed at UCLA. MobEyes is based on wireless-enabled vehicles equipped with video cameras and a variety of sensors to perform event sensing, processing/classification of sensed data, and intervehicle ad hoc message routing. Since it is impossible to directly report the sheer amount of sensed data to the authorities, MobEyes keeps sensed data in mobile node storage; onboard processing capabilities are used to extract features of interest (e.g., license plates). Mobile nodes periodically generate data summaries with extracted features and context information such as timestamps and positioning coordinates; mobile agents (e.g., police patrolling cars), and move and opportunistically harvest summaries as needed from neighbor vehicles. MobEyes adopts VSN custom designed protocols for summary diffusion/harvesting that exploit intrinsic vehicle mobility and simple single-hop intervehicle communications. In that way, MobEyes harvesting agents can create a low-cost opportunistic index to query the distributed sensed data storage, thus enabling us to answer questions such as: which vehicles were in a given place at a given time, which route did a certain vehicle take in a given time interval, and which vehicle collected and stored the data of interest?

The main contributions of this chapter can be summarized as follows:

- We define a vehicular sensing platform and propose MobEyes *vehicular sensing architecture*. We synthesize the existing techniques to build a MobEyes system that satisfies the key design principles, namely disruption tolerance, scalability, and nonintrusiveness.
- We evaluate overhead and overall system stability of MobEyes via extensive simulation. We then test MobEyes in a challenging *homeland defense application* where the police can reconstruct the route followed by a suspect car by simply specifying its plate number.
- We overview the primary *security requirements* in a VSN-based urban monitoring application, and show how they can be addressed by state-of-the-art solutions. We implement these solutions in MobEyes, with the option to enable/disable them at deployment time depending on application security/privacy.

The rest of the chapter is organized as follows. Section 7.2 describes background and related work by positioning the original MobEyes contributions. Section 7.3 presents the overall MobEyes architecture, while Section 7.4 details our original protocols for opportunistic summary diffusion/harvesting. Section 7.5 evaluates our proposed protocols via simulations. Section 7.6 gives a rapid overview of security/privacy issues and related solutions. Finally, Section 7.7 concludes the chapter.

7.2 State of the Art

The idea of embedding sensors in vehicles is quite novel. To our knowledge, the only research project dealing with similar issues is MIT's CarTel [6, 7]. In CarTel users submit their queries about sensed data on a portal hosted on the wired Internet. Then, an intermittently connected database is in charge of dispatching queries to vehicles and of receiving replies when vehicles move in the proximity of open access points to the Internet. Different from CarTel, MobEyes exploits mobile collector agents instead of relying on the wired Internet infrastructure, thus improving robustness. Work relevant to vehicle sensing can, however, be found in two very popular fields: VANETs and "opportunistic" sensor networks. The two following sections have an overview these areas.

7.2.1 VANETs

Recently, researchers have envisioned applications specifically designed for VANET, including: (1) safe cooperative driving where emergency information is diffused to neighbor vehicles and real-time response is required to avoid accidents [3]; (2) entertainment support, e.g., content sharing [1], advertisements [2], and peer-to-peer marketing [8]; and (3) distributed data collection (e.g., parking lot [9] or traffic congestion information [10]). So far, however, most of the published VANET research has focused on routing issues. Several VANET applications (e.g., related to safety or traffic/commercial advertising), call for low delay, high data rate delivery of messages to nodes located close to the sender (e.g., the nodes potentially affected by an intersection crash). Recent research addressed this issue by proposing customized broadcast strategies [3, 11]. However, broadcasting to only nearby nodes does not support well certain advertising applications that wish to reach remote customers. Likewise, homeland defense applications such as the reconstruction of the attackers' escape route require the diffusion of *sensed data* over several hops. This is necessary in order to facilitate the forensic search. Thus, effective dissemination solutions that work over multi-hop must also be investigated [12].

Packet delivery issues in areas with sparse vehicles have stimulated several recent research contributions to investigate carry-and-forward strategies. In [13], authors simulate a straight highway scenario to compare two ideal strategies: pessimistic (i.e., synchronous), where sources send packets to destinations only as soon as a multihop path is available, and optimistic (i.e., carry-and-forward), where intermediate nodes hold packets until a neighbor closer to the destination is detected. Under the implicit assumptions of (1) unbounded message buffers and bandwidth, and (2) of easily predictable mobility patterns as for vehicles on a highway, the latter has achieved a lower delivery delay. However, in more realistic situations, carry-and-forward protocols call for careful design and tuning.

MaxProp [14], part of the UMass DieselNet project [15], has a ranking strategy to determine packet delivery order when node encounters occasionally occur, as well as dropping priorities in the case of full buffers. Precedence is given to packets destined to the other party, then to routing information, to acknowledgments, to packets with small hop-counts, and finally to packets with a high probability of being delivered through the other party. VADD (vehicle-assisted data delivery)[16] rests on the assumption that most node encounters happen in intersection areas. Effective decision strategies are proposed, highly reducing packet delivery failures and delay.

Applications for distributed data collection in VANET call for geographic dissemination strategies that deliver packets to all nodes belonging to target remote areas, despite possibly interrupted paths [10, 17]. MDDV (mobility-centric data dissemination algorithm for vehicular networks)[17] exploits geographic forwarding to the destination region, favoring paths where vehicle density is higher. In MDDV, messages are carried by head vehicles (i.e., best positioned toward the destination with respect to their neighbors). As an alternative, [10] proposes several strategies based on virtual potential fields generated by propagation functions: any node estimates its position in the field and retransmits packets until nodes placed in locations with lower potential values are found; this procedure is repeated until minimael target zones are detected.

7.2.2 Opportunistic Sensor Networking

Traditionally, sensor networks have been deployed in static environments, with application-specific monitoring tasks. Recently, opportunistic sensor networks have emerged, which exploit existing devices and sensors, such as cameras in mobile phones [18–21]. Several of these networks are relevant to our research because they can easily implement opportunistic dissemination protocols [22, 23].

Dartmouth's MetroSense [18, 24] is closely related to MobEyes [5]. Reference [18] describes a three-tier architecture for MetroSense: servers in the wired Internet are in charge of storing/processing sensed data; Internet-connected stationary sensor access points (SAP) act as gateways between servers and mobile sensors (MS); MS move in the field opportunistically delegating tasks to each other, and *muling* [25, 26] data to SAP. MetroSense requires infrastructure support, including Internet-connected servers and remotely deployed SAP. Similarly, Wang et al. proposed data delivery schemes in the delay/fault-tolerant mobile sensor network (DFT-MSN) for human-oriented pervasive information gathering [27]. The trade-off between data delivery ratio/delay and replication overhead is mainly investigated in the case of buffer and energy resource constraints. In contrast, MobEyes does not require any fixed infrastructure by using mobile sinks (or agents) and addresses VANET-specific deployment scenarios: powerful sensing and platforms and distributed index collection.

Application-level protocols for the resolution of queries to sensed data have been proposed in [19, 20]. Contory abstracts the network as a database and can resolve declarative queries; spatial programming hides remote resources, such as nodes, under local variables, thus enabling transparent access; finally, migratory services are components that react to changing context (e.g., the target moving out of range by migrating to other nodes [19]). Dikaiakos et al. [20] presented VITP, a query–response protocol, to obtain traffic-related information from remote areas: the primary idea is that the source specifies the target area when injecting a query in the environment, and nodes in the target area form a virtual ad hoc query server.

Among recent research projects about opportunistic sensing, we mention Intel IrisNet [28] and Microsoft SenseWeb [29]. Both projects investigate the integration of heterogeneous sensing platforms in the Internet via a common data publishing architecture. Finally, we point out CENS's Urban Sensing project [21, 30], a recently started multidisciplinary project addressing *participatory* sensing, where urban monitoring applications receive data from mobile sensors operated by people.

Finally, regarding dissemination of sensed data through peers, we can mention two solutions from the naturalistic environment, namely ZebraNet [22] and SWIM (shared wireless infostation model) [23]. ZebraNet addresses remote wildlife tracking (e.g., zebras in Mpala Research Center in Kenya), by equipping animals with collars that embed wireless communication devices, GPSs, and biometric sensors. As GPS-equipped animals drift within the park, their collars opportunistically exchange sensed data, which must make its way to the base station (the ranger's truck). ZebraNet proposes two dissemination protocols: a flooding-based approach where zebras exchange all the data within their buffers (either locally generated or received from other animals) with neighbors, and a history-based protocol where data is uploaded only to zebras with a good track record of base station encounters. SWIM [23] addresses sparse mobile sensor networks with fixed infostations as collecting points. Sensed data is epidemically disseminated via single-hop flooding to encountered nodes and offloaded when infostations are in reach.

While all the above schemes implement peer-to-peer dissemination, none fits the performance requirements of the MobEyes target scenario. Flooding generates excessive overhead, while history-based protocol is ineffective in the very dynamic VANET where the base station (the agent's vehicle) rapidly moves without following a specific mobility pattern [22]. In addition, let us note that MobEyes nodes do not transmit raw collected data but, thanks to an abundance of onboard computing resources, they locally process data and relay only short summaries. That is a relevant advantage since data collected in VSN (e.g., from video cameras) may be an order-of-magnitude larger than data sensed in naturalistic scenarios.

7.3 MobEyes Architecture

For the sake of clarity, let us present the MobEyes solution using one of its possible practical application scenarios: collecting information from MobEyes-enabled vehicles about criminals that spread poisonous chemicals in a particular section of the city (say, the subway station). We suspect the criminals used vehicles for the attack. Thus, MobEyes will help detect the vehicles and permit tracking and capture. Here, we assume that vehicles are equipped with cameras and chemical detection sensors. Vehicles continuously generate a huge amount of sensed data, store it locally, and periodically produce short *summary chunks* obtained by processing sensed data (e.g., license plate numbers or aggregated chemical readings). Summary chunks are aggregated in *summaries* that are opportunistically disseminated to neighbor vehicles, thus enabling metadata harvesting by the police in order to create a distributed metadata index, useful for forensic purposes such as crime scene reconstruction and criminal tracking.

To support all the above tasks, we have developed MobEyes according to the component-based architecture depicted in Figure 7.1. The key component is the MobEyes diffusion/harvesting processor (MDHP), which is discussed in detail in the next section. MDHP works by opportunistically disseminating/harvesting summaries produced by the MobEyes data processor (MDP), which accesses sensor data via the MobEyes sensor interface (MSI). Since vehicles are not strictly resource constrained, our MobEyes prototype is built on top of the Java Standard Edition (J2SE) virtual machine. MDP is in charge of reading raw sensed data (via MSI), processing it, and generating chunks. Chunks include metadata (vehicle position, timestamp, vehicle ID number, and possible additional contexts such as simultaneous sensor alerts) and features of interest extracted by local filters (see Figure 7.2). For instance, in the above application scenario, MDP includes a filter determining license plate numbers from multimedia flows taken by cameras [31]. Finally, MDP commands the storage of both raw data and chunks in

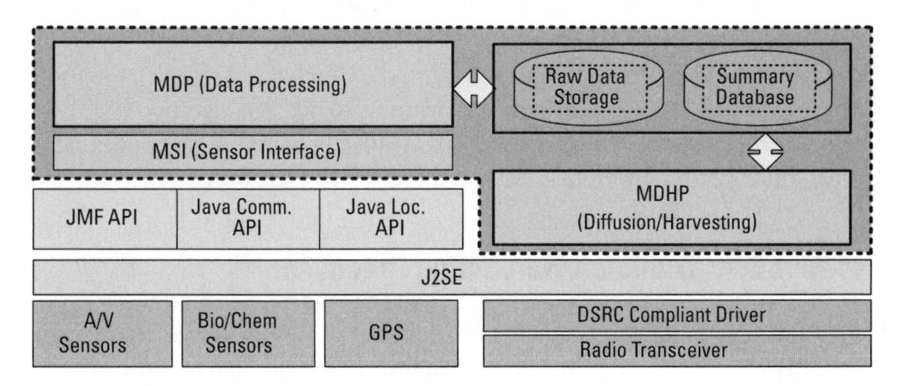

Figure 7.1 MobEyes sensor node architecture.

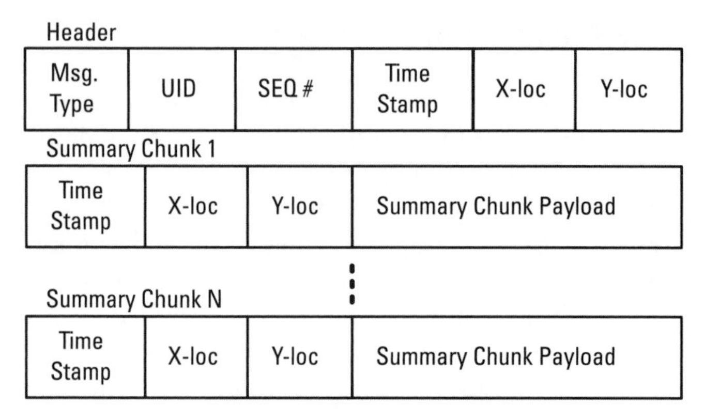

Figure 7.2 Packet format: a single packet contains multiple *chunks*.

two local databases. MDHP disseminates/harvests summaries by packing a set of chunks into a single packet for the sake of effectiveness. Therefore, the generation rate and size of chunks and summaries are relevant to MobEyes performance. The interested reader may find additional details about the design and implementation of the MobEyes prototype in [32].

Developers of MobEyes-based applications can specify the desired generation rate as a function of vehicle speed and expected vehicle density. The chunk size mainly depends on application-specific requirements: in the scenario under consideration, each recognized license plate is represented with 6B, sensed data with 10B (e.g., concentrations of potential toxic agents), timestamp with 2B, and vehicle location with 5B. Then, in our scenario, MDP can pack 65 chunks in a single 1500B summary, even without exploiting any data aggregation or encoding technique. In usual deployment environments chunks are generated every 2–10 seconds and, thus, a single summary can include all the chunks in about a 2–10-minute interval. MSI permits MDHP to access raw sensed data independently of actual sensor implementation, thus simplifying the integration with many different types of sensors. MSI currently implements methods to access camera streaming outputs, serial port I/O streams, and GPS information by only specifying a high-level name for the target sensor. To interface with sensor implementations, MSI exploits well-known standard specifications to achieve high portability and openness: the Java media framework (JMF) API (applications programming interface), the sun communication API, and the JSR179 location API.

7.4 MobEyes Diffusion/Harvesting Processor

Here we review the design principles of MobEyes diffusion/harvesting processor (MDHP) protocols, namely disruption tolerance, scalability, and nonintrusiveness. Private vehicles (regular nodes) opportunistically and autonomously spread summaries of sensed data by exploiting their mobility and occasional encounters. Police agents (authority nodes) proactively build a low-cost distributed index of the mobile storage of sensed data. The main goal of the MDHP process is the creation of a highly distributed and scalable index that allows police agents to place queries to the huge urban monitoring database without ever trying to combine this index in a centralized location.

7.4.1 MDHP Protocol Design Principles

A vehicular sensing platform, built on top of a vehicular ad hoc network, has the following specific characteristics that differentiate it from more established and investigated deployment scenarios. First, it has unique mobility patterns. Vehicles move at a relatively high speed (up to 80 mph) on roads that may have multiple lanes and different speed limits. Instead of random motion patterns, drivers navigate a set of interest points (e.g., home, workplace) by following their preferred paths. The dynamic behavior of mobile nodes (i.e., join/leave/failure) usually results in modifications of the set of participating nodes (called churning). Moreover, there are time-of-the-day effects such that the overall volume of vehicles changes over time (e.g., high density during rush hours or some special event). Thus, the spatial distribution of vehicles is variable, nonuniform and, in some cases, the network can be partitioned. As a result, vehicles may experience disruptions and intermittent connectivity. Second, the network scales up to hundreds of thousands of vehicles because sensing applications primarily target urban environments. Finally, unlike conventional sensor networks where the communication channel is dedicated to sensing nodes, the primary purpose of vehicular communications is for safety navigation and sensing platforms cannot fully utilize the overall available bandwidth.

Under these circumstances, we have decided to consider the following design principles for MDHP protocols:

- *Disruption tolerance:* It is crucial that MDHP protocols must be able to operate even with disruptions (caused by sparse network connectivity, obstacles, and nonuniform vehicle distribution) and with arbitrary delays. High churning of vehicles must be considered; for robustness purposes, data replication is a must.
- *Scalability:* MDHP protocols must be able to scale up to hundreds of thousands nodes (i.e., the number of vehicles potentially interworking in a large city).
- *Nonintrusiveness:* Intrusive protocols may cause severe contention with safety applications and could deter reliable propagation of important messages in a timely fashion. MDHP protocols should not disturb other

safety applications; limiting the use of bandwidth below a certain threshold is imperative.

Given the above motivations and the deriving design principles, simple flooding and probabilistic gossiping cannot be used for MDHP. In fact, they require the network to be fully connected (i.e., nondelay tolerant) and cause the network traffic to scale with the number of nodes in the network (i.e., nonscalable, intrusive). For instance, in epidemic data dissemination (EDD), where data is spread whenever connectivity is available (i.e., data is replicated without any restriction), the size of exchanged data scales with network size; thus, EDD is intrusive and nonscalable. EDD is more suitable for sparse and small-scale wireless networks. Unlike these approaches, we propose to use *mobility-assist* information dissemination and harvesting in MobEyes. Data is replicated via periodic *single-hop* broadcasting (i.e., only the data originator can broadcast its data) for a given period of time. Through the mobility of carriers, the data is delivered to a set of harvesting agents. Mobility-assist dissemination and harvesting per se are delay and disruption tolerant, and, as extensively detailed in the following, single-hop broadcasting-based localized information exchange makes our protocols nonintrusive and scalable.

7.4.2 Summary Diffusion

By following the above guidelines, in MobEyes any regular node periodically advertises a new packet with generated summaries to its current neighbors in order to increase the opportunities for agents to harvest summaries. Clearly, excessive advertising introduces too much overhead (as in EDD), while no advertising at all (i.e., direct contact) introduces unacceptable delays, as agents need to directly contact each individual source of monitoring information to complete the harvesting process. Thus, MobEyes tries to trade off delivery latency with advertisement overhead. As depicted in Figure 7.2, a packet header includes packet type, generator ID, a locally unique sequence number, a packet generation timestamp, and the generator's current position. Each packet is uniquely identified by a <generator ID, sequence number> pair and contains a set of summaries locally generated during a fixed time interval.[1]

Neighbor nodes receiving a packet store it in their local summary databases. Therefore, depending on the mobility and the encounters of regular nodes, packets are opportunistically diffused into the network (*passive* diffusion). MobEyes

1. The optimal interval can be determined by noting that the harvesting time distribution is characterized by average (μ) and standard deviation (ρ). Then, the Chebyshev inequality $P(|x-\mu| \geq k\rho) \leq \frac{1}{k^2}$ allows us to choose k such that we can guarantee the needed *harvesting latency*, thus fixing the period as $\mu + k\rho$.

can be configured to perform either single-hop passive diffusion (only the original source of the data advertises its packet to current single-hop neighbors) or k-hop *passive* diffusion (the packet travels up to k-hop as it is forwarded by j-hop neighbors with $j < k$). Other diffusion strategies could be easily included in MobEyes, for instance, single-hop active diffusion where any node periodically advertises all packets (generated by itself and received from others) in its local summary databases, at the expense of a greater traffic overhead. As detailed in the experimental evaluation section, in a usual urban VANET (node mobility restricted by roads), it is sufficient for MobEyes to exploit the lightweight k-hop passive diffusion strategy with very small k-values to achieve the desired diffusion levels.

Figure 7.3 depicts the case of two sensor nodes, C1 and C2, that encounter other sensor nodes while moving (the radio range is represented as a dotted circle). For ease of explanation, we assume that there is only a single encounter, but in reality any nodes within a dotted circle are considered encounters. In the figure, a black triangle with a timestamp represents an encounter. According to the MobEyes summary diffusion protocol, C1 and C2 periodically advertise a new summary packet $S_{C1,1}$ and $S_{C2,1}$, respectively, where the subscript denotes $\langle ID, Seq.\# \rangle$. At time $T - t_4$, C2 encounters C1, and thus they exchange those packets. As a result, C1 carries $S_{C2,1}$ and C2 carries $S_{C1,1}$.

Summary diffusion is time and location sensitive (spatial–temporal information diffusion). In fact, regular nodes keep track of freshness of summary

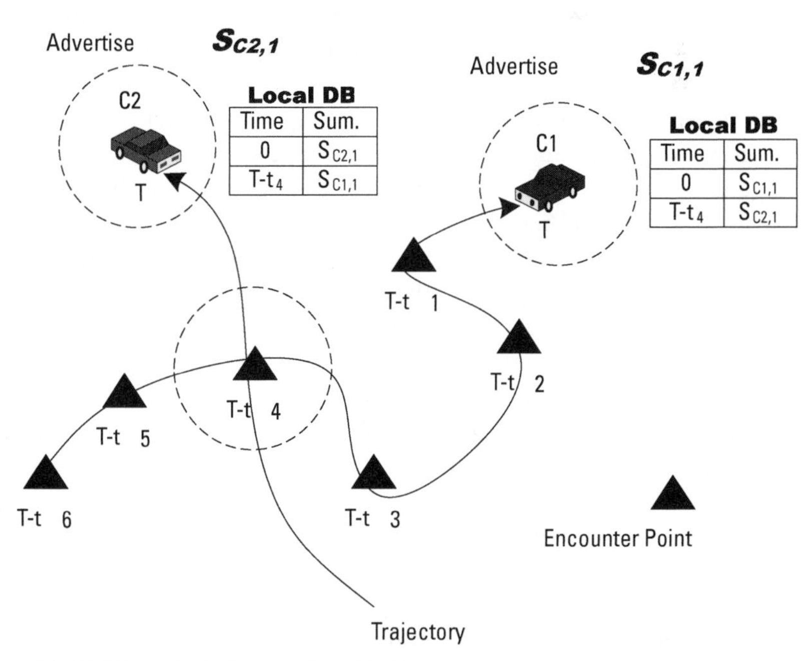

Figure 7.3 MobEyes single-hop passive diffusion.

packets by using a sliding window with a maximum window size (i.e., fixed expiration time). In addition, since a single summary packet may contain multiple summaries, we define the *aggregate* packet sensing location as the average of the sensing locations of all summaries in the packet. When a packet expires or the packet originator moves away more than a threshold distance from the aggregate packet sensing location, the packet is automatically disposed. The expiration time and the maximum distance are system parameters that should be configured depending on urban monitoring application requirements. Let us also briefly note that summaries always include, of course, the time and location where the sample was taken. Upon receiving an advertisement, neighbor nodes keep the encounter information (the advertiser's current position and current timestamp). This also allows MobEyes nodes, when the type of urban monitoring application makes it applicable, to exploit spatial–temporal routing techniques such as last encounter routing [33] and to maintain a georeference service that can be used to proactively access the data. That is obtained as a simple by-product of summary dissemination, without additional costs.

7.4.3 Summary Harvesting

In parallel with diffusion, MobEyes summary harvesting takes place. There are two possible modes of harvesting *diffused* information, namely *on-demand* mode and *proactive* (or background) mode. The on-demand mode is suitable for cases when the police agents react to an emergency call, for example, the earlier mentioned poisonous gas incident. Police agents converge to the outskirts of the area (keeping a safe distance of course) and query vehicles for summaries that correspond to a given time interval and area (i.e., time–space window). The agents can *flood* a query with such information (à la route request in on-demand routing). Each regular node resolves the query and returns its summary to the agents (à la route reply in on-demand routing). So, the *on-demand* strategy is more likely a traditional sensor network–based data harvesting protocol, e.g., directed diffusion [34]—i.e., the reply is *diffused* in the direction of the querier. The main difference in MobEyes would be that a query has a spatiotemporal range. The police agents as a team collect as many summaries of interests as they can.

However, it is not very practical to exploit an on-demand strategy in MobEyes for the following reasons. First, agents should provide a query with the *range* of spatiotemporal information even in the usual cases when they have no precise prior information. An improperly chosen query range may require accessing a large number of vehicles. For instance, for a given chemical attack that happened in a busy street, the police may want to find out *all* the vehicles passing by the scene within the last several hours, resulting in thousands of vehicles. Second, the on-demand scheme is quite similar to conventional data harvesting and requires maintaining a *concast* tree from the query originator. But the number of vehicles is expected to be very large in MobEyes and vehicles are mobile; thus, route management would have relevant implementation costs in terms of overhead. Third, collecting a complete set of summaries would be nontrivial and not always possible due to intermittent connectivity and network partitions. To overcome intermittence, query/response can be opportunistically disseminated in a delay tolerant network style. However, in such a case, the delay will be comparable to *proactive* harvesting, with the additional costof a separate dissemination process. Finally, in *covert* operations, agents may not want to broadcast a query (e.g., in order not to alert the criminals that are currently being pursued). Then, the police must consider physically dispatching agents to the location of interest and collecting summaries via physical contacts. In summary, this on-demand *mechanical* search is extremely costly and potentially very time consuming.

To overcome such issues, we are proposing a *proactive* version of the search, based on distributed index construction. Namely, in each area there are agent vehicles that collect all the summaries as a background process and create a distributed index. In this case, there is no time–space window concern during collection. The only requirement is to collect all the summaries in a particular area. Now, for specific information (e.g., the poisonous gas–level monitoring), the query is directed to the target regular vehicles by exploiting the agents' distributed index. The time–space window concept can be applied to the index to find the vehicles in a particular place and time and then pursue the hot leads. For example, upon receiving a specific query, the agents collectively examine the index, find a match, and decide to inspect in more detail the video files collected by a *limited* number of vehicles. The vehicles can be contacted based on the originator's vehicle ID number stored in each summary. A message is sent to each vehicle requesting it to upload the file at the nearest police access point. Note that the request message can exploit georouting, either exploiting the geolocation service that maps vehicle ID to the current vehicle location or using *last encounter routing* techniques [33, 35]. The latter is particularly convenient here because nodes memorize the time and place of encounters as the time summary exchanges take place.

After the desired summaries have been found, both on-demand and proactive processes require contacting the cars under consideration. However, the proactive approach is much more powerful as it can speed up the search considerably. For instance, if the inspection of the information collected in the crime area indicates a possible escape direction by the terrorists, one can immediately search again the proactively created index for a new time–space window, without having to do another time-consuming collection of summaries from vehicles. On the negative side, maintaining that index is costly, as agent resources must be dedicated to the task.

In the sequel we assume proactive index construction. Thus, the agents collect all summaries indiscriminately. There is no loss of generality, however, since the procedure also allows on-demand index construction for a specific time–space request. In fact, the only difference between the two harvesting schemes is the size of the set being harvested. In the on-demand scheme, the target set is a specific time–space window. In the proactive scheme, the target set is the entire geographic area within agent responsibility; there is no limit on harvesting time, though old records are timed out.

By considering the proactive (or background) harvesting model, the MobEyes police agent collects summaries from regular nodes by periodically querying their neighbors. The goal is to collect all the summaries generated in the specified area. Obviously, a police node is interested in harvesting only summary packets it has not collected so far: to focus only on missing packets, a MobEyes authority node compares its list of summary packets with that of each neighbor (set difference problem). To this end, we use a Bloom filter [36], a space-efficient randomized data structure for representing a set that is mainly used for membership checking.[2] A Bloom filter for representing a set of ω elements, $S = \{s_1, s_2, \ldots, s_\omega\}$, consists of *m bits*, which are initially set to 0. The filter uses ℓ independent random hash functions $h_1,\ldots,h_\ell$ within m bits. By applying these hash functions, the filter records the presence of each element into the m bits by setting ℓ corresponding bits. To check the membership of the element x, it is sufficient to verify whether all $h_i(x)$ are set to 1. A MobEyes police agent uses a Bloom filter to represent its set of already harvested and still valid summary packets. Since each summary has a <unique node ID, sequence number> pair, we use this as input for the hash functions. The MobEyes harvesting procedure consists of the following steps:

1. The agent broadcasts a *harvest* request message with its Bloom filter.
2. Each neighbor prepares a list of *missing* packets based on the received Bloom filter.
3. One of the neighbors returns missing packets to the agent.
4. The agent sends back an acknowledgment with a piggy-backed list of returned packets and, upon listening to or overhearing this, neighbors update their lists of missing packets.
5. Steps 3 and 4 are repeated until all missing packets are sent.

An example of summary harvesting is shown in Figure 7.4. The agent first broadcasts its Bloom filter related to the packets collected so far (P2, P4, P6, P7, P9, and P10) as in Figure 7.4(a). Each neighbor receives the filter and creates a list of missing packets. For example, C3 has P3 and P8 to return, while C4 has P1 and P8. In Figure 7.4(b), C2 is the first node to return missing packets (P1, P3) and the agent sends back an acknowledgment piggy-backed with the list of

2. Readers can find a survey of Bloom filter–based networking applications in [37].

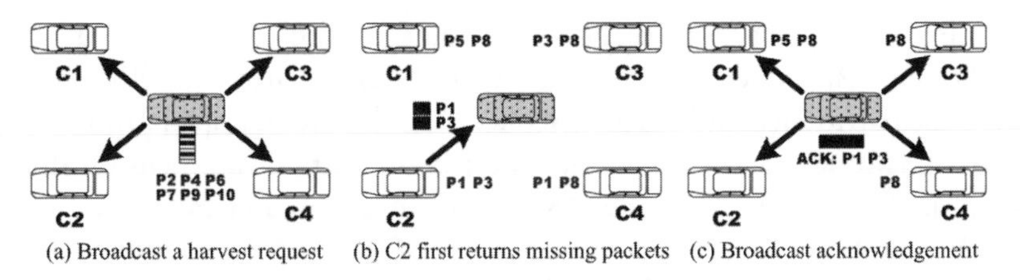

Figure 7.4 (a–c) MobEyes proactive summary harvesting.

received packets. Neighbor nodes overhear the message and update their lists: C3 removes P3 and C4 removes P1 from their lists, as depicted in Figure 7.4(c).

Note that membership checking in a Bloom filter is probabilistic and false positives are possible, even if rare.[3] In Figure 7.4, for example, a false positive on P1 makes C2 return only P3. None of the neighbors can send P1. To deal with this problem, the agent periodically changes the set of hash functions. Suppose that we use m hash functions. Each hash function is a pseudorandom function (PRF) where a PRF takes two arguments, X_k is the key ($k = 1, 2, \ldots, m$) and i is the input value, and produces an output value $o = F_{X_k}(i)$. All nodes are initially given the same set of keys X_k where $k = 1,2,\ldots,m$. For the purpose of periodic changes, the key for a k-th hash function in an nth epoch, X_k^n, can be calculated by hashing the initial value X_k^n n times. The Bloom filter contains the epoch number, which allows the neighbors to find the set of keys for m hash functions. Even with failure, by periodically incrementing the epoch number, the agent can gather the missing packets. Note also that in our application a set changes over time with summaries being inserted and deleted because summaries have spatiotemporal properties. For deletion operations, Fan et al. introduced the idea of counting Bloom filters,where each entry in the Bloom filter is not a single bit but rather a small counter [38]. When an item is inserted, the corresponding counters are incremented; when an item is deleted, the corresponding counters are decremented. For actual filter transfer, instead of sending the full counting Bloom filter, each counter is represented as a single bit (i.e., 1 if its value is greater than 0; 0 otherwise).

For the sake of simplicity, thus far we assumed that there is a single agent working to harvest summaries. Actually, MobEyes can handle concurrent harvesting by multiple agents (possibly several hops apart) that can cooperate by exchanging their Bloom filters among multihop routing paths; thus, this creates a distributed and partially replicated index of the sensed data storage. In particular, whenever an agent harvests a set of j new summary packets, it broadcasts its Bloom filter to other agents, with the benefits in terms of latency and accuracy

3. The false positive probability is $p_f = \left(1-\left(1-\frac{1}{m}\right)^{\ell\omega}\right)^{\ell} \simeq \left(1-\varepsilon^{-\frac{\ell\omega}{m}}\right)$ where m is the total number of bits, ω is the number of elements, and ℓ is the number of hash functions [37].

shown in the following sections. Note that strategically controlling the trajectory of police agents, properly scheduling Bloom filter updates, and efficiently accessing the partitioned and partially replicated index are still open problems and part of our future work. In the following section, instead, we focus on the primary goal of identifying the trade-offs between dissemination and harvesting in a single geographic area, and the dependence of MobEyes performance on various parameters. We also analyze the traffic overhead created by diffusion/harvesting and show that it can scale well to very large node numbers.

7.5 MobEyes Performance Evaluation

We evaluated MobEyes protocols using NS-2, a packet-level network simulator [39]. This section shows the most important results, with the goal of investigating MobEyes performance from the following perspectives.

1. *Analysis Validation.* We simulate MobEyes protocols for summary collection on regular nodes as well as for agent harvesting and show that they confirm our main analytic results.
2. *Effect of k-hop Relay and Multiple Agents.* We examine how MobEyes effectiveness can be increased by leveraging *k*-hop passive diffusion and the deployment of multiple agents.
3. *Summary Diffusion Overhead.* We investigate the trade-off between harvesting delay and the load imposed on the communication channel.
4. *Stability Check.* We verify that the system is stable, even with the high summary generation rate.
5. *Tracking Application.* We prove MobEyes effectiveness in supporting a challenging tracking application, where trajectories of regular nodes are locally reconstructed by a police agent based on harvested summaries.
6. *Border Effects and Turnover.* We show that MobEyes performance does not dramatically change in case of more dynamic mobility models, where nodes are allowed to enter/exit from the simulated area.

Additional experimental results and MobEyes implementation details are available at http://www.lia.deis.unibo.it/Research/MobEyes/.

7.5.1 Simulation Setup

We consider vehicles moving in a fixed region of size 2,400m×2,400m. The default mobility model is real-track (RT), introduced by our colleagues in [40]. RT permits to model realistic vehicle motion in urban environments. In RT nodes move following virtual tracks, representing real accessible streets on an arbitrary loaded road map. For this set of experiments, we used a map of the Westwood area in the vicinity of the UCLA campus, as obtained by the U.S. Census Bureau data for street-level maps [41] (Figure 7.5). At any intersection, each node (i.e., vehicle) randomly selects the next track it will run through; speed is periodically allowed to change (increase or decrease) by a quantity uniformly distributed in the interval $[0, \pm\Delta s]$. Let us remark that MobEyes agents do not exploit any special trajectory or controlled mobility pattern, but move conforming with regular nodes.

Our simulations consider a number of nodes $N = 100, 200, 300$. Vehicles move with an average speed $v = 5, 15, 25$ m/s; to obtain these values, we tuned the minimum speed to $v_m = 1$ m/s and the maximum speed to $v_M = 11, 31, 51$ m/s, respectively. The summary advertisement period of regular nodes and the harvesting request period are kept constant and equal to 3 seconds through all the simulations. We use a Bloom filter with 8,192 bits (1024B) and 10 hash functions. Since we have a large filter size (i.e., 8,192 bits) compared to the number of summaries, the false positive probability is negligible. We note that if the value of this parameter is too large, MobEyes effectiveness is reduced since it is possible that two nodes do not exchange messages, even if they occasionally enter into each other transmission range; this effect is magnified as node speed v increases. The chosen value has been experimentally determined to balance the effectiveness of our protocol and the message overhead, even in the worst case

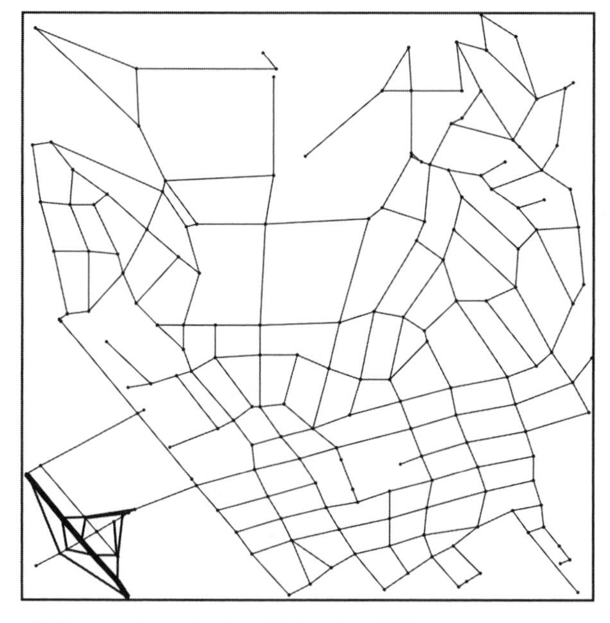

Figure 7.5 Map of Westwood area in the vicinity of the UCLA campus.

(i.e., $v = 25$ m/s). A deeper and more formal investigation of the optimal value of the advertisement period is the object of future work.

Finally, we modeled communications as follows: MAC protocol IEEE 802.11, a transmission band of 2.4 GHz, a bandwidth of 11 Mbps, a nominal radio range equal to 250m, and a two-ray ground propagation model [42]. The values of these parameters have been chosen similar to other work in the field [13, 16]. Where not differently stated, reported results are average values out of 35 repetitions. Other MobEyes configuration parameters are introduced in the following sections, when discussing the related aspects of MobEyes performance.

7.5.2 Stability Check

In the following, we investigate the stability of MobEyes by verifying that continuous summary injections do not influence its performance to a large extent. In particular, we show that the ratio of summaries harvested on longer periods remains acceptable and that the harvesting latency does not grow as time passes. With regard to the results presented so far, here we remove the assumption about the single summary generation epoch at $t = 0$. Nodes generate new summaries with period $T = 120s$ and advertise the last generated summary: let us observe that according to the discussion presented in Section 7.3 this rate represents a practical worst case. For the sake of clarity of presented results, we hold the synchronicity assumption: all nodes simultaneously generate new summaries at intervals multiple of T. We obtained similar performance with differently distributed generation intervals, i.e., Poisson with an average value T, but plots are far more jumbled. The following results are reported for the case of a single harvesting agent, $k = 1$, $N = 100$, $v = 15$ m/s, and nodes moving according to the RT model. Figure 7.6 plots the cumulative distribution of the number of summaries generated and harvested as a function of time (we ran simulations for 6,000s). The graph shows that the harvesting curve tracks the generation curve with a certain delay. Figure 7.7 provides further evidenceof the stability of the system; curves show the harvesting latency for summaries generated during some generation epochs. For the sake of figure clarity, the graph does not exhaustively represent every generation epoch, but only samples one generation epoch every $T * 7 = 840s$ until the end of the simulation time. The different curves show similar trends, without any performance degradation caused by the increase of the number of summaries in the network. These results prove that MobEyes achieves completeness in harvesting generated summaries even in practical worst cases.

We also investigated if higher summary generation rates afflict MobEyes performance. We shortened T from 120s to 6s (with $T = 6s$, the chunk generation rate is 100 ms). Such a generation rate is much greater than the one required for the set of applications addressed by MobEyes. Simulation results prove that Mob-

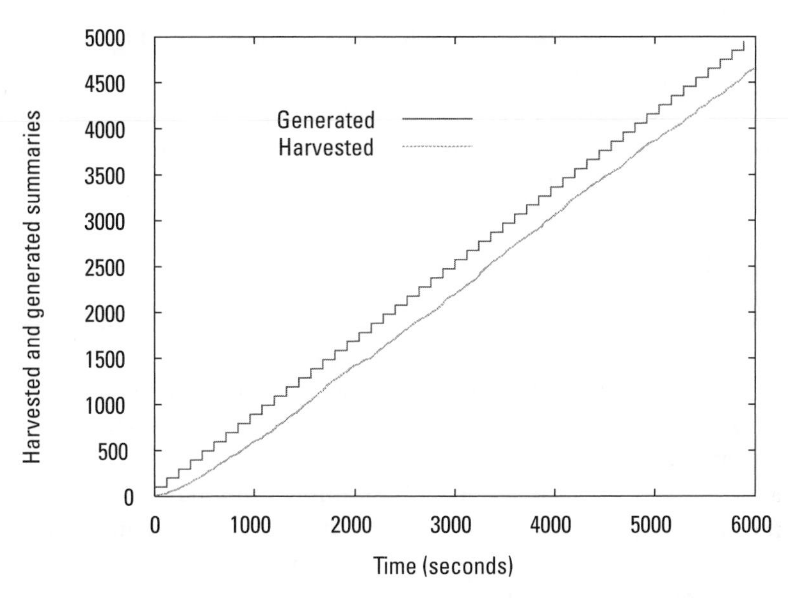

Figure 7.6 Cumulative distribution of generated and harvested summaries over all epochs.

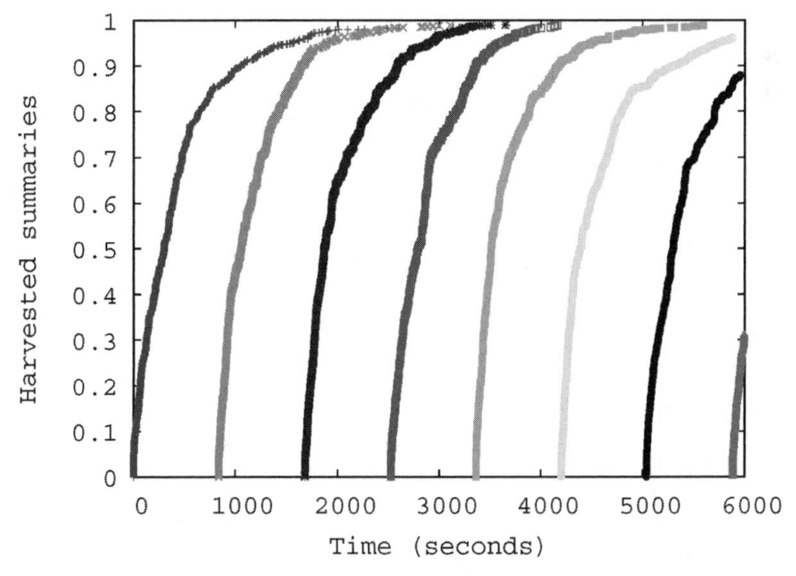

Figure 7.7 Cumulative distribution of harvested summaries per epoch: each curve shows the harvesting latency for summaries generated during some generation epoch. For the sake of clarity, we only show the plots for every seven generation epochs (i.e., every $T * 7 = 840s$).

Eyes performance starts degrading only when we set $T < 30s$. Figure 7.8 shows the harvesting process for two epochs ($0s$ and $2{,}520s$) and compares $T = 120s$ with $T = 6s$. The second case shows that MobEyes performance degrades gracefully as the generation epoch shortens, thus demonstrating the high stability of the system when operating in the usual summary rate conditions.

7.5.3 Tracking Application

In the introduction we sketched some application cases for MobEyes. For the sake of proving its effectiveness in supporting urban monitoring, we also simulated a vehicle tracking application where the agent reconstructs node trajectories exploiting the collected summaries. This is a challenging application, since it requires our system: (1) to monitor a large number of targets (i.e., all participant vehicles), (2) to periodically generate fresh information on these targets, since they are highly mobile, and (3) to deliver to the agent a high share of the generated information. Moreover, since nodes are generally spread all over the area, this application shows that a single agent can maintain a consistent view of a large zone of responsibility. More in detail, as regular cars move in the field, they generate new summaries every $T = 120s$ and continuously advertise the last generated summary. Every summary contains 60 *summary chunks*, which are created every $ChunkPeriod = 2s$ and include the license plate and position of the vehicle nearest to the summary sender at the generating time, tagged with a timestamp. The application exploits the MobEyes diffusion protocol with $k = 1$ to spread the

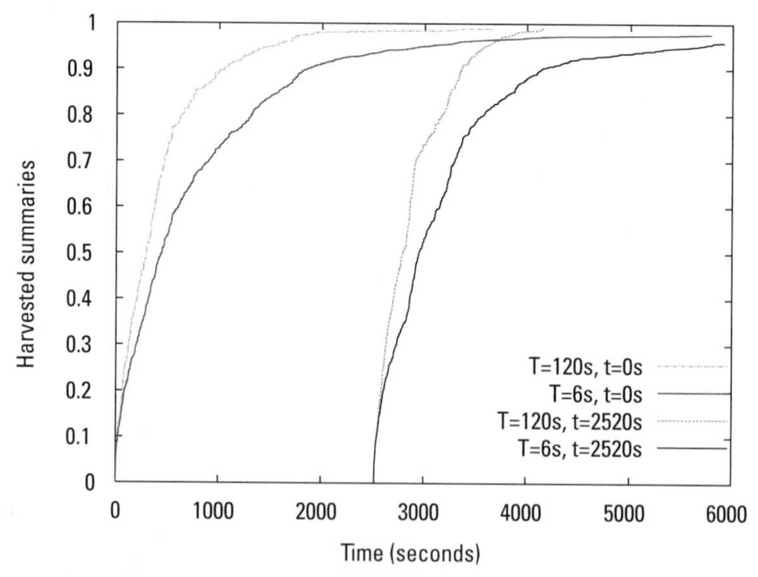

Figure 7.8 Cumulative distribution of harvested summaries per epoch.

summaries and deliver as much information as possible to a single agent scouting the ground. As the agent receives the summaries, it extracts the information about node plates and positions, and tries to reconstruct node trajectories within the area. This is possible by aggregating data related to the same license plate, reported from different summaries.

To determine the effectiveness of MobEyes we decided to evaluate the *average uncovered interval* and *maximum uncovered interval* for each node in the field. Given a set of summary chunks related to the same vehicle and ordered on a time basis, these parameters measure, respectively, the average period for which the agent does not have any record for that vehicle and the longest period. The latter typically represents situations in which a node moves in a zone where vehicle density is low; thus, it cannot be traced by any other participant. We associated the average and maximum uncovered intervals to each simulated node, and present the results in Figure 7.9 (note the logarithmic scale on the Y-axis). Every point in the figure represents the value of the parameter for a different node. We sorted nodes on the X-axis so that they are reported with increasing values of *uncovered interval.* Results are collected along a 6,000*s* simulation. The plot shows that in most cases the average uncovered interval floats between (2.7*s* – 3.5*s*); the maximum uncovered interval shows that even in the worst cases the agent has at least one sample every 200*s* for more than 90% of the participants. A more immediate visualization of the inaccuracy is given in Figure 7.10. This figure shows, for the case of a node with a maximum uncovered interval equal to 200*s* (i.e., locating this node in the lowest 10th percentile), its real trajectory (the unbroken line), and the sample points the agent collected.

7.5.4 Border Effects and Turnover

Usual mobility models [43], such as RWP, MAN, and RT, assume that nodes remain within the simulated area during the entire simulation (in the following, we indicate them as *closed* mobility models). Even if this does not necessary hold for MobEyes applications, we observe that this assumption does not invalidate our findings. First, if we consider a sufficiently large area, on the order of several hundreds *Km* squares, the amount of time that nodes continuously reside within the area is likely very long, namely, for most nodes a closed mobility model. Second, the worst effect of dynamic scenarios takes place when nodes leave a specific area carrying several summaries (locally generated or collected) not harvested by the local agent yet. Nonetheless, we remark that carried information does not vanish as nodes leave, but can be harvested later by remote agents, responsible for the adjacent area the leaving nodes are moving into.

However, to estimate how node entrances/exits impact presented results, we tested MobEyes with a novel mobility model, *open-RT*, which takes these effects into account. In open-RT nodes follow the same patterns of RT, with one

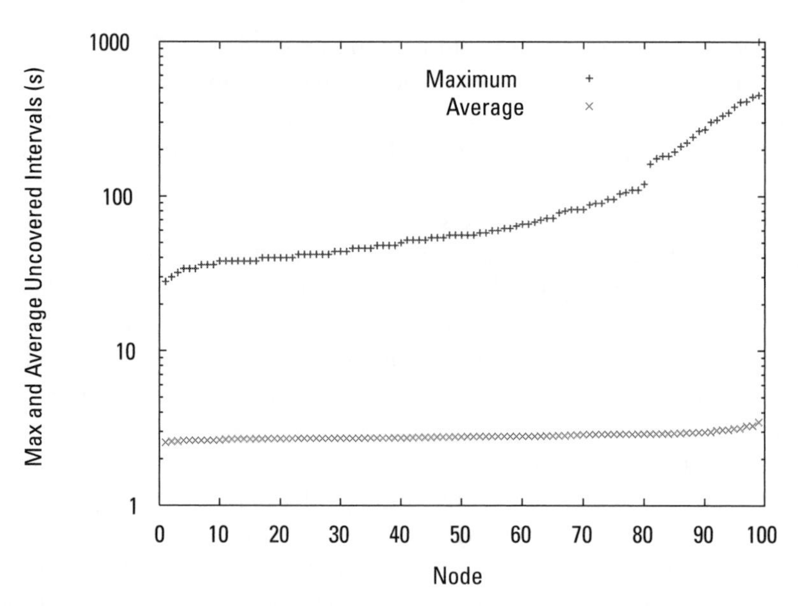

Figure 7.9 Maximum uncovered intervals per node.

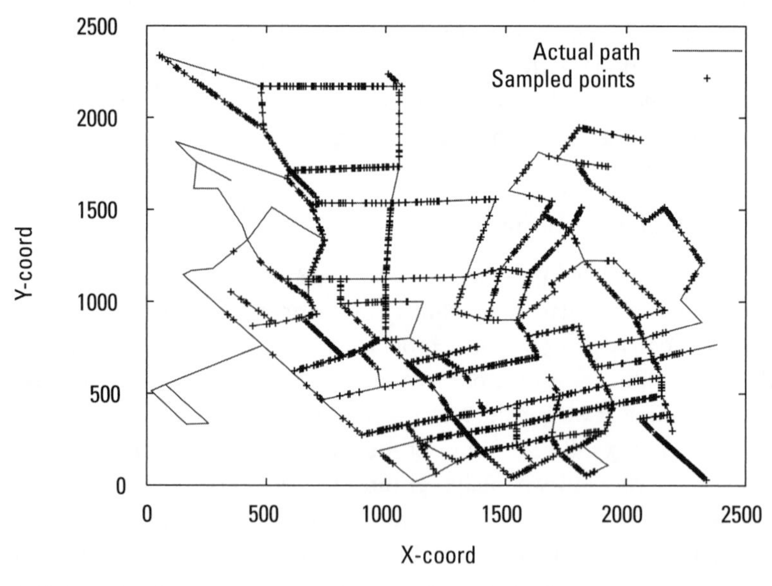

Figure 7.10 Actual node trajectory versus harvested sampled points.

exception: as soon as a node reaches the endpoint of a track, close to the boundary of the area, it suddenly disappears. To maintain unchanged the number of nodes within the area, and obtain results comparable to the ones presented in the previous sections, we assume that the net vehicle flow in/out of the area is null. Thus, any node exiting from the area is immediately replaced with one node entering; the latter is placed at the endpoint of a random road, close to the boundary of the area.

This dynamic effect is better evaluated for long simulation periods and periodic summary generation epochs. Thus, we confirm the settings used in Sections 7.5.2 and 7.5.3; in addition, we consider a single harvesting agent, $k = 1$, $N = 100$, $v = 15$ m/s. Nodes generate new summaries synchronously, and only as long as they remain in the area. To avoid nodes staying within the area only for very short periods, we introduce a constraint on their minimum residing time equal to 10% of the whole simulation. Even with this assumption, more than 550 nodes need to take turns on the simulation area to maintain 100 nodes always present. The agent does not follow the open-RT model, but traditional RT (i.e., it always remains within the area).

Figures 7.11 and 7.12 present results corresponding to those in Sections 7.5.2 and 7.5.3, but obtained with the open-RT model. Significant conclusions can be drawn, especially from Figure 7.11: also under these unfavorable assumptions, the agent is able to collect more than 85% of any generated summary, and in most cases it reaches 90%. By inspecting simulation traces, we could find that missing summaries generally originate by vehicles leaving the area within a short interval from any epoch. In that case, the last generated summary is only advertised for that short interval and cannot spread enough to reach the agent. Let us remark once more that those summaries are not irreparably lost, but will probably be harvested by agents in charge of the adjacent areas. Figure 7.12 shows average and maximum uncovered intervals as obtained with the open-RT model. The quality of the reconstructed trajectories is only slightly degraded, given that the average uncovered interval is below 4*s* for more than 75% of the nodes (and below 10*s* for 90%), and that the 85th percentile of the vehicles can be tracked with a worst-case inaccuracy of 200*s*.

7.6 MobEyes Privacy and Security

MobEyes nodes continually generate and diffuse summaries containing private information (e.g., license plate numbers). Thus, privacy is of critical importance. On the one hand, nonauthorized nodes must not be allowed access to private information, including vehicle location. On the other hand, the harvesting process should not reveal the information that is being sought, since this may tip the attackers and/or cause unnecessary panic in the public. In general, we can summarize the security requirements of MobEyes as follows.

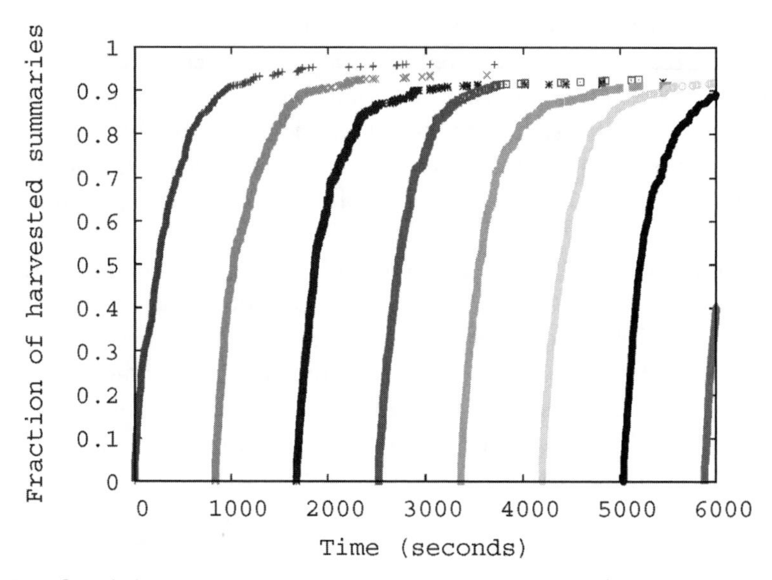

Figure 7.11 Cumulative distribution of harvested summaries per epoch (open-RT).

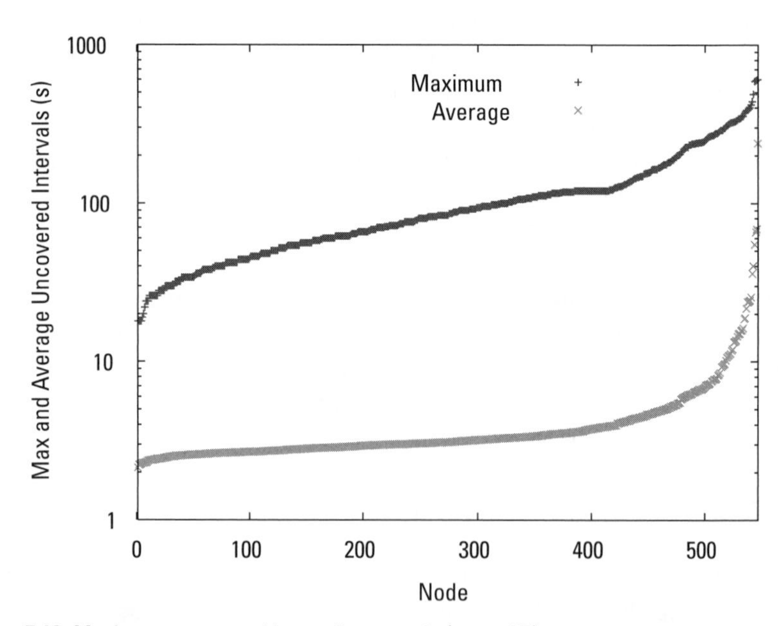

Figure 7.12 Maximum uncovered intervals per node (open-RT).

- *Authentication.* Harvesting agents (authority nodes) must authenticate summary senders and vice versa.
- *Non-repudiation.* A summary originator cannot deny the transmission of a summary (liability issue); in that way, upon request from the agent, the summary source must submit the full file with related sensed data.
- *Privacy.* Only legitimate users (authority nodes) can access summaries. Moreover, summaries must be privately advertised such that the attackers cannot track users.
- *Service Availability.* MobEyes summary diffusion/harvesting should be protected from denial of service (DoS) attacks.
- *Data Integrity.* MobEyes should be able to filter out false summary data injected by attackers.
- *Query Confidentiality.* In some cases (e.g., bioattacks and search for crime suspects), even the nature of the query injected by harvesting agents should not be disclosed, so as to not create unnecessary panic in the population or to avoid tipping the criminals.

One important aspect that sets apart MobEyes forensic sensed data security from conventional VANET security for safe navigation is the real time and criticality of the safe navigation application. For instance, consider a dangerous curve on the road monitored by an e-mirror. If no car is coming, the e-mirror tells the driver to proceed at normal speed, else, it tells her to slow down. An adversary can intercept and replay a message from the mirror and tell the driver that the way is clear, while instead a truck is coming at high speed behind the curve. In this safe drive application, it is mandatory to authenticate alert messages. Thus, in safe navigation applications, message authentication is far more important than privacy. For instance, privacy concerns should not prevent a driver from alerting the vehicles behind her that there is a boulder on the road.

MobEyes has strongly different security requirements. A false report cannot create much damage since it is not acted upon immediately (for example, a wrong set of license plates at the crime scene). There is plenty of time to detect and, if necessary, punish the impostors. On the other hand, drivers that propagate summaries want to be assured that their privacy will not be violated. This major difference in security concerns leads to MobEyes security approaches that are quite different (and in fact much simpler and generally more efficient) than conventional VANET security solutions. Readers can find general security issues for VANET in [44]. For the sake of brevity, in this section we will simply outline several MobEyes security approaches. In MobEyes, we assume the existence of a public key infrastructure (PKI). Since standard PKI mechanisms provide authentication and non-repudiation, we focus on the rest of the MobEyes requirements, namely privacy, service availability, and data consistency, by addressing the following MobEyes-specific attack models and a brief description of possible solutions. Readers can find the details for each solution in the extended version of this chapter [45].

- *Location Tracking.* Periodic broadcasting of identical summaries could facilitate attackers in tracking the route of a vehicle. To change the encrypted summary, one can introduce perturbation of time and position information.
- *Denial of Service (DoS).* Attackers could inject a large number of bogus summaries in order to slow down correct summary harvesting by agents. MobEyes can check summary validity using the PKI and the rate of generating valid summaries can be limited using *rate-limited summary diffusion*, which shares the same idea of the RREQ rate limit in a secure routing protocol [46].
- *False Data Injection.* Attackers could inject fabricated summaries in order to mislead investigations or make the data inconsistent. Statistical methods in conventional sensor networks [47–49] can be used. As noted in [50], mobility of vehicles makes it hard in reality.
- *Query Confidentiality.* Attackers could infer *important* information from the content of police queries. *Private keyword searching*, proposed by Ostrovsky et al. [51], can be used. Secure filters can be distributed to the regular vehicles and agents can harvest the resulting encrypted data.

7.7 Conclusions

In this chapter, we proposed a valid decentralized and opportunistic solution for proactive urban monitoring in VSN. The MobEyes key component is MDHP, which works by disseminating/harvesting summaries of sensed data and uses original opportunistic protocols that exploit intrinsic mobility of regular and authority nodes. One of the reasons for using original dissemination protocols is, for instance, to overcome the intermittent connectivity of urban grids in off-peak hours, which precludes the exploitation of conventional search/propagation techniques based on ad hoc multicast and broadcast. We showed that MDHP protocols are disruption tolerant, scalable, and nonintrusive via both analytic models and extensive simulations. We also illustrated the application of this technique to a typical homeland defense application, namely the reconstruction of routes taken by a suspect vehicle. The reconstruction was possible thanks to the cooperation of the hundreds of nodes that saw the vehicle in question. MobEyes can be configured to achieve the most suitable trade-off between latency/completeness and overhead by properly choosing primarily its k-hop relay scope and the number of harvesting agents. These encouraging results are stimulating further research activities, particularly determining the best trajectory of mobile agents when they collaborate in summary harvesting; establishing the optimal value for the summary advertisement period as a function of node speed/population and of traffic/latency; and finally, exploring hybrid strategies that combine broadcast with epidemic dissemination, with dynamically adapting to urban density conditions and application needs.

References

[1] A. Nandan, et al. "Co-operative Downloading in Vehicular Ad-Hoc Wireless Networks." *IEEE WONS*, St. Moritz, Switzerland, January 2005.

[2] A. Nandan, et al. "AdTorrent: Delivering Location Cognizant Advertisements to Car Networks." *IFIP WONS*, Les Menuires, France, January 2006.

[3] Q. Xu, et al. "Vehicle-to-vehicle Safety Messaging in DSRC." *ACM VANET*, Philadelphia, PA, October 2004.

[4] J. Ott and D. Kutscher. "A Disconnection-Tolerant Transport for Drive-Thru Internet Environments." *IEEE Infocom*, Miami, FL, April 2005.

[5] U. Lee, et al. "Efficient Data Harvesting in Mobile Sensor Platforms." *IEEE PerSeNS*, Pisa, Italy, March 2006.

[6] MIT's CarTel Central. http://cartel.csail.mit.edu/.

[7] B. Hull, et al. "Car-Tel: A Distributed Mobile Sensor Computing System." *ACM SenSys*, Boulder, CO, October–November 2006.

[8] U. Lee, et al. "FleaNet: A Virtual Market Place on Vehicular Networks." *IEEE V2VCOM*, San Francisco, CA, July 2006.

[9] M. Caliskan, D. Graupner, and M. Mauve. "Decentralized discovery of free parking places." *ACM VANET*, Los Angeles, CA, September 2006.

[10] D. Sormani, et al. "Towards Lightweight Information Dissemination in Inter-vehicular Networks." *ACM VANET*, Los Angeles, CA, September 2006.

[11] M. Torrent-Moreno, D. Jiang, and H. Hartenstein. "Broadcast Reception Rates and Effects of Priority Access in 802.11-based Vehicular Ad-hoc Networks." *ACM VANET*, Philadelphia, PA, October 2004.

[12] G. Korkmaz, et al. "Urban Multi-hop Broadcast Protocols for Inter-vehicle Communication Systems." *ACM VANET*, Philadelphia, PA, October 2004.

[13] Z. Da Chen, H. T. Kung, and D. Vlah. "Ad Hoc Relay Wireless Networks overMoving Vehicles on Highways." *ACM MOBIHOC*, Long Beach, CA, October 2001.

[14] J. Burgess, et al. "Max-Prop: Routing for Vehicle-Based Disruption-Tolerant Networks." *IEEE INFOCOM*, Barcelona, Spain, April 2006.

[15] UMass' DieselNet. http://prisms.cs.umass.edu/dome/.

[16] J. Zhao and G. Cao. "VADD: Vehicle-Assisted Data Delivery in Vehicular Ad Hoc Networks." *IEEE INFOCOM*, Barcelona, Spain, April 2006.

[17] H. Wu, et al. "MDDV: A Mobility-Centric Data Dissemination Algorithm for Vehicular Networks." *ACM VANET*, Philadelphia, PA, October 2004.

[18] S. B. Eisenman, et al. "MetroSense Project: People-Centric Sensing at Scale." *ACM WSW*, Boulder, CO, October–November 2006.

[19] O. Riva and C. Borcea. "The Urbanet Revolution: Sensor Power to the People!" *IEEE Pervasive Computing*, 6(2), 2007.

[20] M. D. Dikaiakos, et al. "VITP: An Information Transfer Protocol for Vehicular Computing." *ACM VANET*, Cologne, Germany, September 2005.

[21] J. Burke, et al. "Participatory Sensing." *ACM WSW*, Boulder, CO, October–November 2006.

[22] P. Juang, et al. "Energy-efficient Computing for Wildlife Tracking: Design Tradeoffs and Early Experiences with ZebraNet." *ACM ASPLOS-X*, San Jose, CA, October 2002.

[23] T. Small and Z. J. Haas. "The Shared Wireless Infostation Model—A New Ad Hoc Networking Paradigm (or Where There Is a Whale, There Is a Way)." *ACM MOBIHOC*, Annapolis, MD, June 2003.

[24] University of Dartmouth MetroSense. http://metrosense.cs.dartmouth.edu/.

[25] R. C. Shah, et al. "Data MULEs: Modeling a Three-Tier Architecture for Sparse Sensor Networks." *Elsevier Ad Hoc Networks Journal*, 1(2–3):215–233, September 2003.

[26] Q. Li and D. Rus. "Sending Messages to Mobile Users in Disconnected Ad-Hoc Wireless Networks." *ACM MOBICOM*, Boston, MA, August 2000.

[27] Yu Wang and Hongyi Wu. "DFT-MSN: The Delay/Fault-Tolerant Mobile Sensor Network for Pervasive Information Gathering." *INFOCOM'06*, Barcelona, Spain, April 2006.

[28] P. B. Gibbons, et al. "IrisNet: An Architecture for a Worldwide Sensor Web." *IEEE Pervasive Computing*, 2(4):22–33, October–December 2003.

[29] S. Nath, J. Liu, and F. Zhao. "Challenges in Building a Portal for Sensors World-Wide." *ACM WSW*, Boulder, CO, October–November 2006.

[30] CENS' Urban Sensing. http://research.cens.ucla.edu/projects/2006/Systems/Urban Sensing/.

[31] L. Dlagnekov and S. Belongie. "Recognizing Cars." Technical Report CS2005-0833, UCSD CSE, 2005.

[32] P. Bellavista, et al. "Standard Integration of Sensing and Opportunistic Diffusion for Urban Monitoring in Vehicular Sensor Networks: The MobEyes Architecture." *IEEE ISIE*, Vigo, Spain, June 2007.

[33] M. Grossglauser and M. Vetterli. "Locating Nodes with EASE: Mobility Diffusion of Last Encounters in Ad Hoc Networks." *IEEE INFOCOM*, San Francisco, CA, March–April 2003.

[34] C. Intanagonwiwat, R. Govindan, and D. Estrin. "Directed Diffusion: A Scalable and Robust Communication Paradigm for Sensor Networks." *ACM MOBICOM'00*, Boston, MA, 2000.

[35] J. Li, et al. "A Scalable Location Service for Geographic Ad Hoc Routing." *ACM MOBICOM*, Boston, MA, 2000.

[36] B. H. Bloom. "Space/Time Trade-Offs in Hash Coding with Allowable Errors." *CACM*, 13(7):422–426, July 1970.

[37] A. Broder and M. Mitzenmacher. "Network Applications of Bloom Filters: A Survey." *Internet Mathematics*, 1(4):422–426, 2003.

[38] L. Fan, P. Cao, and J. Almeida. "Summary Cache: A Scalable Wide-Area Web Cache Sharing Protocol." *ACM SIGCOMM*, Vancouver, Canada, August–September 1998.

[39] ns-2 (The Network Simulator). http://www.isi.edu/nsnam/ns/.

[40] B. Zhou, K. Xu, and M. Gerla. "Group and Swarm Mobility Models for Ad Hoc Network Scenarios Using Virtual Tracks." *IEEE MILCOM*, Monterey, CA, October–November 2004.

[41] U.S. Census Bureau. TIGER, TIGER/Line and TIGER-Related Products. www.census.gov/geo/www/tiger/.

[42] T. S. Rappaport. *Wireless Communications: Principles and Practice*. IEEE Press, Piscataway, NJ, 1996.

[43] T. Camp, J. Boleng, and V. Davies. "A Survey of Mobility Models for Ad Hoc Network Research." *Wireless Communications and Mobile Computing (WCMC)*, 2(5):483–502, 2002.

[44] M. Raya and J.-P. Hubaux. "The Security of Vehicular Ad Hoc Networks." *SASN*, Alexandria, VA, November 2005.

[45] U. Lee, et al. "Dissemination and Harvesting of Urban Data Using Vehicular Sensing Platforms." Technical report, UCLA CSD, 2007.

[46] Y.-C. Hu, A. Perrig, and D. B. Johnson. "Ariadne: A Secure On-Demand Routing Protocol for Ad Hoc Networks." *MOBICOM*, Atlanta, GA, September 2002.

[47] S. Tanachaiwiwat and A. Helmy. "Correlation Analysis for Alleviating Effects of Inserted Data inWireless Sensor Networks." *MobiQuitous*, San Diego, CA, July 2005.

[48] F. Ye, et al. "Statistical En-Route Filtering of Injected False Data in Sensor Networks." *INFOCOM*, Hong Kong, March 2004.

[49] S. Zhu, et al. "An Interleaved Hop-by-Hop Authentication Scheme for Filtering False Data Injection in Sensor Networks." *IEEE Symposium on Security and Privacy*, Oakland, CA, May 2004.

[50] P. Golle, D. Greene, and J. Staddon. "Detecting and Correcting Malicious Data in VANETs." *VANET'04*, Philadelphia, PA, October 2004.

[51] R. Ostrovsky and W. Skeith. "Private Searching on Streaming Data." *CRYPTO*, Santa Barbara, CA, August 2005.

8

Modeling and Analysis of Wireless Networked Systems

Phoebus Chen, Songhwai Oh, and Shankar Sastry

8.1 Introduction

The United States Office of Homeland Security identified 17 critical infrastructure sectors in its publication *The National Strategy for Homeland Security* [1]:

- Agriculture;
- Food;
- Water;
- Public Health;
- Emergency Services;
- Government;
- Defense Industrial Base;
- Information and Telecommunications;
- Energy;
- Transportation;
- Banking and Finance;
- Chemical Industry;
- Postal and Shipping;
- Information Technology;

- Telecommunications;
- Banking and Finance;
- Dams;
- Government Facilities;
- National Monuments and Icons.

Many of these sectors involve large, distributed systems and would benefit from distributed sensing and monitoring. For instance, the Agriculture, Food, Water, and Public Health sectors would benefit from monitoring for contaminants and the potential for epidemic outbreaks. Emergency Services can benefit from better situational awareness during natural disasters and terrorist attacks. A distributed sensing and monitoring system can detect attacks on assets in the Defense Industrial Base sector. It can be also used to provide information on how to reroute traffic in the Transportation sector in case our highways and railways are congested or damaged. The Energy sector can benefit from better sensing and prediction of power demands and transmission line failures to prevent cascading blackouts [2]. The Chemical Industry sector can benefit from better monitoring and process control, both for quality control and safe, automated shutdown in case of accidents or sabotage. The Postal and Shipping sector can benefit from better tracking of packages to identify the source of suspicious packages, such as the letters containing anthrax in the 2001 amerithrax case [3].

Many of these distributed sensing and monitoring systems can be implemented by wireless sensor networks (WSNs) to eliminate wiring and reduce costs. As explained in Chapter 2, wireless sensor networks are different from other traditional wireless networks because the nodes have limited energy reserves and processing power. Furthermore, large numbers of them are meant to be deployed for unattended operation over long periods of time. Thus, the ability of the network to self-organize is very important, whether it be to form multihop routing topologies or to fuse, filter, and make inferences on collected sensing data.

Although wireless sensor networks can be built as separate systems to monitor critical infrastructure, ultimately they will become *part* of the critical infrastructure itself. This is because automated/semiautomated reasoning and response will be necessary to process the large amount of information gathered by these sensor networks. The sensor network effectively becomes a part of an integrated system that not only monitors, but also *regulates*, the systems in our critical infrastructure. Humans may be part of this regulation process. For instance, in transportation, automobile drivers can use traffic reports generated from traffic sensor data to avoid congestion caused by a collapsed bridge.

The security of critical infrastructure thus involves the security of the complete *wireless networked system*, which consists of the wireless sensor network, the *decision-making* system that acts on the information from the sensor net-

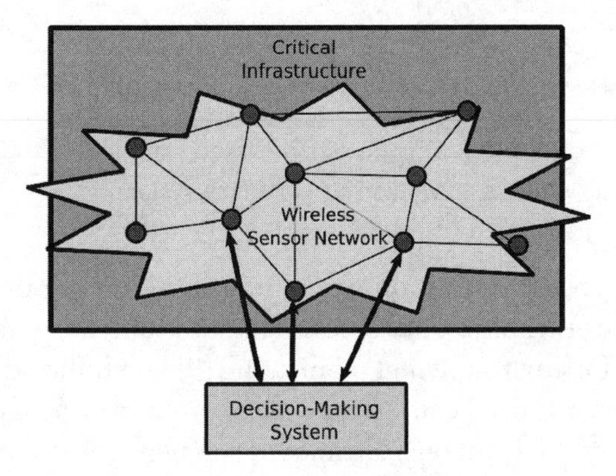

Figure 8.1 The components of a wireless networked system with a wireless sensor network and critical infrastrtucture that we wish to protect.

work, and the original critical infrastructure itself (see Figure 8.1). Traditionally, the security of these components is studied separately (e.g., cybersecurity of the sensor network) but it is necessary to study their interaction to secure the entire wireless networked system. For instance, we would like to understand how partial system compromise will jeopardize the performance of the entire system. This is especially relevant for wireless sensor networks because the sensor nodes are often physically unprotected from attack. The distributed computing systems approach to this problem, known as *Byzantine fault tolerance*, looks at the ability of separate components in the system to come into agreement despite some faulty components giving arbitrary output [4–6]. The control systems approach is to view the entire system as a feedback control loop and ask whether the sensor network can monitor the infrastructure and whether the decision-making system can regulate its behavior when some parts of the system are compromised or unavailable.

This chapter will study the security of critical infrastructure in the context of modeling the entire wireless networked system as a feedback control system. Instead of focusing on the traditional mechanisms of securing the sensor network, such as encryption and authentication of packets on the wireless network [7] and setting up secure routing paths [8], this chapter describes a framework for system-wide security by using models of system components to predict performance of the wireless networked system.

8.2 Wireless Networked Systems for Critical Infrastructure

The design of wireless networked systems to secure critical infrastructure often needs to support *real-time decision making*, whether the decisions are in response to an event/alarm or to help continuously regulate a system. *Real time* refers to

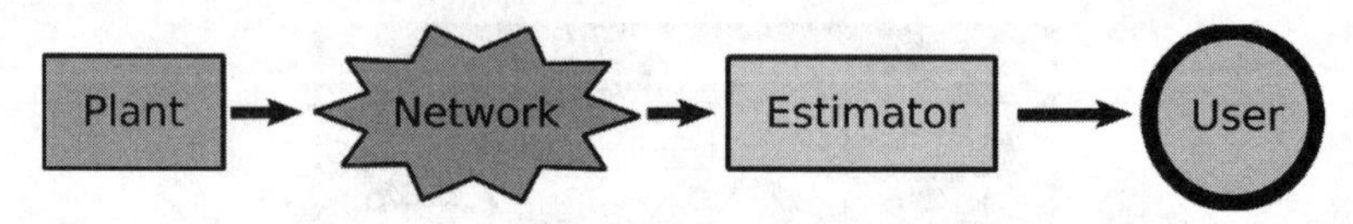

Figure 8.2 A real-time wireless networked measurement system.

having tight latency constraints on communication between the components of the system, on the order of seconds or less. This is in contrast to traditional wireless sensor networks for environmental monitoring [9] and offline data analysis [10], which have latency requirements on the order of hours to send data to a collection point. This chapter will focus on real-time wireless networked systems, particularly wireless *real-time measurement systems* and wireless *real-time control systems.*

8.2.1 Real-Time Measurement Systems

A real-time measurement system consists of networked sensors and processors that perform signal processing, estimation, or inference on the sensor measurements before delivering it to the user with low latency (Figure 8.2). The user is often a human who needs to respond to/act on the data from the system.

Wireless sensor networks can be used as real-time measurement systems if they can deliver data reliably and with low latency. This depends heavily on the networking protocols used in the sensor network. Furthermore, to characterize application performance, we need good models of these networking protocols to predict their reliability and latency. This is the focus of Section 8.3.

8.2.1.1 Examples

Applications of real-time measurement systems to secure the critical infrastructure in various homeland security sectors can be distinguished by their required response times. For instance, operators of video surveillance cameras in office buildings and security alarm systems for banks need to react quickly to intruders, on the order of minutes to dispatch security personnel or lock down critical areas. There are also proposals to use sensor networks to monitor the vital conditions of the injured in large disaster-response scenarios [11] and to monitor the sick and elderly for assisted living at home [12]. In scenarios where a patient's heart rate drops or an old man falls down the stairs, we would like to respond and dispatch emergency medics on the order of minutes.

There are also proposals to use sensor networks as real-time measurement systems to help firefighters navigate through a burning building [13, 14]. The sensor network would keep track of the location of the firefighters and use smoke and temperature sensors to identify the location of the fires and safe evacuation routes, relaying all this information back to the incident commander to coordinate the firefighting efforts. Here, the system response time should be on the order of seconds to match the speed at which the fire can spread and the firefighters can move.

Another real-time measurement system is the *supervisory control and data acquisition* (SCADA) system used in industrial control systems [15]. SCADA system architectures are typically hierarchical, with localized feedback control and wide area monitoring for diagnostics and safety [16], often with a "human in the loop." The diagnostics network needs to relay large amounts of data, and is usually not as sensitive to delay as the safety and control networks. However, diagnostics information can also be used to "close the loop" for equipment shutdown or continuous process improvement, though the actuation is typically event-driven as opposed to time-driven continuous actuation. The safety networks require determinism (guaranteed response time), low delay, and reliable data delivery, though the traffic load may be lower than diagnostics networks. The response time/delay requirements depend on the particular industrial plant, and can range from microseconds to hours, depending on the task [17].

Finally, real-time measurement systems can be used for battlefield awareness in military applications. For instance, sensor networks can be used to track moving targets [18]. The tolerable delay in such a system depends on the distance of the target to critical areas, on the order of seconds to tens of seconds. Sensor networks can also be used to locate snipers from the acoustic trail of bullets and the muzzle blast of guns [19]. Such systems may need time synchronization between sensor nodes on the order of microseconds to compute sniper locations, but they can tolerate communication delays to the user on the order of seconds.

8.2.2 Real-Time Control Systems

From a control systems point of view, a wireless networked system consists of a plant/system to be controlled, sensors, actuators, and computers connected together by a network that is wireless or partially wireless (see Figure 8.3). The distinction

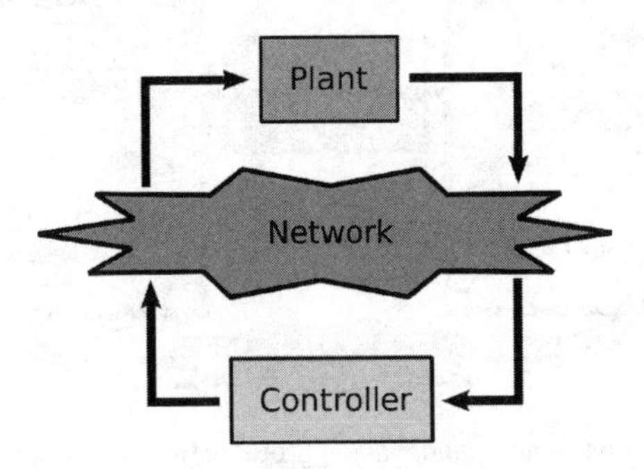

Figure 8.3 A real-time wireless networked control system.

between wireless real-time control systems (or simply, wireless control systems) from real-time measurement systems is that computers, instead of humans, make decisions to regulate the plant. Latency and reliability are still critical network performance parameters, but a theory also needs to be developed to address how to design automatic controllers in these systems. This is the theory of *networked control systems*, presented later in Section 8.2.3.

A wireless networked control system for securing critical infrastructure must be designed to guarantee system performance and stability when the communication medium is lossy and delivers data with variable delay. In some systems, such as distributed heating and ventilation of office buildings, it may be sufficient to guarantee system performance in a statistical sense—for instance, the system works 99.9% of the time. However, it would be foolish to build safety-critical systems over a wireless network without a contingency plan when the wireless channel between the system components is disconnected. Putting in a redundant, wired communication channel may be impossible or too costly. However, we may be able to design a safety-critical control system that continues to perform safely, but with lower performance, without a connected network and performs *better* with a wireless connection. Equivalently, in automated manufacturing, the system can be designed to maintain product *quality* but vary the *yield* depending on the reliability and delay of the wireless network.

Figure 8.4 proposes an abstract scenario and a handshaking protocol for the components of a wireless networked system to coordinate and maintain safety/product quality while adapting to wireless network conditions to maximize performance. In this scenario, the output of system A is being fed as input to system B, with some delay *D*. While system B can measure its immediate inputs and perform local feedback control, it can get better performance with a larger look-ahead horizon by getting direct measurements of the output of system A. System A can measure its output and send it over the wireless network to system B, and furthermore it can adjust the rate of its output in response to how well it can coordinate with system B. If it does not hear an end-to-end ac-

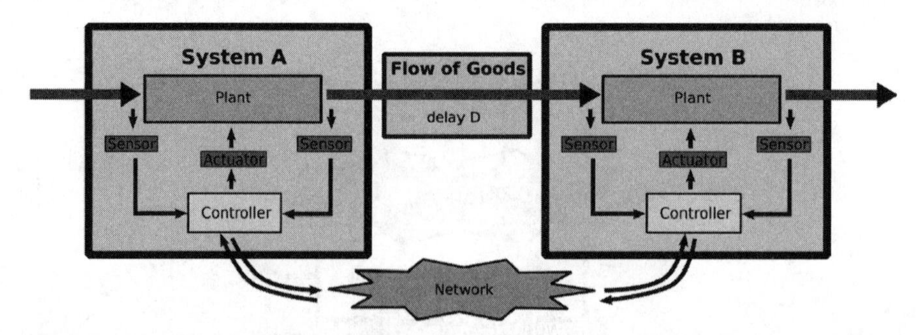

Figure 8.4 A sample scenario and handshaking protocol for coordinating the components of a wireless, networked, safety/quality-critical control system. See text for details.

knowledgment from system B, system A will assume that system B did not receive the measurements at the output of system A, and slow down it's output accordingly. This way, the stability of the system or the quality of the output will not be compromised, but the performance or yield can be tuned based on wireless network conditions.

8.2.2.1 Examples

Real-time control systems are often part of SCADA systems for process control and industrial automation. Traditional applications range from chemical production and petroleum refinery to waste management. Current research is trying to extend these systems to reconfigurable manufacturing systems that can change production capacities to quickly match consumer demands [16, 20]. In these systems, wireless connections between equipment can reduce the cost of reconfiguration.

Real-time control systems can also be used to protect critical infrastructure, such as the use of wireless sensor networks for active/semiactive damping of civil structures [21, 22]. Active damping is the actuation of parts of a structure to dampen vibrations, while semiactive damping changes the dissipation properties of passive dampers in real time. This can be useful for reducing the damage to large buildings and bridges during earthquakes or hurricanes. The latency requirements for such systems are on the order of hundredths of a second. Wireless sensor networks enable existing structures to be retrofitted with active/semiactive dampers without the installation of wires, while simultaneously monitoring them for structural damage [23, 24].

8.2.3 Theory of Networked Control Systems

The theory of *networked control systems* (NCSs) can be used to aid the design and performance evaluation of wireless networked systems. A networked control system

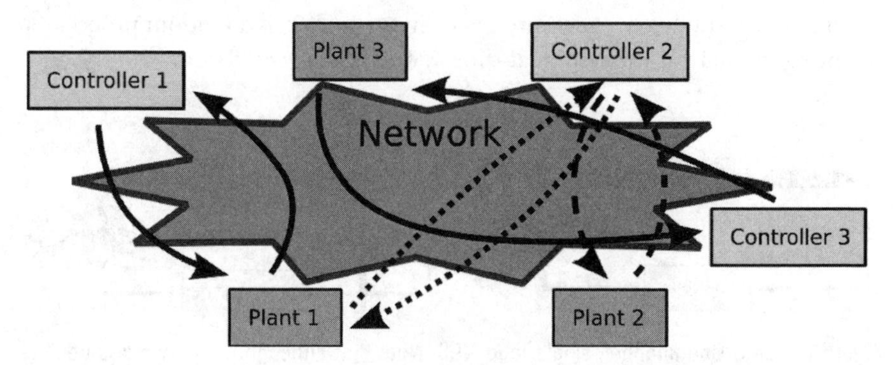

Figure 8.5 The components of a networked control system (NCS). There can be many control loops running over the same network and multiple controllers per plant or multiple plants per controller.

is defined to be a spatially distributed system of controllers, sensors, and actuators that share a band-limited communication network [25] (see Figure 8.5). The study of NCSs focuses on how a shared, lossy communication network with variable delay affects the performance of the closed-loop control system. The literature in this area has largely focused on packet networks, especially since many wired SCADA and distributed control systems are migrating to TCP/IP networks for cost and interoperability [16].

This section will highlight some important contributions to the theory of networked control systems to give a sense of how network conditions translate to system performance. This is a first step toward studying how partial system failure or compromise translates to system performance degradation, which will in turn guide the choice of network architectures and protocols for better system robustness and security. Although there can often be multiple interacting control loops in a networked control system, for clarity this exposition will focus on single-loop NCSs (see Figure 8.6). A more comprehensive overview of NCSs can be found in [25].

The current literature on networked control systems focuses on system stability/performance analysis and controller synthesis, where the network introduces *packet drops, variable sampling, variable delay,* and *measurement/control quantization error*. Much of the current literature analyzes control of a simple plant, such as a discrete-time LTI (linear time-invariant) plant with Gaussian noise and disturbance, to focus on the control issues introduced by using a network.

There are two types of system stability and performance guarantees: *deterministic* and *stochastic*. Deterministic system guarantees often rely on absolute guarantees on network performance—for instance, packets in the network have a bounded delay, or that no more than a fixed number of consecutive packets can be dropped. These network guarantees may be impossible to enforce on a wireless network. The alternative is to use a stochastic characterization of the network to provide stochastic system guarantees. For instance, if $\boldsymbol{x}_k$ is a random process representing the state of the system at time k, one can say the system is:

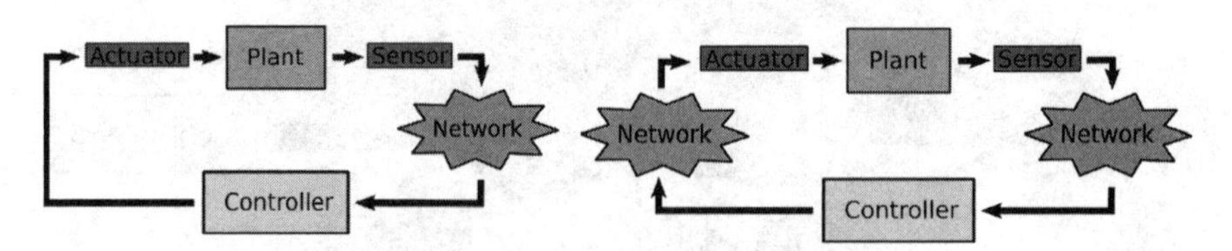

Figure 8.6 (Left) One-channel, single-loop NCS. Note that although the network is depicted between the sensor and the controller, it can also be between the actuator and the controller. This latter situation is less common in practice. (Right) Two-channel, single-loop NCS. (*After*: [1].)

- *almost surely stable* if $\mathbb{P}\left(\sup_{k\in\mathbb{N}} \|\boldsymbol{x}_k\| < \infty\right) = 1$;
- *stable in the 2m-th moment* if $\sup_{k\in\mathbb{N}} \mathbb{E}\|\boldsymbol{x}_k\|^{2m} < \infty$;
- *asymptotically stable in the 2m-th moment* if $\lim_{k\to\infty} \mathbb{E}\|\boldsymbol{x}_k\|^{2m} = 0$;
- *exponentially stable in the 2m-th moment* if there exists α, $\beta > 0$ such that $\mathbb{E}\|\boldsymbol{x}_k\|^{2m} < \alpha\mathbb{E}\|\boldsymbol{x}_0\|^{2m} e^{-\beta k}$ for all $k \in \mathbb{N}$.

If $m = 1$, one can also say the system is (exponentially/asymptotically) *mean-square* stable.

The current literature on NCSs often use a point-to-point channel model as an abstraction for the network to study the impact of channel characteristics (packet drops, variable sampling, variable delay, and measurement/control quantization error) on the closed-loop system. The way we quantify/measure these channel characteristics, the problem of computing *network metrics*, is studied next in Section 8.3.

The architecture of a networked control system depends on whether it is a single-channel network or a two-channel network, and where the network is placed relative the the controller, sensors, actuators, and plant (see Figure 8.6). For instance, if the system has *smart sensors* that include some computational capabilities with the sensor [26], the sensor can transmit a state estimate over the network instead of a raw measurement. This effectively connects part of the controller—the estimator—to the sensor directly without a network, and results in better system performance [27].

The following two examples consider two networked control systems with stochastic packet drops but no variable delay, no variable sampling, and no quantization error on the transmitted data. Although delays are common in networks and are important for control, these simplified systems match a network architecture where packets arriving after a deadline are dropped. The techniques for dealing with time-delayed systems modeled as delayed differential equations (DDE) are covered in [25]. The following examples follow the notation and exposition from [25].

8.2.3.1 Estimator Stability Under Bernoulli Packet Drops

In [28], Schenato et al. give stochastic guarantees on the stability of optimal estimators for a discrete-time LTI system connected by two-channel feedback where packet drops are modeled as a Bernoulli process (Figure 8.6, right). Particularly, there is a threshold probability p_c such that if the probability of dropping a packet $p > p_c$ then the estimation error covariance of the optimal estimator is unbounded. Furthermore, they show that if the controller receives an acknowledgment of whether the previous control packet reached the actuator, what they term "TCP-like protocols," then the separation principle holds. This means that the optimal estimator and the optimal controller can be designed separately.

In fact, the optimal controller is a linear function of the state. On the other hand, if the network does not have acknowledgments, what they term "UDP-like protocols," then the separation principle fails and the optimal controller is in general nonlinear. In [27], Schenato extends this problem to considering estimators that can buffer past measurements to account for out of order packet delivery.

For simplicity, consider just the estimation problem for a discrete-time LTI plant driven by white Gaussian noise:

$$\left.\begin{aligned} \boldsymbol{x}_k + 1 &= A\boldsymbol{x}_k + \boldsymbol{w}_k \\ \boldsymbol{y}_k + 1 &= C\boldsymbol{x}_k + \boldsymbol{v}_k \end{aligned}\right\} \forall k \in \mathbb{N},\ \ \boldsymbol{x}_k, \boldsymbol{w}_k \in \mathbb{R}^n,\ \ \boldsymbol{y}_k,\ \boldsymbol{v}_k \in \mathbb{R}^p \qquad (8.1)$$

The initial state is $\boldsymbol{x}_0 \sim \mathcal{N}(0, \Sigma)$ (i.e., $\boldsymbol{x}_0$ has a Gaussian distribution with mean 0 and covariance Σ) and the Gaussian white noises $\boldsymbol{w}_k \sim \mathcal{N}(0,R_w)$, $R_w \geq 0$ and $\boldsymbol{v}_k \sim \mathcal{N}(0,R_v)$, $R_v \geq 0$ are independent. Let (C, A) be detectable and (A, R_w) be stabilizable.

The plant and the estimator are connected by an erasure channel, meaning the receiver knows when a packet is corrupted and can discard it. The lossy channel is modeled by a stochastic process $\theta_k \in \{0,1\}$, $\forall k \in \mathbb{N}$, which is independent of $\boldsymbol{x}_0$, $\boldsymbol{w}_k$ and $\boldsymbol{v}_k$. Here, $\theta_k = 1$ means a measurement packet reached the destination and $\theta_k = 0$ means it did not.

The goal is to compute the optimal estimate of $\boldsymbol{x}_k$ given all the measurements successfully transmitted up to time $j \leq k$, which is

$$\hat{\boldsymbol{x}}_{k|j} = \mathbb{E}\left[\boldsymbol{x}_k \,|\boldsymbol{y}_l, \forall l \leq j \text{ s.t. } \theta_l = 1\right] \qquad (8.2)$$

This is computed recursively using a time-varying Kalman filter (TVKF). The form of the TVKF and bounds on the threshold probability p_c are provided in [28].

8.2.3.2 H_∞ Controller Synthesis for MJLS

In [29], Seiler and Sengupta study how to synthesize an H_∞ controller for a one-channel feedback NCS with a discrete-time LTI plant (see Figure 8.6, left). Unlike the previous example where packet losses are independent, this problem formulation uses the Gilbert-Elliot channel model where packet losses are modeled by a two-state Markov chain (see Figure 8.7). This is a simple approximation of wireless channel fading where consecutive packets are dropped with high probability. As a result of this channel model, the NCS is modeled as a Markov jump linear system (MJLS). They formulate a semidefinite programming problem using the MJLS model to synthesize an H_∞ controller that makes the closed-loop system exponentially mean-square stable.

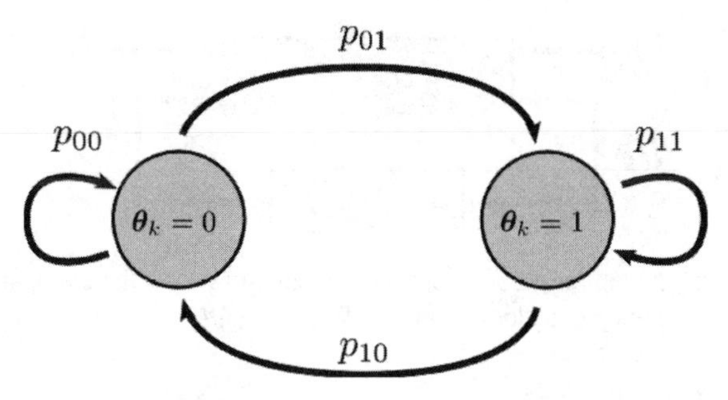

Figure 8.7 The Gilbert-Elliot communication model uses a two-state Markov chain to model correlated packet losses. Note that if we set $p_{00} = p_{10}$ and $p_{01} = p_{11}$,we get back the Bernoulli packet loss model.

First, let the discrete-time LTI plant $\mathcal{P}$ be a Markov jump linear system of the form

$$\begin{bmatrix} x_{k+1} \\ z_k \\ y_k \end{bmatrix} = \begin{bmatrix} A_{1,\theta_k} & B_{1,\theta_k} & B_{2,\theta_k} \\ C_{1,\theta_k} & D_{11,\theta_k} & D_{12,\theta_k} \\ C_{2,\theta_k} & D_{21,\theta_k} & 0 \end{bmatrix} \begin{bmatrix} x_k \\ d_k \\ u_k \end{bmatrix} \tag{8.3}$$

where $\boldsymbol{x}_k \in \mathbb{R}^{n_x}$ is the state, $\boldsymbol{d}_k \in \mathbb{R}^{n_d}$ is the disturbance, $\boldsymbol{u}_k \in \mathbb{R}^{n_u}$ is the control input, $\boldsymbol{z}_k \in \mathbb{R}^{n_z}$ is the error, and $\boldsymbol{y}_k \in \mathbb{R}^{n_y}$ is the measurement. The state matrices are functions of a finite-state discrete-time Markov chain switching at time k to a mode $\theta_k \in \mathcal{N} = \{1,\ldots,N\}$ with transition probabilities p_{ij} between modes i and j.[1] In our notation, each mode θ_k is associated with the set of matrices with θ_k in the second subscript (e.g., A_{1,θ_k}). The switching between modes is governed by whether a packet is delivered through the network ($\theta_k = 1$) or a packet is dropped ($\theta_k = 0$). Hence, there are two modes in our single-loop MJLS, meaning $\mathcal{N} = \{0,1\}$.

Seiler and Sengupta synthesize an H_∞ controller $\mathcal{K}$ of the form

$$\begin{bmatrix} x_{C,k+1} \\ u_k \end{bmatrix} = \begin{bmatrix} A_{C,\theta_k} & B_{C,\theta_k} \\ C_{C,\theta_k} & 0 \end{bmatrix} \begin{bmatrix} x_{C,k} \\ y_k \end{bmatrix} \tag{8.4}$$

Note here that the first subscript C is used to distinguish the controller matrices from the plant matrices. The closed-loop system is denoted $F_L(\mathcal{P},\mathcal{K})$ and depicted in Figure 8.8. For a brief discussion of H_∞ controllers and the procedure for synthesizing the controller, see [29].

1. The term *mode* is used interchangeably with the term *state* to describe θ_k in the Markov chain. This is to avoid confusion with the term *state* used to describe $\boldsymbol{x}_k$.

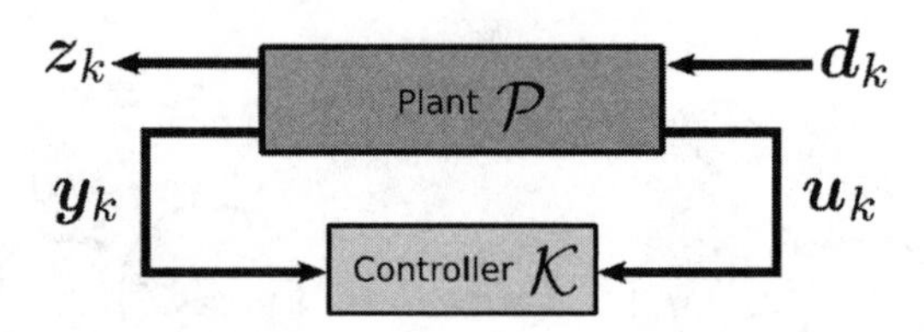

Figure 8.8 Closing the loop for the H_∞ control problem. Here, $\mathcal{P}$ is the plant, and $\mathcal{K}$ is the controller, and the closed loop system is $F_L(\mathcal{P}, \mathcal{K})$. (*After:* [2].)

8.2.3.3 Discussion

The current networked control system theory is developed by abstracting the network as a simple point-to-point channel and studying how packet loss, variable delay, variable sampling, and quantization error affect the closed-loop control system. In order to apply this theory to wireless networked systems, we need to develop network models where we can extract the relevant parameters for this abstraction. We term these parameters *network metrics,* and they serve as an *interface* for tuning the NCS to network conditions and tuning the network to meet the requirements of the NCS.

The network metrics allow us to draw conclusions about system performance from network performance, which in turn is directly affected by network security attacks. At the most basic level, a security attack can be modeled as a subset of network nodes being compromised, resulting in packets being dropped and a degradation in the stability and performance of the system. Although rudimentary, this approach allows us to quantify the security of a system by combining existing the NCS theory with the network models developed in the following section.

8.3 Wireless Network Metrics

In communications, the typical metrics used to measure the quality of a network include:

- *Throughput/goodput,* or rate, the amount of application data traversing through the network per unit time;
- *Latency* of a particular piece of data;
- *Fairness* of access to the network if there are multiple senders and/or receivers;
- *Robustness/reliability* of the end-to-end connections in a network to changing network conditions;
- *Resilience to failure and attack,* particularly node failure and node compromise—for instance, avoiding the formation of routing loops;

- *Availability* of the network—for instance, the network is not partitioned;
- *Maintenance overhead/complexity* of the algorithms, measured not only in terms of the memory and processing requirements but also whether it is state free and easy to troubleshoot and implement correctly;
- *Power consumption/network lifetime,* which is related to traffic distribution across nodes, particularly if they run on batteries.

Oftentimes, it is important to know how one metric is a function of another metric in a network. For instance, a network can often increase throughput at the cost of increased energy consumption and latency. This relationship between two metrics itself can be a metric of network performance.

From a control system point of view, if one models the network as a black box, the three most basic metrics that directly affect system performance are *reliability, latency,* and *data rate.* Here, reliability can be measured by the probability of end-to-end packet delivery (or simply, the *end-to-end connectivity* of the network). To model the network properly for the control system, we need to know how reliability is a function of latency and data rate. In the study below, the metric is the end-to-end connectivity of the network as a function of latency, given a fixed data rate. We will refer to this metric as the *connectivity metric.*

8.3.1 Choice of Network Models

To compute our connectivity metric, we will model the wireless network at the network and data link (MAC) layers of the seven-layer OSI model (see Figure 8.9). We select implementations of these OSI layers that enable the wireless network to provide high reliability while being easy to model.

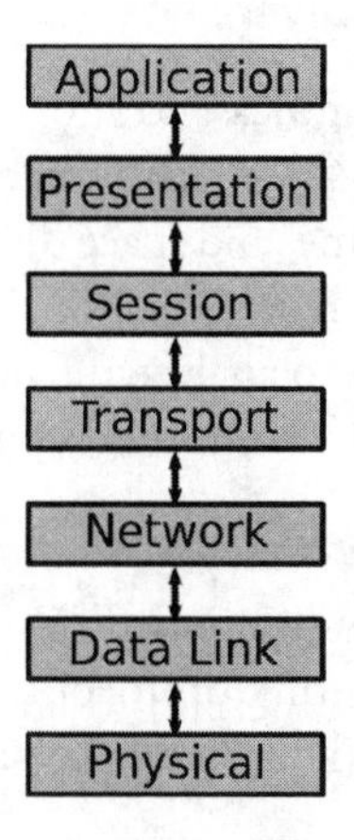

Figure 8.9 The seven-layer Open Systems Interconnect (OSI) model of networks. For more details, see a textbook on communication networks, such as [3].

At the data link layer, we will model *time division multiple access* (TDMA) media access schemes, where nodes are scheduled to transmit during time slots. The other popular media access scheme used by the simple sensor network radios, *carrier sense multiple access* (CSMA), is not studied here because of the difficulty of modeling random collisions when multiple nodes simultaneously try to access the medium. The variants of TDMA schemes (e.g., LEACH [30], BMA [31], LMAC [32], TRAMA [33], UPD over TSMP [34]) essentially differ in the protocols for scheduling the nodes. There are also variants that are hybrid CSMA/TDMA schemes to improve performance while avoiding some of the complexity of optimal scheduling in TDMA, such as scheduled channel polling (SCP) [35] and Z-MAC [36], but for modeling simplicity we stick to the most basic TDMA model.

At the network layer, we will model multipath routing from a single source to a single sink. There are many types of multipath routing. *Multipath source routing,* such as braided multipath [37], sends packets along multiple paths, specifying the entire routing path from the source to the sink for each packet. *Deterministic mesh routing,* such as unicast path diversity[2] [34], does not specify an end-to-end path for each packet, but instead sets up a mesh of paths for the packets that all flow to the sink. A node in the mesh tries to transmit a packet on its outgoing links in a deterministic order. *Stochastic mesh routing,* such as ARRIVE [38], is similar except that nodes in the mesh will stochastically select a link for transmission. Finally, *flooding* or some form of *controlled flooding,* such as trickle [39], sends multiple copies of a packet down multiple paths to the sink.

We will model two TDMA multipath routing schemes—unicast path diversity and directed staged flooding—to compute the connectivity metric and the traffic distribution throughout the network. The following models are from [40].

8.3.2 Modeling Unicast Path Diversity

Dust Networks, Inc., proposed unicast path diversity (UPD) over time synchronized mesh protocol (TSMP) [34] for reliable networking in sensor networks. The algorithm exploits frequency, time, and space diversity to achieve what they claim is over 99.9% typical network reliability [41]. We use a general mesh TDMA Markov chain (MTMC) model to analyze the performance of UPD over TSMP (hereafter referred to simply as UPD) for incorporation into a control system.

8.3.2.1 Modeling Characteristics

We model UPD as a frequency-hopping TDMA scheme with multipath routing. UPD is a many-to-one routing protocol (i.e., there is one sink in the network). Each node in the network has multiple parents and the routing graph has

2. The name *unicast path diversity* is not explicitly mentioned in the reference, but this is the name of the routing protocol that is described.

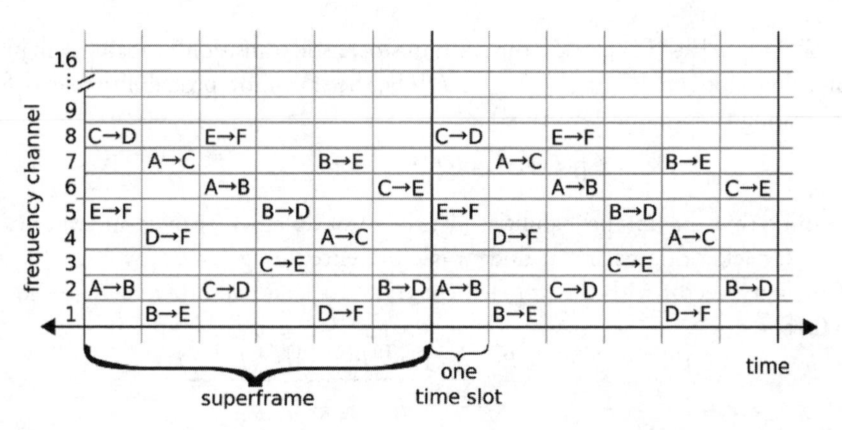

Figure 8.10 Example of a UPD schedule with superframes and time slots. Here, only 8 of the 16 frequency channels are used.

no cycles. The links selected for routing are bidirectional, and hence every link transmission can be acknowledged. If a packet transmission is not acknowledged, it is queued in the node for retransmission. As for scheduling, time is divided into time slots, and grouped into superframes (see Figure 8.10). At each time slot, pairs of nodes are scheduled for transmitting a packet on different frequencies. The superframe containing the schedule of transmissions is repeated over time. Our model uses frequency hopping to justify the assumption that links are independent over retransmissions.

To construct a model of mesh TDMA routing, we assume knowledge of the routing topology, schedule, and all the link probabilities. Furthermore, we study single-packet transmission in the network and do not analyze the effects of queuing.

8.3.2.2 Mesh TDMA Markov Chain Model

Let us represent the routing topology as a graph $G = (\mathcal{V},\mathcal{E})$, and name a node in the network as $i \in \mathcal{V} = 1,\ldots,\mathcal{N}$, and a link in the network as $l \in \mathcal{E} \subset \{(i,j) \mid i,j \in \mathcal{V}\}$, where $l = (i,j)$ is a link for transmitting packets from node i to node j. Time t will be measured in units of time slots, and let T denote the number of time slots in a superframe. The link success probability for link $l = (i,j)$ at time slot t is denoted, $p_l^{(t)}$ or $p_{ij}^{(t)}$. We set $p_l^{(t)} = 0$ when link l is not scheduled to transmit at time t.

For a packet originating from a source node a routed to a sink node b, we wish to compute $p_{net}^{(t_d)}$, the probability the packet reaches b at or before time t_d has elapsed. This is done by a time-varying, discrete-time Markov chain.

8.3.2.3 Mesh TDMA Markov Chain Model

Let the set of states in the Markov chain be the nodes in the network, $\mathcal{V}$. The transition probability from state i to state j at time t is simply $p_{ij}^{(t)}$, with $p_{ii}^{(t)} = 1 - \Sigma_{j \neq i} p_{ij}^{(t)}$.

Let $P^{(t)} = [p_{ij}^{(t)}]^T \in [0,1]^{N \times N}$ be the column stochastic transition probability matrix for a time slot and $P^{(\underline{T})} = P^{(T)}P^{(T-1)} \ldots P^{(1)}$ be the transition probability matrix for a repeating superframe.[3] Assume

$$P^{(T+h)} = P^{(cT+h)}, \ \forall c,h \in \mathbb{Z}_+ \tag{8.5}$$

meaning that the link probabilities in a time slot do not vary over superframes.

A packet originating at node a is represented by $\mathbf{p}^{(0)} = \mathbf{e}^{[a]}$, where $\mathbf{e}^{[a]}$ is an elementary vector with the ath element equal to 1 and all other elements equal to 0. Then,

$$\mathbf{p}^{(t_d)} = P^{(t_d)} \ldots P^{(2T+1)} \underbrace{P^{(2T)}P^{(2T-1)} \ldots P^{(T+1)}}_{P^{(\underline{T})}} \cdot \underbrace{P^{(T)}P^{(T-1)} \ldots P^{(1)}}_{P^{(\underline{T})}} \mathbf{p}^{(0)} \tag{8.6}$$

represents the probability distribution of the packet over the nodes at time t_d.

The sink node b is an absorbing state in the Markov chain, meaning there are no transitions out of that state. This means $p_{net}^{(t_d)} = \mathbf{p}_b^{(t_d)}$, the bth element of the vector $\mathbf{p}^{(t_d)}$. One of the characteristics of a good routing schedule is that $p_{net}^{(td)} \xrightarrow{t_d \to \infty} 1$, meaning the packet will eventually reach the sink. This condition is satisfied when the MTMC model has only one recurrent class consisting of the sink (see [42]).

8.3.2.4 MTMC Examples and Discussion

An example of a small UPD routing schedule is given in Figure 8.11, where p_{ij} is the link probability for link (i, j) and $\bar{p}_{ij} = 1 - p_{ij}$. We get the transition probability matrices,

$$P^{(1)} = \begin{bmatrix} \bar{p}_{12} & 0 & 0 & 0 \\ p_{12} & 1 & 0 & 0 \\ 0 & 0 & \bar{p}_{34} & 0 \\ 0 & 0 & p_{34} & 1 \end{bmatrix} \quad P^{(2)} = \begin{bmatrix} \bar{p}_{14} & 0 & 0 & 0 \\ 0 & \bar{p}_{23} & 0 & 0 \\ 0 & p_{23} & 1 & 0 \\ p_{14} & 0 & 0 & 1 \end{bmatrix}$$

$$P^{(3)} = \begin{bmatrix} \bar{p}_{13} & 0 & 0 & 0 \\ 0 & \bar{p}_{24} & 0 & 0 \\ p_{13} & 0 & 1 & 0 \\ 0 & p_{24} & 0 & 1 \end{bmatrix} \tag{8.7}$$

$$\mathbf{p}^{(0)} = [1 \ \ 0 \ \ 0 \ \ 0]^T \quad P^{(\underline{3})} = P^{(3)}P^{(2)}P^{(1)}$$

3. [0, 1] denotes the closed interval between 0 and 1. $[\cdot]^T$ denotes the transpose of a matrix, while $P^{(T)}$ denotes the transition matrix at time T.

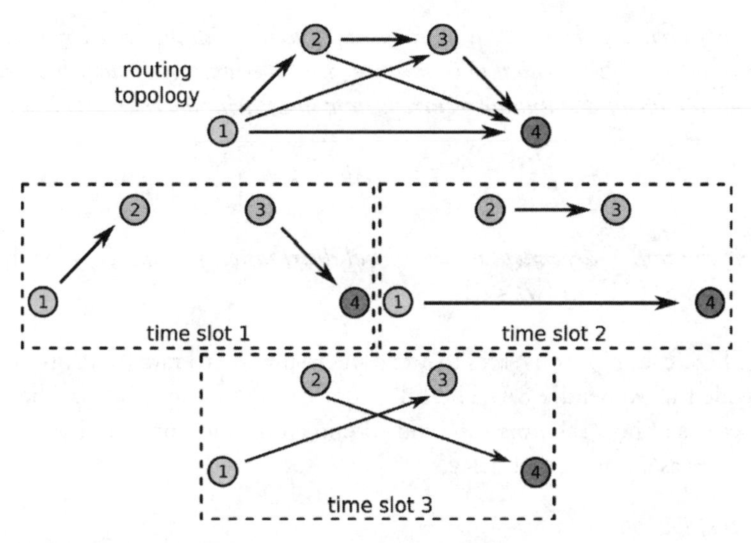

Figure 8.11 Multipath routing example corresponding to (8.7).

The MTMC model can be modified to represent routing topologies and schedules not used by UPD. For instance, UPD avoids creating cycles in the routing graph, as one would want from a good routing algorithm. The MTMC model, however, can model routing cycles that may arise when the network malfunctions. We can still calculate $p_{net}^{(t_d)}$, and we still have $p_{net}^{(t_d)} \xrightarrow{t_d \to \infty} 1$ if no recurrent classes besides the sink are added to the Markov chain. The MTMC model can also be extended to represent mesh networks with multiple sink nodes (e.g., two Internet gateways to a sensor network). In this case, if we let $\mathcal{B}$ be the set of sinks, $p_{net}^{(t_d)} = \sum_{i \in \mathcal{B}} \mathbf{p}_i^{(t_d)}$. Finally, if we wish to model schedules that never retransmit packets, we simply remove the requirement in our MTMC model that $p_{ii}^{(t)} = 1 - \sum_{j \neq i} p_{ij}^{(t)}$, instead replacing it with $p_{ii}^{(t)} = 0$. To ensure that the transition probability matrix $P^{(t)}$ is a column stochastic matrix, we add a dummy state $N+1$ to represent a packet being lost after transmission. Now, $P^{(t)} = \left[p_{ij}^{(t)}\right]^T \in [0,1]^{N+1 \times N+1}$, where $p_{i(N+1)}^{(t)} = 1 - \sum_{j \neq i} p_{ij}^{(t)}, p_{(N+1)i}^{(t)} = 0$ for all $i \neq N+1$, and $p_{(N+1)(N+1)}^{(t)} = 1$.

8.3.2.5 MTMC Analysis

Network-Wide Rate of Convergence for $p_{net}^{(t_d)}$

Besides calculating $p_{net}^{(t_d)}$ for one node transmitting to the sink, we would like to calculate the rate of convergence of $p_{net}^{(t_d)} \to 1$ for the *entire* network from $P^{(\mathcal{T})}$. This may be a useful metric for designing routing schedules to optimize the performance of the network.

Theorem 8.1 (MTMC $p_{net}^{(t_d)}$ converges exponentially to 1.) *Let* $P^{(\mathcal{T})} \in [0,1]^{N \times N}$ *be a diagonalizable, column stochastic matrix with* $\lim_{k \to \infty} (P^{(\mathcal{T})})^k \mathbf{p} = \mathbf{e}^{[b]}$ *for all probability vectors* $\mathbf{p}$. *Here,* $\mathbf{e}^{[b]}$ *is an elementary vector with the bth element equal to 1 and all other elements equal to 0, meaning that the routing topology has a unique sink node b which is the unique recurrent state in the Markov chain. Then,*

$$p_{net}^{(t_d)} \geq 1 - C(\rho_*)k, \quad k = \left\lfloor \frac{t_d}{T} \right\rfloor \tag{8.8}$$

for some constant C dependent on the initial distribution $\mathbf{p}^{[0]}$ *and* $\rho_* = \max\{|\lambda|:\lambda$ *is an eigenvalue of* $P^{(T)}$ *and* $|\lambda| < 1\}$.

Therefore, $p_{net}^{(t_d)}$ converges to 1 exponentially with a rate ρ_*. A proof sketch is provided in Appendix 8A. The full proof can be found in [40]. The rate ρ_* gives a sense of how the *worst-case* end-to-end connection probability in the network varies as a function of delay.

Traffic Distribution

To identify hot spots in the network, we compute the probability that the packet visits a node *i at or before* time *t*. This can be done by making *i* an absorbing state in the MTMC model and finding $\mathbf{p}_i^{(t)}$ on the new model.

In other words, $\forall t \in \mathbb{N}, \forall j \in \mathcal{V}$, let

$$\widetilde{P}_{ji}^{(t)} = 0$$

$$\widetilde{P}_{ii}^{(t)} = 1$$

$$\widetilde{P}_{mn}^{(t)} = P_{mn}^{(t)} \quad \forall m,\ n \in \mathcal{V},\ n \neq i$$

(see Figure 8.12). The resulting model has two absorbing states, *b* and *i*. $\alpha_i^{(t)} = \widetilde{\mathbf{p}}_i^{(t)} = \widetilde{P}^{(t)} \ldots \widetilde{P}^{(1)} \mathbf{p}^{(0)}$ is the probability that the packet visits node *i* in the original

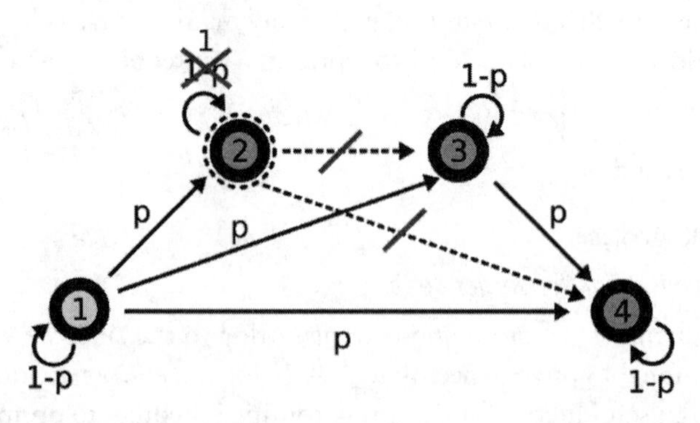

Figure 8.12 Illustration of how to create absorbing states in the Markov chain to calculate the probability that a packet sent from node 1 to node 4 passes through node 2 by time *t*, using the routing topology of Figure 8.11.

model at or before time t, while $\alpha_b^{(t)} = \tilde{\mathbf{p}}_b^{(t)}$ is the probability that the packet arrives at the sink at or before time t through an alternate path disjoint with node i.

To find $\alpha_i = \lim_{t\to\infty} \tilde{\mathbf{p}}_i^{(t)}$, the probability the packet ever visits node i, we solve a system of equations for the probability that any state $j \neq i$ is absorbed into state i.

Theorem 8.2 (Absorption Probability Equations [42]) *For a given Markov chain, choose an absorbing state i. Then, the probabilities α_{ji} of reaching state i starting from j are the unique solution to the equations*

$$
\begin{aligned}
&\alpha_{ii} = 1 \\
&\alpha_{ji} = 0 \quad \textit{for all absorbing } j \neq i \\
&\alpha_{ji} = \sum_{k=1}^{N} p_{jk}\alpha_{ki} \quad \textit{for all transient } j
\end{aligned}
\tag{8.9}
$$

If a packet is transmitted from a source node a, then $\alpha_i = \alpha_{ai}$.

8.3.3 Modeling Directed Staged Flooding

We propose a simple constrained flooding scheme called *directed staged flooding* (DSF) for one-to-many and one-to-one routing, focusing on the latter. Unlike UPD, DSF provides increased end-to-end connectivity with less latency by multicasting packets instead of using acknowledgments and retransmissions. We use a directed staged flooding Markov chain (DSFMC) model to find $p_{net}^{(t_d)}$. As with UPD, we build the model assuming we are provided with a routing schedule, the way nodes are grouped into stages (discussed later), and all the link probabilities. An algorithm for grouping nodes into stages was proposed by Dubois-Ferriere in [43] for two similar routing protocols, *ERS-best and ERS-any anypath routing*, where a single copy of a packet is routed with multicasting from each node on the path. We leave the development of an algorithm to construct a specific routing schedule for DSF for future work.

8.3.3.1 Modeling Characteristics

Like UPD, DSF also assumes that the nodes follow a TDMA routing schedule. During a transmission each node transmits to a subset of its neighboring nodes. Furthermore, we group the nodes along the end-to-end transmission path such that a packet is modeled as being passed between groups of nodes, and we call each group of nodes a *stage*. Figure 8.13 illustrates this on a *wide path* topology between a source and destination where the nodes lie on a regular grid and each stage, except the first and last, consists of a column of 3 nodes.

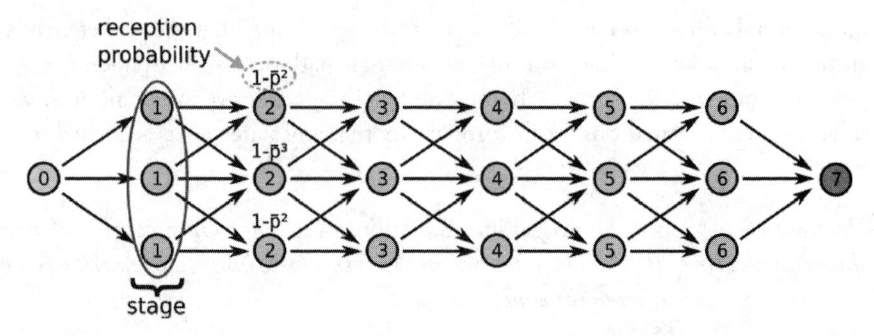

Figure 8.13 Directed staged flooding example on a wide path topology containing stages with a path width of 3. (See more details in Section 8.3.3.)

DSF does not use acknowledgments to signal a node to retransmit a packet on a failed link. Thus, with careful scheduling consecutive packets will not queue in the network if there is only a single source transmitting to a single sink.

Our DSFMC model of DSF requires the *sets* of link transmissions between *distinct* pairs of stages to be independent. Like UPD, DSF uses frequency hopping over time to help justify this assumption. However, the model allows the link transmissions between the *same* pair of stages to be correlated. This mirrors reality because on any single multicast transmission, all the receiving nodes are listening on the same channel.

Our DSFMC model also assumes that all nodes in one stage transmit their copy of the packet before the nodes in the next stage transmit their copy of the packet. Furthermore, the transmissions of nodes within a stage will interfere with each other, so they must be scheduled in separate time slots. We make this assumption because most sensor network nodes have only one radio and can only listen to one channel at a time.

In DSF routing schedules, a node can be shared between multiple stages (see Figure 8.14). Like UPD, we assume that the links in the routing topology for DSF do not form a cycle. Complications arise when sharing nodes between stages because unlike flooding, staged flooding puts on the constraints that a packet can only be transmitted from a node if it received the packet *prior* to the

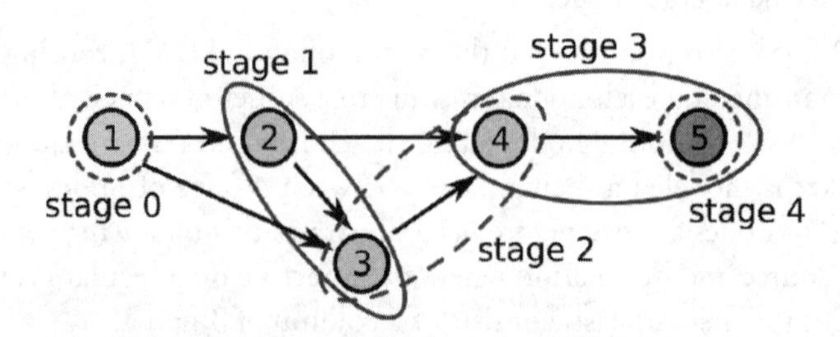

Figure 8.14 Directed staged flooding example corresponding to (8.13).

time another node in its stage first transmits. This is necessary for the DSFMC model developed below to hold. To enforce this condition, packets may carry with them a field indicating during which stage they were last transmitted. Last of all, we assume that if a node i is shared between stages k and k+1, then node i will retain the packet after transmission in stage k so it "receives the packet" with probability 1 in stage $k + 1$.

8.3.3.2 Directed Staged Flooding Markov Chain Model

As before, we represent the routing topology as a graph $G = (\mathcal{V},\mathcal{E})$ and denote a node in the network as $i \in \mathcal{V} = 1,\ldots,N$ and a link in the network as $l \in \mathcal{E} \subset \{(i,j)|\ i,j \in \mathcal{V}\}$ where $l = (i,j)$ is a link for transmitting packets from node i to node j. Because each link is used only once when transmitting a single packet, the link success probability for link $l = (i,j)$ is treated as being time invariant and is denoted p_l or p_{ij}.

Unlike the MTMC model, in the DSFMC model a state in the Markov chain at a stage represents the *set* of nodes in the stage that successfully received a copy of the packet. The transition probabilities between the states depend on the joint probability of successful link transmissions between stages. We state the DSFMC model for the special case where the links are all independent. For the general model, see [40].

Directed Staged Flooding Markov Chain Model

Let's assume we have a routing topology with $K + 1$ stages $0,\ldots,K$. Each stage k has N_k nodes, and the set of 2^{N_k} possible states in stage k is represented by the set of numbers $\mathcal{S}^{(k)} = \{0,\ldots,2^{N_k} - 1\}$. Let $\mathcal{K}^{(k)}$ be the set of nodes in stage k and for each state $\sigma^{(k)} \in \mathcal{S}^{(k)}$, let $\mathcal{R}_\sigma^{(k)} \subset \mathcal{K}^{(k)}$ be the set of nodes that have received a copy of the packet and $\mathcal{U}_\sigma^{(k)} = \mathcal{K}^{(k)} \setminus \mathcal{R}_\sigma^{(k)}$ be the set of nodes that have not received a copy of the packet (see Figure 8.15). Let $0^{(k)}$ denote the state where no nodes received a copy of the packet in stage k.

The conditional probability of the next state $\mathbf{X}^{(k+1)}$ being state $\sigma^{(k+1)}$ given that the current state $\mathbf{X}^{(k)}$ is $\sigma^{(k)}$ can be expressed as

$$\mathbb{P}\left(\mathbf{X}^{(k+1)} = \sigma^{(k+1)} \middle| \mathbf{X}^{(k)} = 0^{(k)}\right) = \begin{cases} 1 : \sigma^{(k+1)} = 0^{(k+1)} \\ 0 : \text{otherwise} \end{cases}$$

$$\text{if } \sigma^{(k)} \neq 0^{(k)}$$

$$\mathbb{P}\left(\mathbf{X}^{(k+1)} = \sigma^{(k+1)} \middle| \mathbf{X}^{(k)} = \sigma^{(k)}\right) = \tag{8.10}$$

$$\left(\prod_{\substack{u \in \mathcal{U}_\sigma^{(k+1)} \\ i \in \mathcal{R}_\sigma^{(k)}}} (1 - p_{iu})\right) \prod_{r \in \mathcal{R}_\sigma^{(k+1)}} \left(1 - \prod_{i \in \mathcal{R}_\sigma^{(k)}} (1 - p_{ir})\right)$$

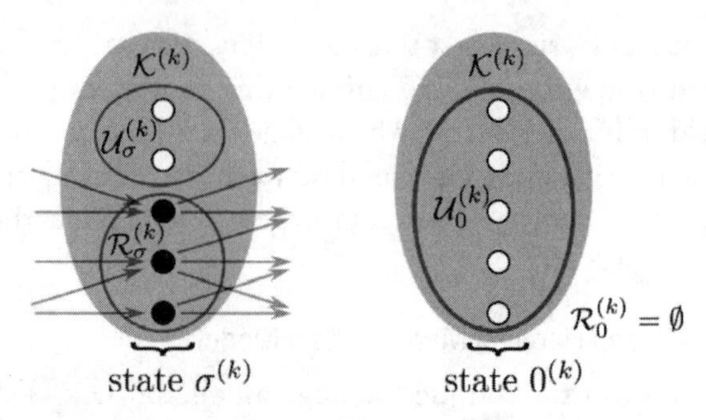

Figure 8.15 Mapping of states to nodes that received a packet in the DSFMC model. On the left is an example of a state $\sigma^{(k)}$ and on the right is the state $0^{(k)}$ where no packets have been received.

The transition probability matrices between stage k and $k+1$ are $P^{(k+1)} \in [0,1]^{N_{k+1}\times N_k}$, where the entry in position $(\sigma^{(k+1)},\sigma^{(k)})$ of the matrix is $\mathbb{P}(\mathbf{X}^{(k+1)} = \sigma^{(k+1)}|(\mathbf{X}^{(k)} = \sigma^{(k)})$.

The initial state $\mathbf{X}^{(0)}$ is the state $\sigma^{(0)}$ corresponding to $\mathcal{R}_\sigma^{(0)} = \{a\}$, where a is the node sending the initial packet. Then, the probability distribution $\mathbf{p}^{(k)} \in [0,1]^{N_k}$ of the state at stage k is

$$\mathbf{p}^{(k)} = \underbrace{P^{(k)} \cdots P^{(2)} P^{(1)}}_{P^{(\underline{k})}} \mathbf{p}^{(0)} \tag{8.11}$$

We can obtain the probability that a copy of the packet is at a node i at time t directly from our model by translating t to k from the relation $t = \sum_{i=0}^{k-1} N_i$ and looking at $\sum_{\{\sigma^{(k)} | i\in\mathcal{R}_\sigma^{(k)}\}} \mathbb{P}(\mathbf{p}^{(k)} = \sigma^{(k)})$. In the case where the last stage contains only the sink node and only the nodes in stage K–1 transmit to the sink, if b is the state in stage K where the sink receives a copy of the packet, we have

$$p_{net}^{(t_d)} = \begin{cases} 0 & : \; t_d \le \sum_{i=0}^{K-2} N_i \\ \mathbf{p}_b^{(K)} & : \; t_d \ge \sum_{i=0}^{K-1} N_i \end{cases} \tag{8.12}$$

and $0 \le p_{net}^{(t_d)} \le \mathbf{p}_b^{(K)}$ when $\sum_{i=0}^{K-2} N_i < t_d < \sum_{i=0}^{K-1} N_i$.

Finally, note that except in the special case where there exists a path from the source to the destination with all link probabilities equal to 1, $p_{net}^{(t_d)} < 1$ for all t_d. All copies of a packet can be lost in the network because we do not use acknowledgments and retransmissions to guarantee a copy of the packet has been delivered.

8.3.3.3 DSFMC Examples and Discussion

As an example, let's consider the stages with path width 3 in Figure 8.13. Assume the links are independent, that each link has the same transmission success

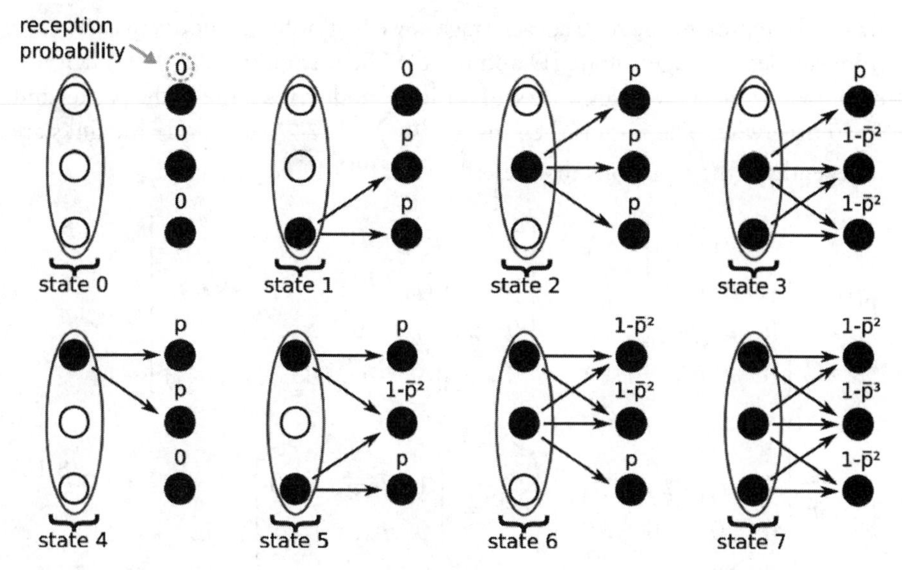

Figure 8.16 Markov chain states for the routing topology in Figure 8.13, excluding the states for the source and the destination.

probability p, and let $\bar{p} = 1 - p$. Then, the probability that a node in stage k +1 receives a copy of the packet, given the state of stage k, is 1 minus the product of incoming link failure probabilities, as shown in Figure 8.16. The transition probability between states can be obtained by applying (8.10). Figure 8.17 illustrates the transitions out of state 7.

In the example in Figure 8.14, the dimensions of the state probability distribution vector vary with time, and also some of the nodes are shared between

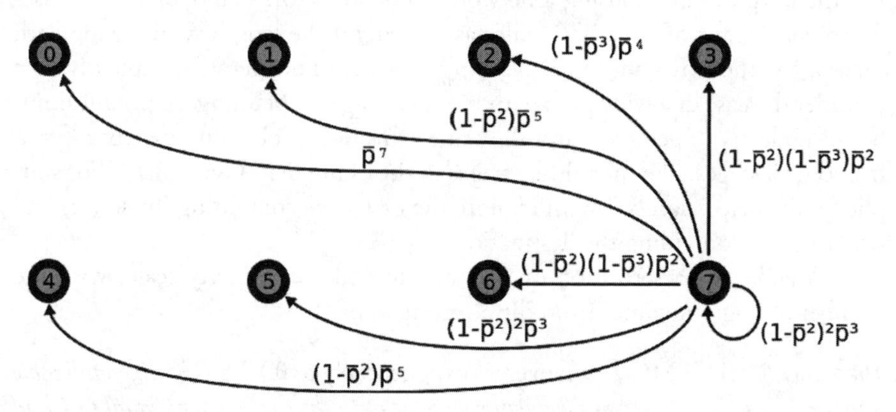

Figure 8.17 Markov chain transition diagram for a stage of path width 3 in the routing topology in Figure 8.13. Here, only the outgoing transitions and associated transition probabilities from state 7 are shown.

stages. To represent the state at each stage k, we first order the nodes in each stage from smallest to largest node ID and reindex them from $0,\ldots,N_k-1$. Then, for each node with a new index n we set $i_n=1$ if the node has a copy of the packet and $i_n=0$ otherwise. The state is then just $\sigma^{(k)}=\sum_{n=0}^{N_k-1} i_n 2^n$. Assuming the links are independent, the equations that describe the DSFMC model are

$$P^{(1)}=\begin{bmatrix}1 & \bar{p}_{12}\bar{p}_{13}\\ 0 & p_{12}\bar{p}_{13}\\ 0 & \bar{p}_{12}p_{13}\\ 0 & p_{12}p_{13}\end{bmatrix}\quad P^{(2)}=\begin{bmatrix}1 & \bar{p}_{23}\bar{p}_{24} & 0 & 0\\ 0 & p_{23}\bar{p}_{24} & \bar{p}_{34} & \bar{p}_{24}\bar{p}_{34}\\ 0 & \bar{p}_{23}p_{24} & 0 & 0\\ 0 & p_{23}p_{24} & p_{34} & (1-\bar{p}_{24}\bar{p}_{34})\end{bmatrix}$$

$$P^{(3)}=\begin{bmatrix}1 & \bar{p}_{34} & 0 & 0\\ 0 & p_{34} & \bar{p}_{45} & \bar{p}_{45}\\ 0 & 0 & 0 & 0\\ 0 & 0 & p_{45} & p_{45}\end{bmatrix}\quad P^{(4)}=\begin{bmatrix}1 & \bar{p}_{45} & 0 & 0\\ 0 & p_{45} & 1 & 1\end{bmatrix} \tag{8.13}$$

$$\mathbf{p}^{(0)}=\begin{bmatrix}1 & 0\end{bmatrix}^T \quad P^{(\underline{4})}=P^{(4)}P^{(3)}P^{(2)}P^{(1)}$$

where p_{ij} is indexed by the original node IDs and again $\bar{p}_{ij}=1-p_{ij}$. As mentioned in Section 8.3.3, we assume that if a node i in stage k has a copy of the packet and node i is also in stage $k+1$, then node i will have a copy of the packet in stage $k+1$ with probability 1.

8.3.3.4 DSFMC Analysis

$p_{net}^{(t_d)}$ for Wide Paths with Repeated Stages

For the purposes of choosing a network topology before deployment, it is useful to get a grasp of how $p_{net}^{(t_d)}$ scales as we extend the length K of a wide path topology without having to calculate $p_{net}^{(t_d)}$ for each new network explicitly. We consider the case of a wide path with repeated stages containing a constant number of nodes N_{stage} per stage and the same transition probability matrix $P^{(k)}=P$ between all stages, like the middle stages in the example in Figure 8.13. For simplicity, the discussion below will ignore the first stage containing the source and the last stage containing the destination.

A good characterization of how end-to-end connectivity scales with the number of stages K comes from the eigenvalues of P.

Theorem 8.3 (DSFMC $p_{net}^{(t_d)}$ converges exponentially to 0.) *Let P be diagonalizable and $\lim_{K\to\infty} P^K\mathbf{p}^{(0)}=\mathbf{e}^{[1]}$, an elementary vector with the first element equal to 1 and all other elements equal to 0. $\mathbf{e}^{[1]}$ represents the state 0 where* no *nodes received a copy of the packet. Then*

$$p_{net}^{(t_d)} \leq C\left(\rho_*\right)^K, t_d = KN_{stage} \tag{8.14}$$

for some constant C dependent on the initial distribution $\mathbf{p}^{(0)}$ *and* $\rho_* = \max\{|\lambda|:\lambda$ *is an eigenvalue of P and* $|\lambda| < 1\}$.

The proof of this is similar to that for Theorem 8.1 and can be found in [40]. While this relation is an upper bound, ρ_* is the dominant decay rate for large K because all the eigenvectors of P with eigenvalue magnitudes less than 1 decay exponentially with K. In practice, a good routing topology has ρ_* very close to 1. When choosing a routing topology for wide paths, one can use ρ_* with repeated stages of different widths to quickly compare the gain in reliability at the cost of extra latency.

Traffic Distribution

To calculate the probability that a copy of the packet visits a node i at or before time t, $\alpha_i^{(t)}$, we first remove all the outgoing edges of i, and add a "self-transmission" link of probability 1 from node i to itself over all time slots. Then, we compute $\alpha_i^{(t)} = \sum_{\{\sigma^{(k)}|i \in \mathcal{R}_\sigma^{(k)}\}} \mathbb{P}\left(\widetilde{\mathbf{p}}^{(k)} = \sigma^{(k)}\right)$, where $\widetilde{\mathbf{p}}^{(\mathbf{k})}$ is the state probability distribution on the modified routing schedule and topology.

8.3.4 Using the Network Metrics

Earlier, we mentioned that a direct use for the connectivity metric $p_{net}^{(t_d)}$ is to check the feasibility of running a wireless networked system over the network. Below, we use the metric to compare UPD and DSF. The connectivity metric can also be used as an optimization objective for scheduling network access and for optimal controllers running over the network.

8.3.4.1 UPD and DSF Comparisons

We wish to compare UPD and DSF using end-to-end connectivity as a function of latency, $p_{net}^{(t_d)}$, as the primary metric. This is effectively a comparison to see when retransmissions in UPD are better than "preemptive retransmissions" by multicast in DSF. We chose the example of routing on a wide path grid, where the width of the path is the number of rows and the length of the path is the number of columns. Here, every node in one column of a grid (a stage in DSF) can route to every other node in the next column with equal, independent link probabilities p_l. The routing schedule for directed staged flooding and unicast path diversity routing is described in Figure 8.18 for paths of width 3. Also, in all our comparisons, we assume that the time to send an acknowledgment for UPD is negligible and ACKs can be sent back in the same time slot as the original transmission.

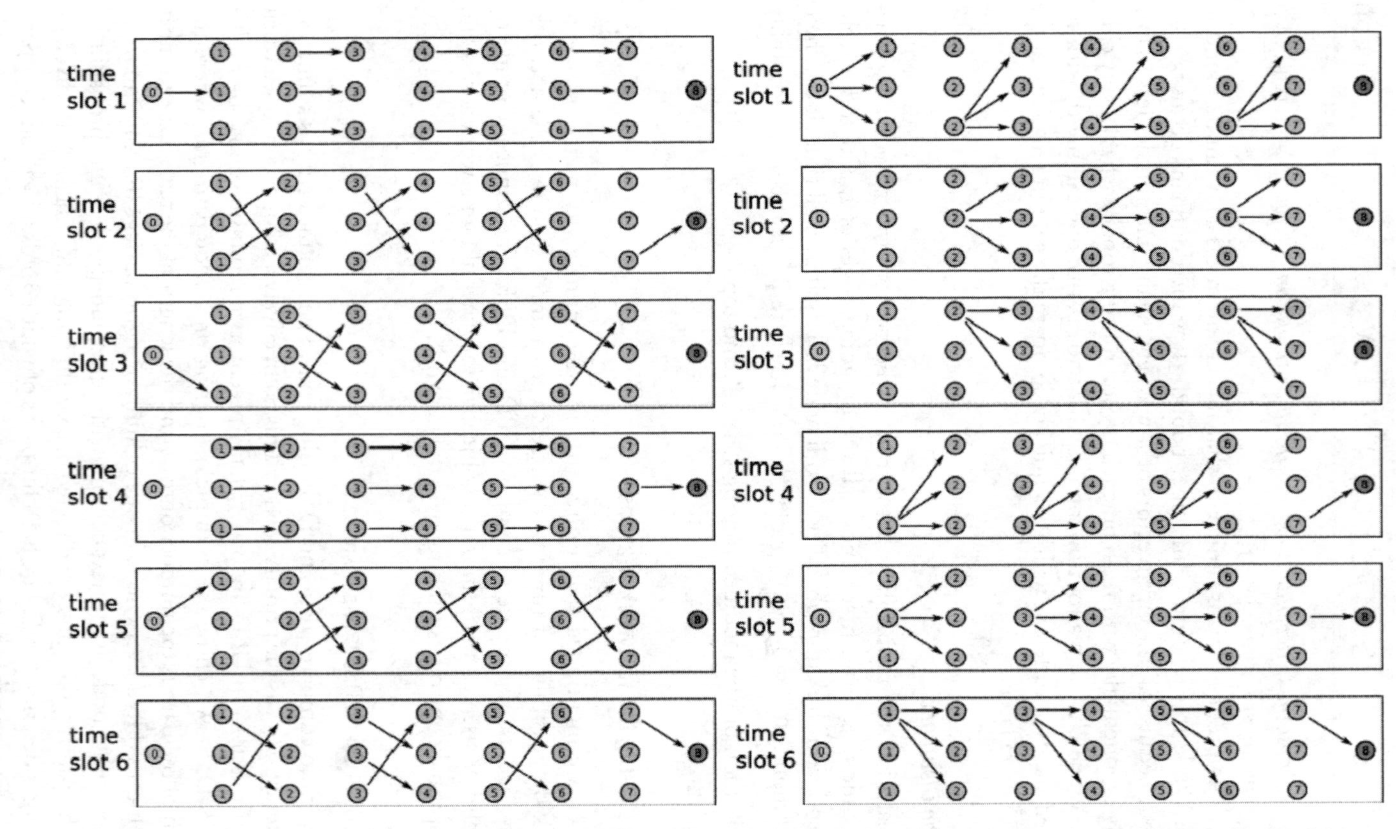

Figure 8.18 (Left) UPD and (right) DSF schedules for routing on a grid of width 3, used in the calculations for the graphs in Section 8.3.4.

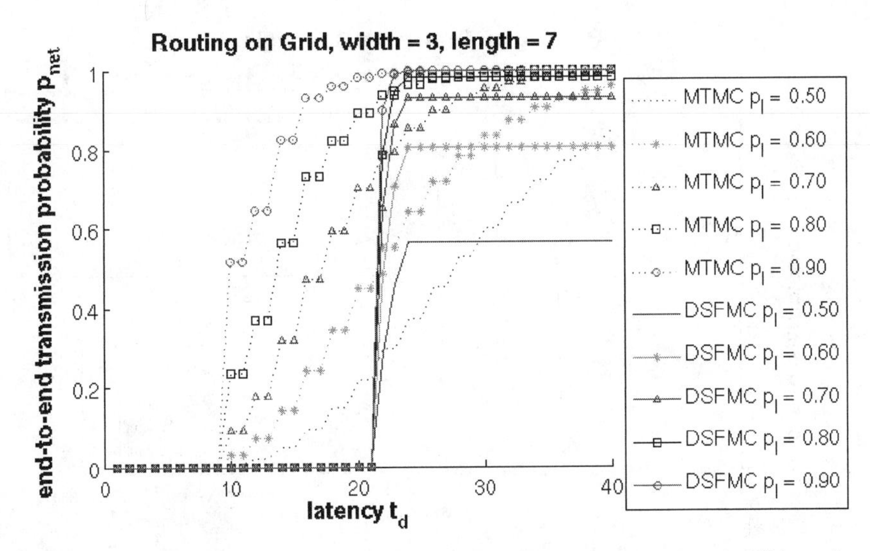

Figure 8.19 End-to-end connectivity as a function of latency for varying link probabilities using the routing schedules described in Figure 8.18.

Figure 8.19 compares $p_{net}^{(t_d)}$ of the two routing schemes under a range of different link probabilities.[4] UPD has the potential to deliver packets from the source to the sink in a shorter period of time, but the packet delivery time has a larger variance. Also, because $\lim_{t\to\infty} p_{net}^{(t_d)} = 1$ for UPD and p_{net} for DSF is a fixed value less than 1 after the last stage transmits (assuming $p_l \neq 1$), UPD can always provide better end-to-end connectivity at high latencies t_d.

Figure 8.20 shows that the final end-to-end connectivity p_{net} for DSF is higher for wider paths at the cost of larger latency. Also, the figure illustrates the limitations of our MTMC model—the model is unable to capture the benefit of diversity from using multiple paths instead of a single path because it assumes that all links are independent. Hence, retransmission on the same link is just as good as transmitting on a different link. What is modeled is that wider paths require more time slots to schedule transmissions from the nodes in the last column to the sink, and hence Figure 8.20 shows that UPD on wide paths with a smaller width tend to perform better.

8.3.4.2 Scheduling and Adaptive Control

Optimal scheduling on a general communication graph, for instance, to maximize throughput, has been shown to be an NP-hard problem [44]. Nonetheless, the importance of the problem spurred many research efforts to find suboptimal

4. Note that in this and subsequent plots, we perform the DSFMC calculations at the time granularity of time slots, not stages, unlike the description of (8.12) in Section 8.3.3.

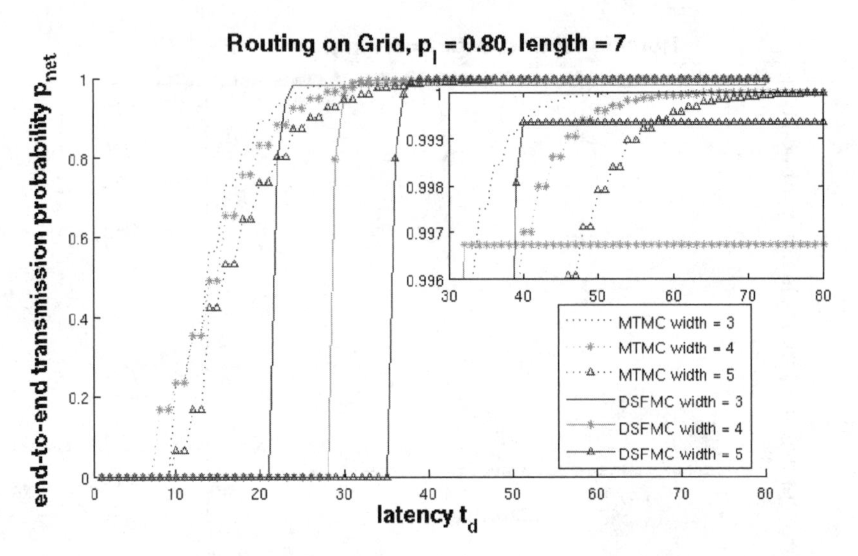

Figure 8.20 End-to-end connectivity as a function of latency for varying path widths using the routing schedules described in Figure 8.18, with magnification of the plot for p_{net} near 1.

schedules that are computationally tractable. Developing new network metrics may not reduce the computational complexity of finding an optimal schedule, but it will allow us to design scheduling algorithms that directly improve the overall performance of the wireless networked system.

For instance, we can set the objective of our optimization problem to yield a higher end-to-end connectivity for a given latency. Then, we can add other constraints to the schedule that correspond to desirable characteristics. For example, the connectivity metric is highly dependent on link probability estimates, so we would like a schedule that is robust to link estimation errors. We also would

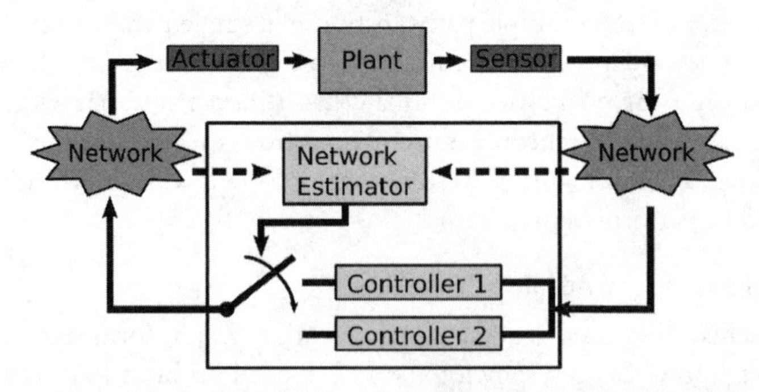

Figure 8.21 An example of a switching controller that adapts to varying network conditions. The network estimator is part of the full system controller. The dashed lines indicate the monitoring of network conditions (e.g., in directly through checking time stamps and sequence numbers of delivered packets).

like a schedule that can be negotiated by nodes in a distributed manner through "prices," and converges in a reasonable amount of time.

Network metrics can also be used to adjust the wireless networked system to varying network conditions. For instance, we can monitor the packet success rate to determine when to switch between controllers (see Figure 8.21), using a more aggressive controller when the network has a high packet delivery success rate with low latency. In fact, the H_∞ controller proposed by Seiler and Sengupta in Section 8.2.3 is an example of such a switching controller. The network is modeled as being in two states—high probability of packet delivery ($\theta_k = 1$) and low probability of packet delivery ($\theta_k = 0$)—with corresponding controller matrices described by (8.4).

8.3.5 Limitations of the Models and Future Work

The proposed MTMC and DSFMC models abstract many of the essential characteristics of the network for the wireless networked system, but they can be improved. One of the underlying assumptions necessary for the model is that links between stages are independent over time. In reality, the quality of wireless links is often correlated over time because of short-term channel fading or interference. While exact modeling of the correlation between links is difficult, a simple network state model similar to the Gilbert-Elliot channel model for a point-to-point link may more closely match the conditions in a real wireless network deployment.

Another important area for improvement is to try to incorporate queuing into the model. Network queuing theory is a notoriously difficult problem to model precisely, but an interesting area for further study is the development of guidelines on the data rates and number of independent data streams that make queuing less likely to occur in the network. It is also interesting to study how to give priority to different data streams when packets are queued at a node in the network so as to maximize the performance of the entire system or maintain stability. Understanding queuing can guide the design of a transport-layer congestion control scheme that performs well in the context of a networked control system.

8.4 Conclusions

This chapter provides a framework for analyzing the performance and security of wireless networked systems for protecting critical infrastructure. The trend in industry is to integrate wireless sensor networks with critical infrastructure both for monitoring and self-regulation. The security of systems integrated with sensor networks requires a framework and model for translating component performance/compromise/availability to system performance. We model the wireless networked system as a networked control system or a networked real time measurement system. Here, the first step is to define the appropriate network metrics that can be translated to system performance using analysis tools such as the existing networked control system theory. This in turn will help us identify the most vulnerable/important components in the system. From this framework, we can then ask what additional network metrics are necessary for characterizing system security.

This chapter has only scratched the surface of wireless networked system security by considering only the simple attack of dropping packets at compromised nodes. Not much has been said about the *dynamics* of the system under attack—how the entire system reacts over time to different attacks. This is particularly relevant for attacks on network routing, which can influence the delay of packets. Furthermore, if the attacker can *alter* the contents of some packets in addition to delaying and dropping them, we will need to reexamine and combine networked control system theory with *game theory*, where we model the attacker as an adversary, to understand the impact on system performance.

A layered security approach is necessary to properly secure a wireless networked system for critical infrastructure. At the bottom layer is physical security on the individual sensor nodes to prevent the extraction of keys by probing the hardware. The next layer is cryptography, authentication, and integrity on messages sent over the wireless channel. The layer above that is authentication for network services, such as joining nodes to the network or establishing routing paths. And finally, robustness and security of the entire system need to be considered to allow it to continue operating under partial system compromise. Only by considering security at all layers of the system can we hope to properly protect our critical infrastructure in the future.

Acknowledgments

The authors were supported under ARO's MURI program "Urban Target Recognition by Ad-hoc Networks of Imaging Sensors and Low-cost, Nonimaging Sensors," under contract number W911NF-06-1-0076.

References

[1] Department of Homeland Security, "National infrastructure protection plan," http://www.dhs.gov/xlibrary/assets/NIPP_Plan.pdf, 2006.

[2] North American Electric Reliability Council, "Technical analysis of the August 14, 2003, blackout: What happened, why, and what did we learn?" ftp://www.nerc.com/pub/sys/all_updl/docs/blackout/NERC_Final_Blackout_Report_07_13_04.pdf, Princeton, NJ, July 2004.

[3] L. Cole, *The Anthrax Letters,* Washington, D.C.: Joseph Henry Press, 2003.

[4] M. Barborak, A. Dahbura, and M. Malek, "The consensus problem in fault-tolerant computing," *ACM Computing Surveys,* vol. 25, no. 2, pp. 171–220, 1993.

[5] L. Lamport, R. Shostak, and M. Pease, "The Byzantine generals problem," *ACM Transactions on Programming Languages and Systems,* vol. 4, no. 3, pp. 382–401, 1982.

[6] M. Castro and B. Liskov, "Practical byzantine fault tolerance and proactive recovery," *ACM Transaction on Computer System,* vol. 20, no. 4, pp. 398–461, 2002.

[7] *Part 15.4: Wireless Medium Access Control (MAC) and Physical Layer (PHY) Specifications for Low-Rate Wireless Personal Area Networks (LR-WPANs),* LAN/MAN Standards Committee of the IEEE Computer Society, New York, October 2003, 802.15.4 Standard.

[8] Y.-C. Hu and A. Perrig, "A survey of secure wireless ad hoc routing," *IEEE Security & Privacy Magazine,* vol. 2, no. 3, pp. 28–39, 2004.

[9] R. Szewczyk et al. "An analysis of a large scale habitat monitoring application," *SenSys '04: Proceedsing of the 2nd International Conference on Embedded Networked Sensor Systems,* New York, 2004, pp. 214–226, http://doi.acm.org/10.1145/1031495.1031521.

[10] G. Tolle et al. "A macroscope in the redwoods," *SenSys '05: Proceedsing of the 3rd International Conference on Embedded Networked Sensor Systems,* New York: ACM Press, 2005, pp. 51–63. http://doi.acm.org/10.1145/1098918.1098925.

[11] V. Shnayder et al. "Sensor networks for medical care," Division of Engineering and Applied Sciences, Harvard University, Tech. Rep. TR-08-05, April 2005. http://www.eecs.harvard.edu/~mdw/papers/codeblue-techrept05.pdf.

[12] M. Becker, et al. "Approaching ambient intelligent home care systems," *Pervasive Health Conference and Workshops,* pp. 1–10, 2006.

[13] J. Wilson et al. "Design of monocular head-mounted displays for increased indoor firefighting safety and efficiency," *Helmet- and Head-Mounted Displays X: Technologies and Applications,* vol. 5800, no. 1, pp. 103–114, 2005. http://link.aip.org/link/?PSI/5800/103/1.

[14] D. Steingart et al. "Augmented cognition for fire emergency response: An iterative user study," *Proceedings of the 11th International Conference on Human-Computer Interaction (HCI),* July 2005.

[15] D. J. Gaushell and H. T. Darlington, "Supervisory control and data acquisition," *Proceedings of the IEEE,* vol. 75, pp. 1645–1658, 1987.

[16] J. R. Moyne and D. M. Tilbury, "The emergence of industrial control networks for manufacturing control, diagnostics, and safety data," *Proceedings of the IEEE,* vol. 95, pp. 29–47, 2007.

[17] J. Marcuse, B. Menz, and J. R. Payne, "Servers in SCADA applications," *IEEE Transactions on Industry Applications,* vol. 33, pp. 1295–1299, 1997.

[18] S. Oh, et al. "Tracking and coordination of multiple agents using sensor networks: System design, algorithms and experiments," *Proceedings of the IEEE,* vol. 95, pp. 234–254, 2007.

[19] Á. Lédeczi et al. "Countersniper system for urban warfare," *ACM Transaction Senon Sensor Network,* vol. 1, no. 2, pp. 153–177, 2005.

[20] M. G. Mehrabi, A. G. Ulsoy, and Y. Koren, "Reconfigurable manufacturing systems: Key to future manufacturing," *Journal of Intelligent Manufacturing,* vol. 11, pp. 403–419, August 2000, 10.1023/A:1008930403506. http://dx.doi.org/10.1023/A:1008930403506.

[21] K. D. Frampton, "Distributed group-based vibration control with a networked embedded system," *Smart Materials and Structures,* vol. 14, no. 2, pp. 307–314, April 2005.

[22] S. Seth, J. P. Lynch, and D. M. Tilbury, "Wirelessly networked distributed controllers for real-time control of civil structures," *Proceedings of the American Control Conference,* vol. 4, pp. 2946–2952, June 8–10, 2005.

[23] S. Kim et al. "Health monitoring of civil infrastructures using wireless sensor networks," *Proceedings of the 6th International Conference on Information Processing in Sensor Networks (IPSN),* ACM Press, pp. 254–263, April 2007.

[24] N. Xu et al. "A wireless sensor network for structural monitoring," *SenSys '04: Proceedings of the 2nd International Conference on Embedded Networked Sensor Systems,* New York: ACM Press, pp. 13–24, 2004. http://doi.acm.org/10.1145/1031495.1031498.

[25] J. P. Hespanha, P. Naghshtabrizi, and Y. Xu, "A survey of recent results in networked control systems," *Proceedings of the IEEE,* vol. 95, pp. 138–162, 2007.

[26] Y. Xu and J. P. Hespanha, "Estimation under uncontrolled and controlled communications in networked control systems," *44th IEEE Conference on Decision and Control (CDC '05),* pp. 842–847, 2005.

[27] L. Schenato, "Optimal estimation in networked control systems subject to random delay and packet loss," *Proceedings of the 45th IEEE Conference on Decision and Control,* December 2006.

[28] L. Schenato et al. "Foundations of control and estimation over lossy networks," *Proceedings of the IEEE,* vol. 95, pp. 163–187, 2007.

[29] P. Seiler and R. Sengupta, "An H_∞ approach to networked control," *IEEE Transactions on Automatic Control,* vol. 50, pp. 356–364, 2005.

[30] W. R. Heinzelman, A. Chandrakasan, and H. Balakrishnan, "Energy-efficient communication protocol for wireless microsensor networks," *Proceedings of the 33rd Annual Hawaii International Conference on System Sciences,* vol. 2, p. 10, 2000.

[31] J. Li and G. Y. Lazarou, "A bit-map-assisted energy-efficient MAC scheme for wireless sensor networks," *3rd International Symposium on Information Processing in Sensor Networks (IPSN),* pp. 55–60, 2004.

[32] L. F. W. van Hoesel and P. J. M. Havinga, "A lightweight medium access protocol (LMAC) for wireless sensor networks," *Proceedings of the International Workshop on Networked Sensing Systems (INSS),* June 2004.

[33] V. Rajendran, K. Obraczka, and J. J. Garcia-Luna-Aceves, "Energy-efficient collision-free medium access control for wireless sensor networks," *SenSys '03: Proceedings of the 1st International Conference on Embedded Networked Sensor Systems,* New York: ACM Press, pp. 181–192, 2003.

[34] Dust Networks, Inc., "Technical overview of time synchronized mesh protocol (TSMP)," http://www.dustnetworks.com/docs/TSMP_Whitepaper.pdf, 2006.

[35] W. Ye, F. Silva, and J. Heidemann, "Ultra-low duty cycle MAC with scheduled channel polling," *SenSys '06: Proceedings of the 4th International Conference on Embedded Networked Sensor Systems,* New York: ACM Press, pp. 321–334, 2006.

[36] I. Rhee, et al. "Z-MAC: A hybrid MAC for wireless sensor networks," *SenSys '05: Proceed-*

ings of the 3rd International Conference on Embedded Networked Sensor Systems, New York: ACM Press, pp. 90–101, 2005.

[37] D. Ganesan, et al. "Highly-resilient, energy-efficient multipath routing in wireless sensor networks," *SIGMOBILE Moblie Computing and Communications Review,* vol. 5, no. 4, pp. 11–25, 2001.

[38] C. Karlof, Y. Li, and J. Polastre, "ARRIVE: Algorithm for robust routing in volatile environments," University of California at Berkeley, Tech. Rep. UCB/CSD-03-1233, May 2002.

[39] P. Levis, et al. "Trickle: A self-regulating algorithm for code propagation and maintenance in wireless sensor networks," *NSDI '04: Proceedings of the 1st Conference on Symposium on Networked Systems Design and Implementation,* Berkeley, CA: USENIX Association, pp. 2–2, 2004.

[40] P. Chen and S. Shankar Sastry, "Latency and connectivity analysis tools for wireless mesh networks," EECS Department, University of California, Berkeley, Tech. Rep. UCB/EECS-2007-87, June 2007. http://www.eecs.berkeley.edu/Pubs/TechRpts/2007/EECS-2007-87.html.

[41] Dust Networks, Inc., *SmartMesh-XT M2030 Product Specification,* http://www.dustnetworks.com/docs/M2030.pdf, 2006, datasheet.

[42] D. P. Bertsekas and J. N. Tsitsiklis, *Introduction to Probability,* Belmont, MA: Athena Scientific, 2002.

[43] H. Dubois-Ferriere, "Anypath routing," Ph.D. dissertation, EPFL, Lausanne, 2006. http://library.epfl.ch/theses/?nr=3636.

[44] S. Ramanathan, "A unified framework and algorithm for channel assignment in wireless networks," *Wireless Networks,* vol. 5, pp. 81–94, March 1999, 10.1023/A:1019126406181. http://dx.doi.org/10.1023/A:1019126406181.

[45] J. S. Rosenthal, "Convergence rates of Markov chains," *SIAM Review,* vol. 37, no. 3, pp. 387–405, 1995.

Appendix 8A Proof Sketch for Theorems 8.1 and 8.3

The proofs of Theorems 8.1 and 8.3 depend on the convergence rate of the second largest eigenvalue of an indecomposable, aperiodic stochastic matrix. The formula for bounding the explicit rate of convergence is reproduced from [45] here.

Theorem 8A.1 (Part of Fact 3 from [45]) *Suppose* P *satisfies* $\rho_* < 1$ *and the state space* $\mathcal{S}$ *is finite. Then, there is a unique stationary distribution* π *on* $\mathcal{S}$ *and, given an initial distribution* $\mathbf{p}^{(0)}$ *and point* $i \in \mathcal{S}$, *there is a constant* $C_i > 0$ *such that*

$$\left|\mathbf{p}_i^{(k)} - \pi_i\right| \leq C_i k^{J-1} \left(\rho_*\right)^{k-J+1}$$

where J *is the size of the largest Jordan block of* P. *It follows immediately that*

$$\left\|\mathbf{p}^{(k)} - \pi\right\|_{var} \leq Ck^{J-1}(\rho_*)^{k-J+1} \tag{8A.1}$$

where $C = \frac{1}{2}\sum C_i$. *In particular, if P is diagonalizable (so that J = 1), then*

$$\begin{aligned}\left\|\mathbf{p}_i^{(k)} - \pi_i\right\|_{var} &\leq \sum_{m=1}^{n-1} |a_m \mathbf{v}_m(i)||\lambda_m|^k \\ &\leq \left(\sum_{m=1}^{n-1} |a_m \mathbf{v}_m(i)|\right)(\rho_*)^k\end{aligned}$$

where $\mathbf{v}_0,\ldots,\mathbf{v}_{n-1}$ *are a basis of right eigenvectors corresponding to* $\lambda_0,\ldots,\lambda_{n-1}$, *respectively, and where* a_m *are the (unique) complex coefficients satisfying*

$$\mathbf{p}^{(0)} = a_0\mathbf{v}_0 + a_1\mathbf{v}_1 + \ldots + a_{n-1}\mathbf{v}_{n-1}$$

Here, $\mathbf{v}_m(i)$ *denotes the ith coordinate of the vector* $\mathbf{v}_m$.

The details of the proofs involve showing that the transition probability matrices satisfy the conditions for applying Theorem 8A.1, and then translating the results of Theorem 8A.1 to bounds on the end-to-end connection probability as a function of time. The full proofs are provided in [40].

9

Large Systems Modeling and Simulation

Alfonso Farina, Antonio Graziano, Luciana Ortenzi, and Emidio Spinogatti

9.1 Introduction

This chapter addresses the application of modeling & simulation (M&S) to the analysis of large systems in the homeland protection domain. Throughout this Chapter the term homeland protection (HP) will be used to identify the overarching mission covered by both homeland security and homeland defense. The use of a more general word such as "protection" refers to the involvement of all stakeholders in this critical national security mission, which needs to be unbiased and unhampered by national organizational title references (i.e., Department of Defense, Department of Homeland Security). Homeland protection addresses the broad civilian and military effort produced by a country to protect its territory—including citizens, assets, and activities that are vital and fundamental for its growth and prosperity—against internal and external hazards and to reduce its vulnerability to attacks, whatever their origin, as well as natural disasters. HP is therefore a wide and complex domain: systems in this domain are large, to mean that size and scope of such systems are conspicuous and that system boundaries may not be easy to identify; systems are integrated, to mean that it is generally not sufficient to study each subsystem in isolation; systems are different in purpose and require a multidisciplinary approach for their analysis; and their interconnections may range from a loose coupling at the information level to a tight coupling at the physical level.

System analysis is a well-established engineering discipline whose aim is to characterize systems and their properties, to predict and evaluate system performance, to support the design of solutions matching given performance requirements, and to assess alternative design options using methods spanning from analytic formulas to computer simulations. M&S is a well-known technique that supports system analysis and complements or replaces purely mathematical approaches; this is especially true in the HP domain where analytic techniques are not appropriate to provide an end-to-end characterization of systems. However M&S poses new challenges due to the broadness and size of systems. M&S of large integrated systems is debated in the specialized literature, but the emphasis is mostly on architectures, standards, software, and simulation technicalities; yet a formal methodology needs to be developed to drive the M&S effort of large systems and assist engineers in what today is still a practice of art.

The M&S of an integrated system in the framework of HP has been dealt with by the authors with a multilevel approach in which trade-offs are realized at each level between field of view and modeling accuracy. The present results originate from the initiative promoted by Finmeccanica and SELEX Sistemi Integrati to strengthen collaboration and cooperation throughout the Finmeccanica Group and to address the analysis and performance evaluation of large integrated systems; Finmeccanica Companies, with their expertise, products, and systems, cover almost entirely the homeland protection domain and therefore this effort is extremely purposeful. A proper working group, which includes representatives from all the Finmeccanica Companies and also experts from the academic world, has been set up since 2005 and is actively working on the analysis of large systems for homeland protection [1].

The chapter is organized as follows. Section 9.2 illustrates the HP domain and highlights some of the characteristics of systems in this specific domain. Section 9.3 briefly reviews the history of system theory and references new emerging trends in complexity theory. In Section 9.4 the role of modeling and simulation is discussed. Two classical approaches to the modeling and simulation of a large system are described in Section 9.5. The multilevel approach to simulation is introduced in Section 9.6. Section 9.7 describes two study cases that have been analyzed in the context of maritime border control: the first shows a performance simulation of the whole system from detection of intruders to final intervention, while the second provides a deeper analysis of the detection, classification, and identification performance that may be achieved by a suite of heterogeneous sensors in different meteorological conditions. Concluding remarks are given in Section 9.8.

9.2 The Challenge of Homeland Protection

The diagram of Figure 9.1 depicts the scope of the homeland protection domain: the two main subdomains are homeland defense (HD) and homeland security (HS).

HD includes the typical duties and support systems of military joint forces and single armed forces. Usually HD systems are strictly military; are employed

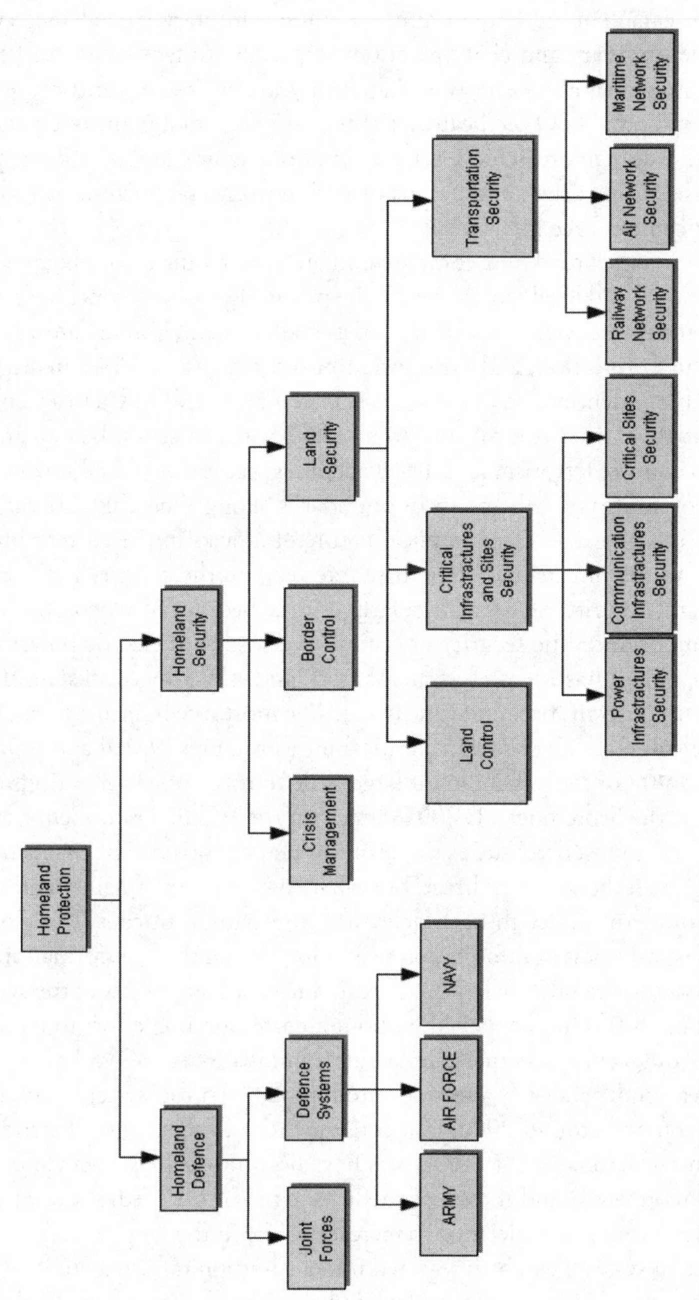

Figure 9.1 Homeland protection domain.

by military personnel only; satisfy specific technical requirements, operational needs, and environmental scenarios; and in most cases are designed to face only military threats. The new trend aims to employ military surveillance systems in combined military and civil operations, especially to face terrorism [2]. The military domain has also been swept in recent years by the net-centric operations (NCO) paradigm; NCO predicates a tighter coupling among forces, especially in the cognitive domain, to achieve synchronization, agility, and decision superiority and it is a strong driver in the transformation from a platform-centric force to a network-centric force [3].

HS is a very broad and complex domain that requires coordinated action among national and local governments, the private sector, and concerned citizens across a country; it covers issues such as crisis management, border control, critical infrastructure protection, and transportation security [4, 5]. Crisis management is the ability to identify and assess a crisis, plan a response, and act to resolve the crisis situation. Border control aims to build a smart protection belt all around a country to counter terrorism and illegal activities; yet it is not resolute due to the difficulty of controlling the country boundaries along their full and variegated extension, the non-necessarily physical nature of attacks in the current information age, and the threats that often arise internally to the country itself. HS also includes land security that is particularly critical because of its complexity and strategic importance; the security of critical assets, such as electric power plants, communication infrastructures, strategic areas, and railway networks, must be ensured continuously in space and time [6–8]. The most recent terrorist attacks have shown the vulnerability of national critical infrastructures [9, 10] and have made the world aware of the possibility of large-scale terrorist offensive actions against civil society: the September 11, 2001, attack on the World Trade Center in New York City is the most dramatic example of this new terrorism. The main emphasis has been put on the terrorist threat, but what emerges is the fragility and vulnerability of modern society to both deliberate and natural threats. The globalization, the pervasiveness of information technologies, and the transformation of the industrial sector and civil society have created new vulnerabilities in the system as a whole, but all this has happened without a corresponding effort to increase its robustness and security. As an example, single infrastructure networks have grown over the years independently, creating autonomous "vertical" systems with limited points of contact; around 2000, as a consequence of the change of trend in the sociotechno scenario, the infrastructures have begun to share services and thus to create interconnected and interdependent systems. In the medium term the degree of interconnection is deemed to increase, while in the long term the expectation is that all systems will be integrated into a common infrastructure.

Nowadays challenge is to understand this new scenario and to address its global implications by approaching large integrated systems in a more systematic and unified way.

9.3 Definitions and Background

Before addressing in more detail the topic of M&S applied to the analysis of large systems, it is useful to briefly review the evolution of system theory and, more recently, the definition of new paradigms and the introduction of complex systems theory.

A *system* may be defined as a group of interacting, interrelated, or interdependent parts forming a complex whole. This definition is very general and applies to systems from a wide array of scientific disciplines (e.g., in biology where a system is an organism as a whole, especially with regards to its processes and functions, or in the engineering field where the term system encompasses anything from a collection of interacting mechanical or electrical components to a complex communication network with hundreds of nodes and links).

The analysis of a "simple" system is generally performed resorting to the classical *system theory,* which models the system as a "black box" represented by a transfer function that describes the relationship between the inputs and the outputs [11, 12]. Classical system theory typically studies characteristics of systems such as stability, observability, and controllability [13, 14]. Significant developments in system theory began in the second half of the nineteenth century and, among its founders, the main contributions were given by H. Poincarè (1854–1912), with his studies on nonlinear dynamic systems; J. von Neumann (1903–1957), who developed the basic concepts of cellular automata; A. Lyapunov (1857–1918), who formulated several theorems about system stability; and—moving from a very different perspective—the psychiatrist W. R. Ashby (1903–1973), who studied self-organizing complex systems, such as the human brain and learning systems. Other fundamental contributions to system theory came in the second half of the twentieth century, from D. Luenberger, who developed the observability theory [15]; R. Kalman, who developed the well-known Kalman-Bucy recursive filtering technique [16]; and L. Zadeh, who formulated fuzzy logic [17–19]. In more recent times the Santa Fe Institute, founded—among others—by M. Gell-Mann at the beginning of the 1980s, started to study complex systems and to establish a complexity theory [20–22], as, for instance, systems comprising several unattended surveillance sensors, randomly spread on the territory, which self-organize their distribution and their behavior when a target is detected [23, 24].

To address the study of large systems in a structured way, it would be useful to have a well-established theory providing a categorization of systems in classes, identifying properties and characteristics of each class of systems, and defining a set of class-specific techniques and methods by which system analysis may be performed. Currently there is no such theory and no general system categorization is worldwide and cross-discipline accepted, even though a lot of technical papers have been written on the subject and definitions given in specific disciplines. An interesting distinction can be found in [25] with regard to complicated and complex systems: in a *complicated* system the components and their connections are equally important, the rules by which components interact produce predictable responses, and so the system response is fully determined; in a *complex* system, the components, which are called *agents*, exhibit a self-organizing collective behavior, which is very difficult to anticipate from the knowledge of each agent's behavior since it does not result from the existence of a central control or a specific law. As a result, if we know that a system is complicated, linear, and determined, we can expect to be able to control and predict its outcomes, while if we know that a system is complex and adaptive we have to expect novel, creative, and emergent outcomes. However, the edge between the two system classes is not a sharp one, definitions are qualitative, and no standard way is proposed to measure how complicated or complex a system is.

9.4 The Role of Modeling and Simulation

M&S plays an important role in the analysis of large systems. A model is a simplified representation of a "system" at some particular point in time or space intended to capture the relevant and essential aspects of the real system; a simulation is the dynamic application of the model in such a way that it operates on time or space to reduce it and promote understanding of the system; thus, especially in the case of large systems, M&S enables one to perceive the interactions that would not otherwise be apparent because of their separation in time or space [26]. The behavior of large integrated systems may not be fully understood by studying its components separately because the interactions between them are a determinant to produce a global system outcome; the decomposition of a large system in subsystems does not capture the propagation of effects when the system components are interconnected. The performance evaluation of large integrated systems is not an easy task and cannot be achieved only by means of analytical methods, which, on the other hand, can be applied effectively to its single components, for instance, a radar. In fact, if we consider again a radar, parameters such as detection and false alarm probabilities are adequate to represent its performance and can be determined analytically; yet in a large system these radar parameters concur to determine the overall system performance or effectiveness, which may be expressed, for example, in terms of reaction time against an incoming intruder, but this overall measure can be determined analytically only in the most simplified scenarios.

M&S represents the only viable solution to address the overall performance evaluation of large systems, yet M&S applied to a large system exacerbates the following drawbacks:

- Large modeling and software effort: the whole system is large and therefore many models need to be defined and developed; large systems for HP also operate in an environment that is extremely rich in details and diversified, so that the simulation in the virtual environment requires the use of sophisticated software tools, which may require intensive software development.
- Exploration of a large multidimensional trade-off space: the trade-off space of a large system is extremely wide and therefore the evaluation of multiple configurations of the system over multiple evolutions of the scenarios through simulation can be a formidable task.
- Assessment of system properties, such as stability, via simulation is unfeasible for a large system.
- Validation and verification of simulation models: validation checks if an "adequate model" has been chosen to represent the real system, it cannot be conducted in a strictly formal way and requires consolidated knowledge and experience; verification answers the question: "is the model implemented correctly in the computer?" It is a process very similar to the debugging of software programs and therefore requires a very big effort.

The simulation of integrated systems has been approached recently in the technical literature in a rather disjointed way and using very specific cutting-edge technologies. A lot of papers have been published on the DIS (distributed and interactive simulation) standard and HLA (high-level architecture) to address the issue of interoperable and distributed computer simulation systems [27]. In the mid-1990s N. Negroponte, A. Kay, and others introduced a simulation technology based on an "Intelligent Software Agent," whose theoretical concepts derive from artificial intelligence [28, 29]. In 1996 J. Epstein and R. Axtell proposed the concept of ABMS (agent-based modeling and simulation) [30]; this technique, usually based on fuzzy logic and genetic algorithms, is suited, in particular, for the analysis of complex systems because it can understand and predict emerging behaviors thanks to the learning capability of the models used; for example, it may be potentially useful for the modeling of terrorist asymmetric threats.

Since from the analysis of the technical literature no clear indication emerges on the methodology to adopt for the modeling and simulation of large systems, in the following we describe two classical reductionist approaches that can be effective in the analysis and in the performance evaluation of large systems. Then we introduce a simulator architecture, which addresses the issue of system modeling and simulation in a novel, integrated multilayer perspective.

9.5 Reductionist Approaches to the Modeling of Large Systems

The proposed approaches are based on the classical "divide and conquer" concept, which is well known and widespread in many technical and nontechnical fields; it relies on the decomposition of a large integrated whole into more manageable independent parts. The two approaches are horizontal and vertical decomposition.

9.5.1 Horizontal Decomposition

Horizontal decomposition approaches the study of the system by identifying separate system layers; layers may refer to subsystems or specific functionalities and several layers contribute to end-to-end performance. The study may focus on a subset of layers (e.g., those that are considered more critical or whose performance we are interested in); layers that typically may require a specific study are, for example, the sensor layer to evaluate detection and classification performance, or the man in the loop layer to characterize the effect the human element introduces in the system. Typical layers include:

- Sensors;
- Effectors and weapons;
- Command and control;
- Communication networks;
- Man in the loop or human element.

Sensors are responsible for the acquisition of data from the real world and may be classified according to different criteria, for example, by sensing principle (e.g., electromagnetic, acoustic, chemical), by domain (e.g., maritime, underwater, space, and urban and indoor spaces), or by platform type (aircraft, helicopter, and so forth). Analysis of the sensor layer is important to evaluate the detection and classification capability of the system, because often it heavily conditions the overall system performance. Effectors interact with the environment and are capable of acting on real-world objects; effectors include warning and dissuasive devices and firing weapons that produce a disabling or kill effect. Effectors may also be classified by domain, by platform, by specialization, and so forth. Command and control (C2) is defined in NATO publications [31] as "the authority, responsibilities and activities of military commanders in the direction and co-ordination of military forces and in the implementation of orders related to the execution of operations." The C2 layer therefore includes the operational doctrine and tactics and all the processing that manages and reduces data and information and supports the decision level. Data and information are exchanged, processed, or used in the C2 layer by several typical components including:

- Static databases (e.g., geographical data and maps and own forces data);
- Dynamic archives collecting quasi-real-time data from the sensors layer;
- Geographical information systems (GIS), which process and display georeferenced data;
- Data fusion and display functions (HMI—human–machine interface);
- Functions at the planning, tasking, and execution levels.

The communication network provides the required connectivity among the nodes of the system. The communication layer is extremely diversified, with the coexistence of digital and analog techniques, different physical means (e.g., copper, fiber optic, wireless, satellite), different technologies and standards (e.g., GSM—global system mobile, UMTS—universal mobile telephone system, WiMax—worldwide interoperability for microwave access), and different frequencies. The human element is certainly the most difficult layer to model and analyze due to the many intangible factors that may influence behavior, including emotions, morale, and stress.

The environment in which the integrated system operates is the real world that is not modeled as a system layer, but can be characterized using a set of scenarios; in the case of a land scenario, it is further characterized by type of terrain, presence and density of infrastructures, meteorological conditions, unintentional and intentional disturbances, and so forth. Scenarios include many actors—some active (e.g., threats, effectors), some passive (e.g., civilians), some neutral (e.g., nongovernmental organizations, Red Cross)—and their actions. This short account cannot be exhaustive, but it only aims to introduce the reader to the complexity of integrated systems analysis, its manifold aspects and numerous possible perspectives, and to highlight how demanding the performance evaluation of an integrated system is.

9.5.2 Vertical Decomposition

The *vertical decomposition* approach identifies end-to-end critical paths along which the system may be described during the evolution of a given scenario. Given a mission and the corresponding MoE (measure of effectiveness) and a description of the integrated system, vertical decomposition can be typically performed as follows:

- Identification of a critical system path relative to the mission;
- Decomposition of the path into a sequence of significant events;
- Formulation of the MoE in terms of the performance associated with the single events (MoP, measures of performance);

- Analysis of each event and quantification of the corresponding MoP;
- Synthesis of the results obtained for each single event and quantification of the MoE.

Vertical decomposition is effective when few critical paths dominate and characterize system behavior or in the preliminary phases of system design and risk reduction when the feasibility of the most stringent requirements must be assessed.

Horizontal and vertical decomposition are two effective ways by which performance and preliminary behavior of large systems may be captured; other strategies are possible (e.g., combining the horizontal and vertical approaches in a hybrid way or focusing on specific aspects of the systems). Of course, these approaches are inadequate to capture all those situations in which the external world triggers different paths at the same time and these paths are not independent; a further level of analysis where everything is working together is generally needed to complement local analysis and fully characterize the system.

9.6 Simulation Architecture for Performance Evaluation

The aim of this section is to present the general architecture, depicted in Figure 9.2, which has been conceived to address the issue of modeling and simulation of an integrated system in a structured and efficient way.

The purpose of the architecture is to integrate in a single frame models and tools having different granularity and that may refer to a specific function, component, or aspect of the system or address the whole system. The architecture is structured in levels that realize different trade-offs between field of view and accuracy of the modeling and simulation; a global field of view of the whole system is necessary to capture the overall system behavior and a high level of modeling accuracy is necessary to analyze in depth each component of the integrated system and to perform sensitivity analysis. It does not seem practical and in any way efficient to represent a large system with extremely detailed models; time is also an issue and results must be available in a short time following the evolution of concepts and system architectures. The underlying idea is then to derive and validate coarser grain models, to be used at the end-to-end level, from the fine grain–level models and to narrow conveniently the field of view if the need arises for a higher level of modeling accuracy. In particular, three levels have been identified: (1) the end-to-end level, characterized by global field of view and coarse grain accuracy, (2) the medium grain level, characterized by intermediate field of view and medium grain accuracy, and (3) the fine grain level, generally characterized by narrow field of view and fine grain accuracy. In the following a brief description of the levels and typical tools available at each level is given.

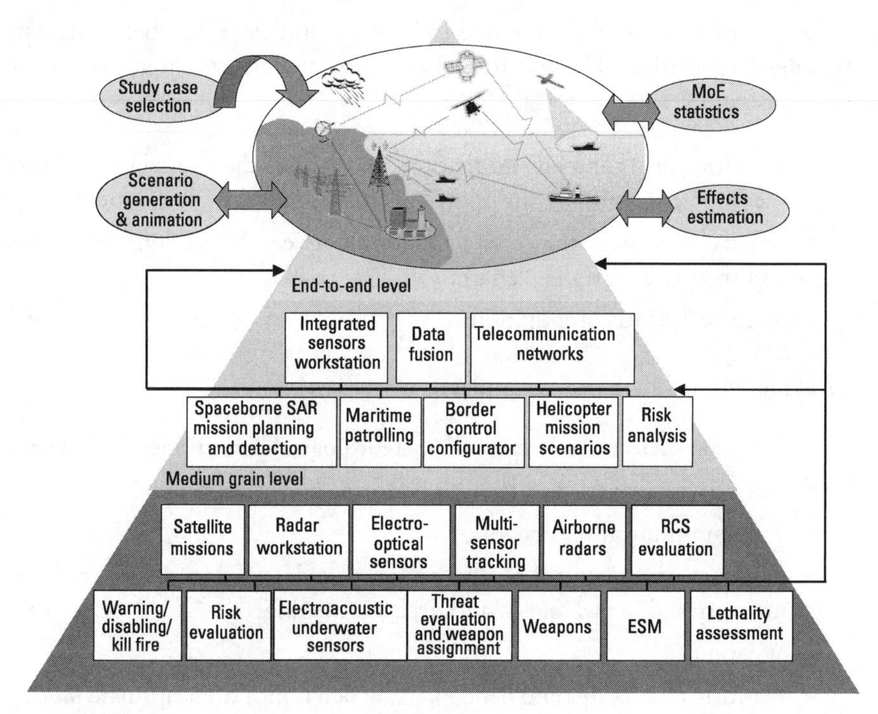

Figure 9.2 Simulator architecture for integrated system performance evaluation.

9.6.1 End-to-End Level

The *end-to-end level* aims to achieve an overall representation and a global view of the system. The end-to-end approach models the system as a network of system components, each viewed as an interacting black box, and enables the simulation of system behavior in a selected scenario.

At the end-to-end level it is of paramount importance to identify the "right complexity" of the modeling (i.e., the minimum model complexity that is sufficient to preserve consistency in the representation of the real world and in the estimation of performance results). There is no simple rule to determine such a right complexity, yet the objectives of the simulation, the type and number of features of the selected scenarios, and engineering expertise provide an indication of the right complexity to use in the system modeling. Right complexity can also be managed by restricting models to work within specific bounds and considering more general behavior at the medium grain level. As a consequence, an important aspect of this level is that models are able to accept input from the lower levels; input is generally obtained off-line, after that specific bounds have been established, and is used to set up the models to achieve a specific behavior.

The input of the simulator is a set of parameters for the selected scenario, and the outputs are the system measures of effectiveness (MoE), which are evaluated by means of the Monte Carlo method [32] over a sufficient number of statistically independent trials. The end-to-end simulator includes the following functions (see Figure 9.2):

- Scenario generation and animation, which runs the scenario and allows the interaction among the operative entities and the environment;
- Effects assessment that closes the loop between the actions of entities and their effect on the scenario;
- Statistical MoE computation.

Typical models included in the simulator are relative to:

- Natural environment (sea state, meteorological conditions, day/night);
- Sensors (heterogeneous, fixed or platform-borne sensors);
- Telecommunication networks;
- Command and control centers performing data fusion, sensor management, situation awareness, and decision making;
- Weapons;
- Platforms (i.e., ships and helicopters, which typically encapsulate models of other components like sensors and weapons).

9.6.2 Medium Grain Accuracy Level

This level includes models relative to subsystems or system components and analysis tools developed to study their behavior under specific constraints. Models at this level use the same scenarios defined at the end-to-end level to estimate metrics and aspects of interest and return inputs for the end-to-end simulation. Exchange of results is also performed horizontally among the tools of this level. These models bridge the gap between the end-to-end coarse grain and the fine grain levels, because models at the latter level are often much too complex to be used directly in the end-to-end simulator; reuse of fine grain models is difficult also because they have often been developed over many years and lack the required portability and customizability.

In the following a short review of typical medium grain–level tools, with a brief description of their inputs and outputs, is reported.

9.6.2.1 Integrated Sensor Workstation

This tool evaluates the performance of a network of heterogeneous sensors such as land-based, naval, submarine, or airborne sensors. Typical output is provided in terms of target probability of detection and accuracy of dynamic target state estimation as a function of the performance of each sensor.

9.6.2.2 Data Fusion

This toolset comprises a set of techniques whose aim is to process and combine the available data in order to infer about the quantities of interest. Target class, type, identity, and intent are examples of quantities of interests in homeland protection applications. Several tools implement processing of signal and data pertaining to conventional and high-resolution radars and sensors with the aim of noncooperative target classification, discrimination, and identification; typical techniques are multidimensional processing for JEM (jet engine modulation) [33, 34] and HERM (helicopter rotor modulation) [34, 35] recognition, SAR and ISAR image reconstruction and analysis, and high-resolution range profiling [36]. At a higher level of abstraction the data fusion toolset deals with data that is often incomplete, contradictory, imprecise, and uncertain [37, 38] and provides algorithms for the representation of the information in different forms, the extraction of the relevant information and the combination of pieces of information coming from different sources. Typical techniques are based on "reasoning under uncertainty" theories such as Bayes, fuzzy, Dempster-Shafer [39], Smets [40, 41] and Dezert-Smarandache [42].

9.6.2.3 Telecommunication Networks

This suite of tools supports the design and analysis of the telecommunication network as part of the integrated system [43]. Typical inputs are the number and the type of traffic sources and sinks (e.g., radars, IR (infrared) cameras, EW (electronic warfare) sensors, ships, aircraft, C2 centers, harbors, and airports), their geographic distribution, and the type of coverage required. Typical outputs are the performance of the telecommunication network in terms of throughput, load, delays, BER (bit error rate), percentage of packet drop, and queue size. Examples of tools include NS2, an open source event network simulator developed at Berkeley University [44], and tools for the visualization and the topological analysis of networks.

9.6.2.4 Spaceborne SAR Mission Planning and Detection

This tool evaluates the potential revisit time of a satellite constellation on a specific region of the globe and the detection capability of a spaceborne SAR (synthetic aperture radar) as a function of relevant parameters such as incidence angle, polarization, wind direction and intensity, and number of looks [45].

9.6.2.5 Maritime Patrolling

This tool comprises a set of analytical formulas that provide the probability of detecting and classifying vessels in maritime scenarios; one example is linear barrier patrolling where the target is moving perpendicularly to the barrier line and one or multiple searchers are moving back and forth on a straight path along the barrier line. Results typically are given as a function of the number of searchers, the size of the search area, the speed, and the search radius of the onboard sensors.

9.6.2.6 Border Control Configurator

This tool optimizes the placement of a set of system resources (e.g., sensors, effectors) so that the given system configuration provides the best reaction against an incoming threat. The tool may also be used to perform a trade-off analysis of the configuration itself in terms of the number, position, and characteristics of resources.

9.6.2.7 Helicopter Mission Scenarios

This tool allows the modeling and analysis of the temporal evolution of maritime operational scenarios, where helicopters—as part of the maritime task force and working as nodes of a network-centric warfare tactical net—play a significant role in the search, tracking, identification, and engagement of targets.

9.6.2.8 Risk Analysis

This tool implements the probabilistic risk analysis method to a set of assets exposed to a threat. The risk may be computed according to the canonical threat-vulnerability–consequence model [46] using input parameters that represent the elements in the defined scenario (number, value, position and type of assets, threat weapon range, and attack probability). The vulnerability, without and in the presence of countermeasures, is another input which may be derived from other tools, for instance, the border control configurator. The tool computes the residual risk and the reduction of the risk deriving from the adoption of different sets of countermeasures, providing the user with a basis upon which the effectiveness of countermeasures to protect the assets may be estimated.

9.6.3 Fine Grain Accuracy Level

Models and tools at this level are very detailed and accurate, are generally confined to components or specific aspects of the system, and are representative of consolidated knowledge and expertise in the field. They are characterized by a large set of inputs and parameters so that the model is applicable in a wide set of scenarios; they may be used to derive performance tables that can be fed into the end-to-end models. Sensitivity analysis is also performed at this level to ascertain which factors significantly affect the performance of the component and therefore it provides valuable feedback in the development of the end-to-end models. Typical fine grain tools are briefly reviewed in the following.

9.6.3.1 Satellite Missions

This tool allows configuring, planning, and simulating a multisatellite Earth observation mission via the definition of the space segment resources, including the most important and critical satellite subsystems and payloads, the ground segment resources (e.g., ground stations, processing centers), and the communication network. It may be used to estimate access capabilities and revisit times for different satellite configurations.

9.6.3.2 Radar Workstation (RWS)

The RWS is a collection of models and tools for performance prediction and validation of radars ranging from active mechanically scanning 2D/3D to multifunctional phased array and passive radars [47, 48]. Typical outcomes are measures of performance in clear, clutter, and ECM (electronic countermeasures) conditions, such as: range calculation, radar elevation coverage diagrams, and accuracy of target position estimation. Additional tools provide analysis of anomalous propagation and multi-path reflection. RWS also includes software packages for analysis and test of adaptive ECCM (electronic counter-countermeasures) algorithms.

9.6.3.3 Electro-Optical Sensors

This tool evaluates the detection, recognition, and identification performance of an IR and a TV camera, and takes into account factors such as target signature, sensor characteristics, and the atmosphere attenuation model [49–54]. The typical output provides a probability curve versus sensor–target range.

9.6.3.4 Multisensor Tracking (MST)

This tool evaluates tracking performance in complex scenarios comprising several fixed and mobile sensors and sea/air/ground targets, including pedestrians and sea-skimmer and ballistic missiles. Cramer-Rao lower bounds (CRLB), which pose a lower bound on the best theoretical accuracy that may be achieved by an estimator, are evaluated to assess algorithm performance.

9.6.3.5 Airborne Radars

This tool evaluates the detection performance of airborne radars against air, sea, and ground targets. The radar is characterized by several parameters (e.g., transmission power, waveform) and takes into account factors such as target RCS, sea and land clutter, and target-to-radar relative kinematics.

9.6.3.6 RCS Evaluation

This software tool predicts the electromagnetic monostatic and bistatic radar cross sections (RCS) of air, ground, and naval targets defined via a suitable CAD (computer aided design) model. Different electromagnetic prediction codes are used to deal with distinct frequency bands, such as UHF (ultra high frequency) and X.

9.6.3.7 Warning/Disabling/Kill Fire Against Ships

This tool evaluates the probability of warning/disabling/kill fire against ship targets by taking into account factors such as target–weapon relative kinematics, ship layout, gun characteristics, meteorological conditions, and accuracy of tracking radar. Measures of performance include collateral damage probability.

9.6.3.8 Risk Evaluation

This tool supports the decision-maker providing a security risk analysis, supplying assistance at all the stages of a security policy implementation.

9.6.3.9 Electroacoustic Underwater Sensors

This tool evaluates detection and classification performance of electroacoustic underwater sensors, both active and passive. The sensor is characterized by several parameters and takes in account target spectrum and sea depth.

9.6.3.10 TEWA (Threat Evaluation and Weapon Assignment)

This tool allows us to experiment with different TEWA logics, which determine the type of mission and the effector to be assigned against threatening targets in the scenario; moreover the tool allows defining items such as the scenario, characteristics, availability, and location of system resources and targets' behavior.

9.6.3.11 Weapons

This tool evaluates the performance of various types of weapons, including surface-to-air and surface-to-surface missiles and guns.

9.6.3.12 ESM

This tool evaluates the performance of ESM (electronic support measures) sensors: typical outputs of the tool are the probability of intercept and classification of an emitting communication source and measurement accuracies.

9.6.3.13 Lethality Assessment

This tool, on the basis of the target vulnerability model and the munition lethality model, evaluates the lethality of the munition against the target.

9.7 Study Cases

This section describes two study cases that have been conducted using the architecture proposed in the previous section. The first study case considers an end-to-end simulation of a maritime border control system that comprises several and heterogeneous sensors and performs detection, tracking, classification, and identification of vessels; evaluation of threat level; and finally selection of an adequate reaction against the threat: this study case shows how the fine grain levels of the architecture contribute to the end-to-end simulation. The second study case focuses once again on the detection, classification, identification, and threat evaluation performance achieved by a suite of heterogeneous sensors and shows how the results provided by the specific tools of the fine grain level of the simulation architecture may be effectively integrated.

9.7.1 End-to-End Simulation of a Maritime Border Control System

9.7.1.1 Model of the Simulated System

The scheme in Figure 9.3 depicts a maritime border control system that is composed of a land-based surveillance asset, an airborne platform, a C2 center, and effectors.

9.7.1.2 Land-Based Asset

The land-based asset comprises a radar, an IR sensor, and an AIS (automatic identification system) device [55,56]. The radar is the typical sensor employed in a vessel traffic management (VTM) system; it operates in X-band, it is elevated in order to increase its optical horizon, and it provides, with respect to a medium-sized vessel, the full coverage of territorial waters (about 22 km). For instance, a vessel characterized by a 10 sqm RCS and height of 2m can be detected with a probability of 90% (and false alarm probability, $P_{fa} = 10^{-6}$) at the range of about 25 km, if the radar antenna is sited on a 40-m-high tower. IR sensors usually support the classification process and are activated by radar cueing; they operate in the bandwidths corresponding to the wavelengths $3 \div 5$ μm or $8 \div 12$ μm according to climate and operational conditions. The AIS is a system operating in the VHF maritime band where shore-based stations and shipboard transponders exchange information about identity, position, route, cargo of the ship, and additional navigation data. AIS is mandatory on all vessels meeting or exceeding

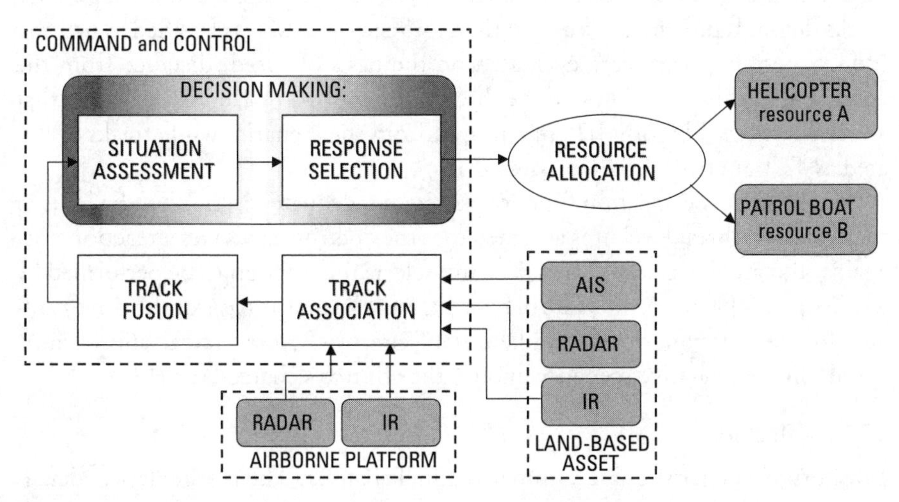

Figure 9.3 Maritime border control system.

300 gross tons, therefore, the AIS allows us to distinguish between cooperating and noncooperating vessels and provides identification data of cooperating vessels. Its maximum range is about 40 nautical miles and is usually higher than the VTM radar maximum range.

9.7.1.3 Airborne Platform

The airborne platform is equipped with a radar and an IR sensor. The airborne radar operates in X-band and its maximum range is also dependent on sea state and platform height.

9.7.1.4 C2 Center

The C2 center receives track data from the airborne platform and the land-based asset; it performs track-to-track correlation and fusion and combines track data with IR classification and AIS data. The C2 center then evaluates the threat level (situation assessment) of the track and selects an adequate reaction (response selection).

The situation assessment function associates each track to one of the following threat levels (TL):

- Neutral track (TL0);
- Suspect track (TL1);
- Threatening track (TL2);

where neutral tracks pose no threat, threatening tracks represent a threat that must be contrasted, and suspect tracks are ambiguous tracks for which no final decision has yet been made. The track threat level is determined by AIS data, position, and kinematic parameters of the track. It is assumed that cooperative tracks do not represent a threat and therefore they are always classified as neutral. Noncooperative targets are evaluated on the basis of speed, distance from the coast, and heading according to the thresholds reported in Table 9.1. Tracks that are evaluated as TL0 or TL2 are removed from the scenario, while tracks evaluated as TL1 are periodically reassessed.

The response selection function selects an adequate reaction on the basis of the evaluated threat level; it is assumed that the doctrine envisages a reaction only against threatening tracks. The C2 center selects the reaction to be performed by a helicopter (effector A) or a patrol boat (effector B) on the basis of the time available for the reaction, on the availability of effectors, on the target approaching speed, on the operative scenario, and on the effector's characteristics.

9.7.1.5 Effectors

Effectors' characteristics are summarized in Table 9.2. The mathematical details to compute the reaction time are omitted for the sake of brevity.

Table 9.1
Thresholds Employed for Threat Level Evaluation

Speed	LOW ($V < 12$ knots) MEDIUM ($12 \le V < 30$ knots) HIGH ($V > 30$ knots)
Distance	FAR (out of warning zone) WZ (inside the warning zone) OFF (inside the off-limit zone)
Heading	IN (approaching the coast) OUT (moving away from the coast)

Table 9.2
Effectors' Characteristics

	Effector A HELICOPTER	**Effector B PATROL BOAT**
Number of effectors	$N_A = 3$	$N_B = 7$
Speed	$V_A = 300$ km/h	$V_B = 60$ km/h
Inspection time	300 sec	300 sec
Readiness*	300 sec	300 sec

*It represents the delay introduced by the effector to start the mission from mission notice.

It is important to highlight that the models of the sensors and processing functions are entirely derived from the tools of the medium and fine grain accuracy levels (e.g., RWS, electro-optical sensor modeling, data fusion).

9.7.1.6 Scenario

The scenario involves vessels such as high-speed rigid hull inflatable boats (RHIB), boats carrying illegal immigrants, fishing boats, and oil tankers. The parameters characterizing each vessel class are summarized in Table 9.3 in terms of length, speed, radar cross section (RCS), IR[1] and electromagnetic[2] (E.M.) signature, and a yes/no cooperation tag. A scenario snapshot depicting the initial positions (black dots) of the vessels is presented in Figure 9.4. The line r_0 is the edge of the off-limit region, which is at the distance of 20 km from the coast and is comprehensive of territorial waters; the line r_1, at the distance of 50 km from

1. IR signature is provided in terms of critical size and temperature difference with respect to background.
2. E.M. signature accounts for onboard radio frequency emitters, together with an indication of the percentage of time an emitter is transmitting.

Table 9.3
Classes of Vessels and Their Characteristics

Class	Length	Speed	RCS	IR Signature	E.M. Signature	Emission Time	Cooperation
RHIB	2.5 ÷ 12.5 m	0 ÷ 40 knots	3 m^2	$\Delta T = 6K$ $d_c = 4.5m$	Wireless phones*	50%	N
Immigrant boat	10 ÷ 25 m	0 ÷ 24 knots	10 m^2	$\Delta T = 6K$ $d_c = 6.5m$	Wireless phones	20%	N
Fishing boat	6 ÷ 18 m	0 ÷ 12 knots	100 m^2	$\Delta T = 6K$ $d_c = 5m$	Wireless phones, VHF radio, navigation radar	80%	Y/N
Oil tanker	90 ÷ 150 m	0 ÷ 20 knots	1000 m^2	$\Delta T = 10K$ $d_c = 120m$	Wireless phones, VHF radio, navigation radar	80%	Y/N
Generic boat	20 ÷ 35 m	0 ÷ 25 knots	20 m^2	$\Delta T = 6K$ $d_c = 6.5m$	Wireless phones, VHF radio, navigation radar	60%	Y/N

*Wireless phones: it includes mobile and GSM cellular phones, walkie-talkies, and satellite phones.

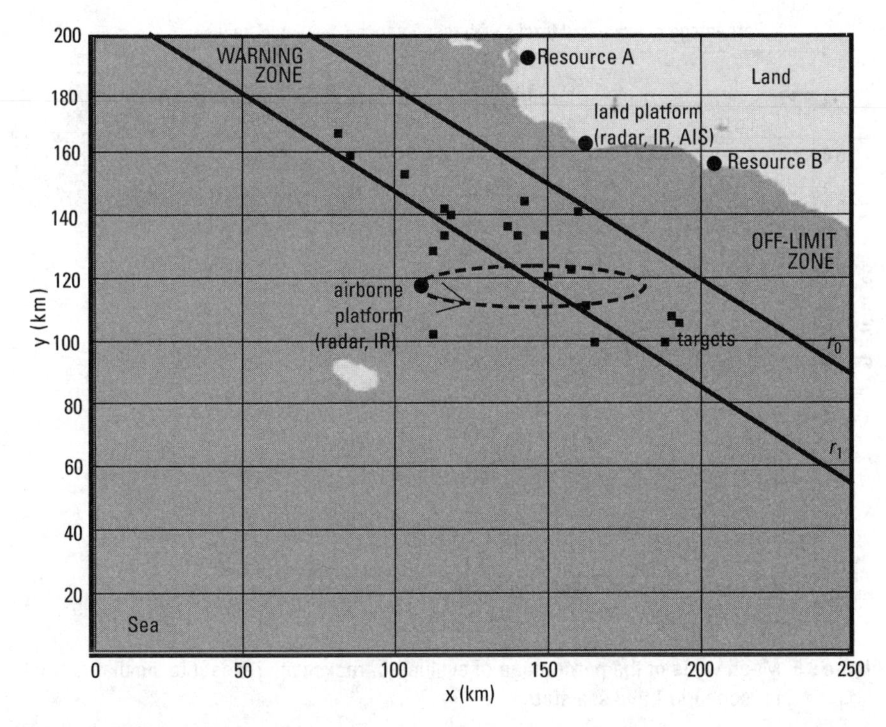

Figure 9.4 Reference scenario.

the coast, is the limit of the warning zone. As represented in Figure 9.4, the airborne platform moves along an elliptical trajectory located at the limit of the warning zone, with a speed of 90 km/h. In particular, two scenarios, each comprising 20 vessels belonging to all of the above-listed classes, have been simulated; each vessel is characterized by a specific speed, heading, and threat level. In the first scenario all vessels are noncooperative, while in the second one an oil tanker and two fishing boats are simulated as cooperative. All the vessels are distributed off the coastline in the territorial, contiguous, and international waters. Meteorological conditions, including sea state, wind speed, and wave height, according to the Beaufort scale [57] may be specified; in this case the simulation has been performed with sea state 0 (absence of clutter and no wind).

9.7.1.7 Simulation Results

Several measures of performance have been evaluated to characterize the performance of the system. Figure 9.5 plots the mean value of the number of tracks evaluated by the C2 center, with respect to the simulated tracks, as a function of observation time for scenario 1 and sea state 0. The number of tracks depends on the number of targets that is in radar coverage. The pseudo-periodicity of the plot depends on the fact that the airborne radar moves along an elliptical trajectory;

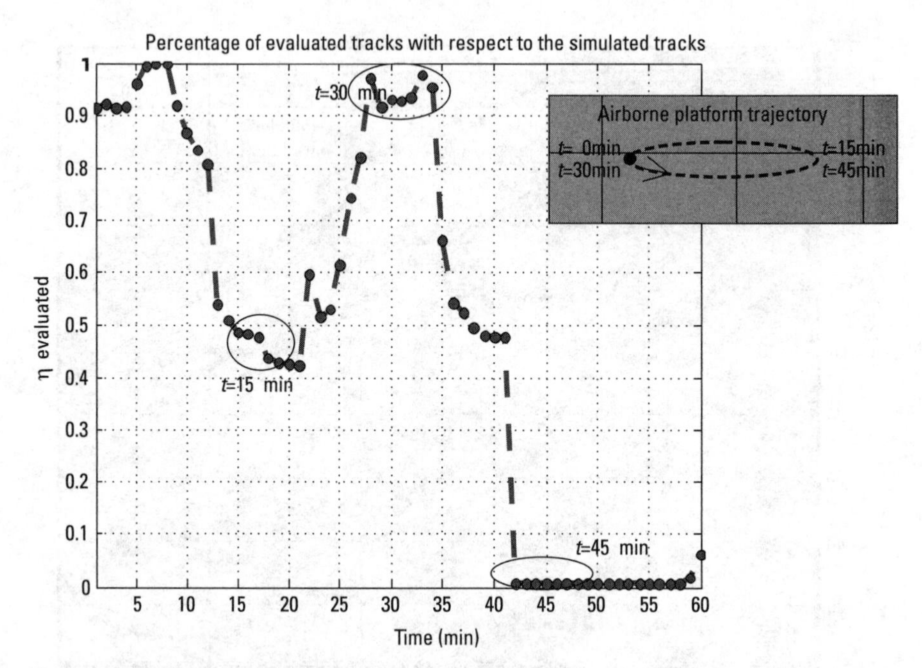

Figure 9.5 Mean value of the percentage of evaluated tracks with respect to simulated tracks for scenario 1 and sea state 0.

after 45 minutes most vessels have changed position and several tracks can be lost. Figure 9.6 reports the mean values of the probabilities $\mathbf{P}_{FI}$, $\mathbf{P}_{MI}$, and $\mathbf{P}_{MA}$, which assess the performance at threat evaluation level and are defined as follows:

- Probability of *false intervention* ($\mathbf{P}_{FI}$): the probability, *Pr*(TL2/TL0), to identify a track as threatening (TL2) given it is a neutral one (TL0); this produces an unnecessary intervention;
- Probability of *missing intervention* ($\mathbf{P}_{MI}$): the probability, *Pr*(TL0/TL2), to identify a track as neutral (TL0) given it is threatening (TL2); this results in a lack of intervention;
- Probability of *missing assessment* ($\mathbf{P}_{MA}$): the probability, *Pr*(TL1/TL0,TL2), that the *track* is not definitively identified during the assessment phase.

The probability $\mathbf{P}_{FI}$ is computed with respect to the twelve neutral tracks of the scenario, the probability $\mathbf{P}_{MI}$ is computed with respect to the eight TL2 tracks, and the probability $\mathbf{P}_{MA}$ with respect to all tracks. The $\mathbf{P}_{MA}$ value is always the highest probability because the selected scenario does not provide enough information to obtain a definitive assessment.

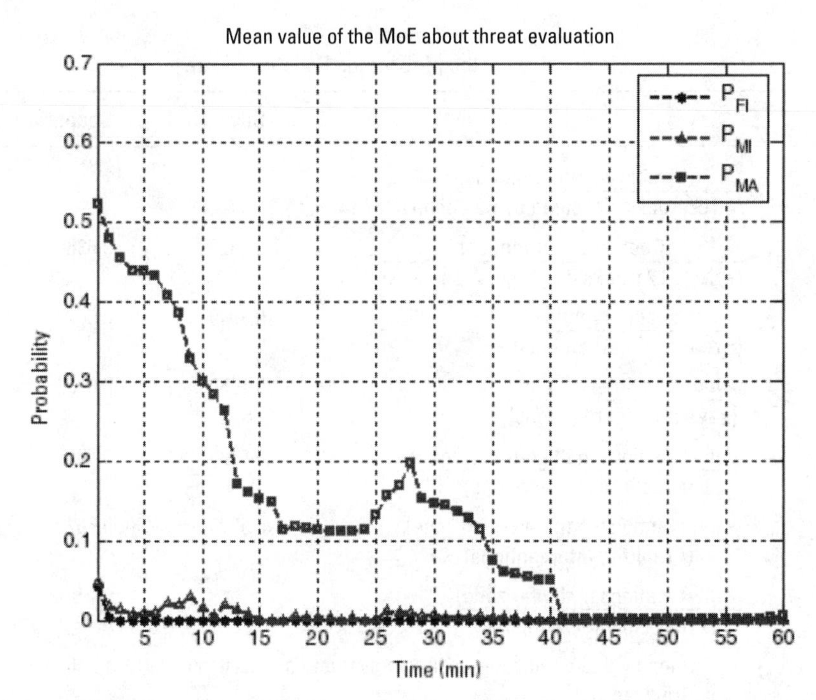

Figure 9.6 MoE mean values for the threat evaluation phase for scenario 1 and sea state 0.

Mean values of the metrics pertinent to the reaction phase are reported in Table 9.4 for scenario 1 and scenario 2. The table shows that effectors are always available to ensure an adequate reaction ($\eta_1 = 1$) and that about 54% of the threatening tracks detected by the system are engaged ($\eta_2 \approx 0.54$). The parameters η_{correct} and $\eta_{\text{incorrect}}$ show that 10% of the interventions are wrong, causing an unnecessary use of the effectors. The values of η_{OL}, η_{WZ}, and η_{FAR} show the distribution of interventions in the various zones.

Finally, Table 9.4 reports the mean value and the standard deviation of the evaluation time for the threatening tracks (TL2) and the mean number of TL0 and TL2 tracks that after an hour of observation have not been assigned a final threat level. From the analysis of the table, it is evident that, on average, after 15 minutes all the threatening tracks (TL2) are evaluated and after an hour of observation, 10% of the threatening tracks are still waiting for their evaluation.

9.7.2 Multisensor Fusion for Naval Threat

9.7.2.1 Model of the Simulated System

The integrated system, depicted in Figure 9.7, comprises several sensors: four land-based, one submarine, and a C2 center.

Table 9.4
Mean Values of the MoE for the Reaction Phase

		Scenario 1	Scenario 2
η_1:	(# realized interventions) / (# TL2 tracks classified by C2 center)	1	1
η_2:	(# correct interventions) / (# true TL2 tracks detected by sensors)	0.5429	0.5368
$\eta_{correct}$:	(# correct interventions) / (# realized interventions)	0.8940	0.9020
$\eta_{incorrect}$:	(# incorrect interventions) / (# realized interventions)	0.1060	0.0980
η_{OL}:	(# interventions in the off-limit zone) / (# realized interventions)	0.4380	0.4380
η_{WZ}:	(# interventions in the warning zone) / (# realized interventions)	0.5620	0.5620
η_{FZ}:	(# interventions in the far zone) / (# realized interventions)	0	0
$E\{\Delta\tau_{TL2}\} \rightarrow$	mean value of time necessary to evaluate a track as TL2	15.66 min	15.36 min
$std\{\Delta\tau_{TL2}\} \rightarrow$	standard deviation of time necessary to evaluate a track as TL2	8.137 min	6.68 min
$TL2_{remaining}$:	mean number of TL0 tracks that after an hour of scenario evolution are still in the assessment phase	0.1000	0.0900
$TL0_{remaining}$:	mean number of TL2 tracks that after an hour of scenario evolution are still in the assessment phase	0.0400	0.0500

The land-based sensors are a VTM-like radar, an IR sensor and an AIS device, which constitute a sensor suite similar to the one encountered in the previous study case, and a C-ESM (communication electronic support measure) sensor. The radar, sited on the coast at the height of 40m, contributes to target detection with probability P_d^R and to target recognition by means of kinematical parameters estimation, with probability P_{ric}^R. The IR camera, sited close to the radar, contributes to both detection and recognition of the target, respectively, with probabilities P_d^{IR} and P_{ric}^{IR}. The AIS contributes to target identification of all targets equipped with a transponder. The C-ESM provides passive surveillance through the spectral analysis of radio frequency emissions in the bandwidth employed by telecommunication devices. By the analysis of characteristics such as the carrier, bandwidth, modulation, code, and emission duration, an emitter

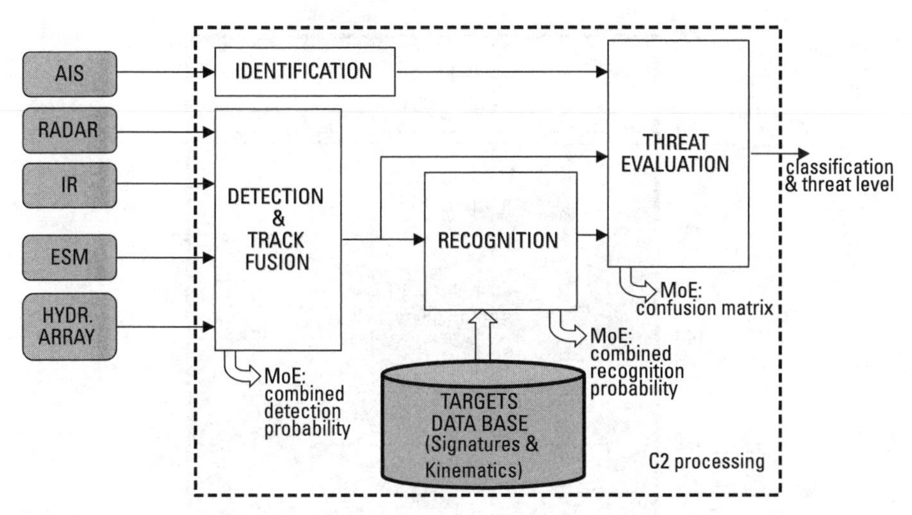

Figure 9.7 Multisensor system.

can be intercepted and recognized with a certain probability and its direction of arrival can be estimated with a certain accuracy. In this kind of application the C-ESM sensor coverage is typically limited by line of sight; for instance, the C-ESM sensitivity allows to intercept a typical VHF naval radio with very high probability at a range of 65 km or more, so long as the emitter and receiver are in line of sight. Therefore the C-ESM is sited 20 km inland at a height of 300m above sea level (see Figure 9.8) to increase its optical horizon.

The submarine sensor is a hydrophonic linear array, sited 20 km off the coast at a depth of 200m (see Figure 9.8). It contributes to target detection with probability P_d^{hydr} and to target recognition by means of the analysis of its acoustic signature, with probability P_{ric}^{hydr}. The maximum range of the hydrophonic array strongly depends on the kind of engine of the target (which determines its acoustic signature) and on the sea state. For instance, a diesel engine can be detected with 90% probability ($P_{fa} = 10^{-4}$) also at ranges beyond 100 km in case of sea state 0; the maximum range reduces to less than 40 km in case of sea state 4.

The C2 center receives data from each sensor and provides a global probability of detection P_d^{comb}, a global probability of recognition P_{ric}^{comb}, and a threat level for each target.

9.7.2.2 Scenario

The scenario comprises one target that may be a high speed RHIB, an immigrant boat, or a generic boat. The parameters characterizing each target are the same as provided in Table 9.3 in terms of target length, speed, RCS, IR, and electromagnetic signature and cooperation tag. The E.M. signature accounts for onboard radiofrequency emitters and a certain percentage of emission time is assumed. The

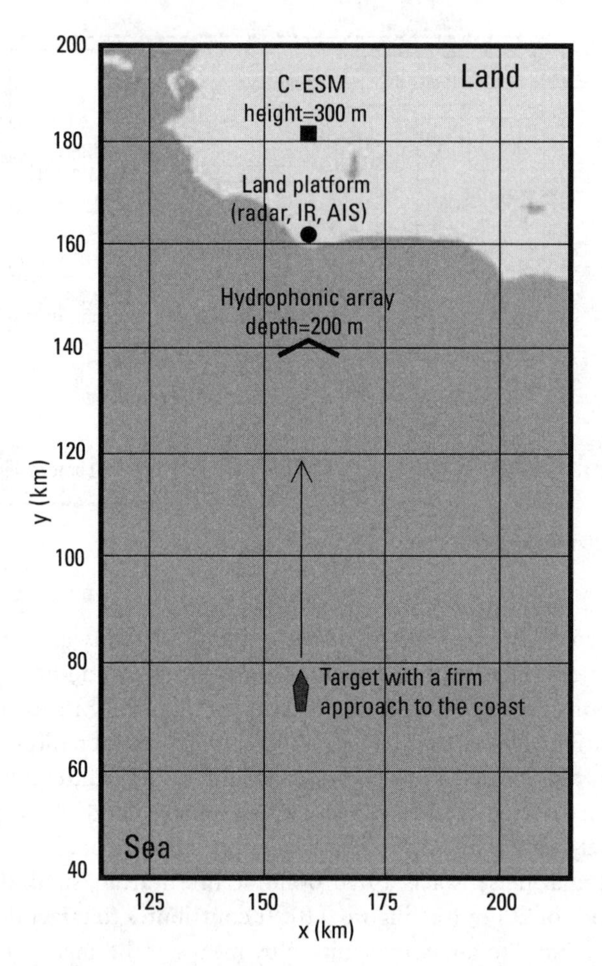

Figure 9.8 Sensors involved in the detection and recognition of naval targets.

simulation has been performed with sea state 0 (in case of absence of clutter, no wind) and presence of clutter with sea state 4 (wind speed: 8 m/s, wave height: 1 m, according to the Beaufort scale [57]).

9.7.2.3 Detection

The detection performance of hydrophonic array, radar, IR, and C-ESM sensors is combined to obtain the overall detection probability, which is higher than the detection probability of each single sensor; the basic combination scheme is provided in Figure 9.9. The global detection probability is:

$$P_d^{comb} = 1 - \prod_{k=1}^{Nsensors} (1 - P_{ric,k}) \quad (9.1)$$

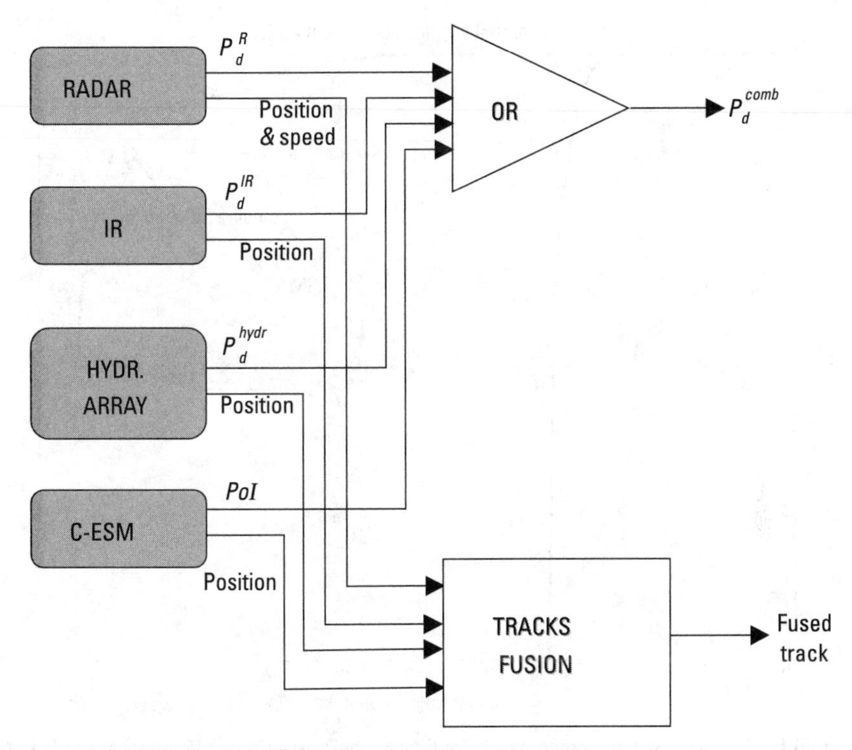

Figure 9.9 Block diagram for the combination of sensor detections.

Detection probabilities are provided in Figure 9.10 for the immigrant boat in case of sea state 0 and, in Figure 9.11, for the RHIB in case of sea state 4 versus distance from the coast. The C-ESM detection probability is quantified by means of a Monte Carlo simulation with 200 independent trials accounting for the percentage of emission in time. In the first case (sea state 0), it is evident that the main contribution to detection is given by the hydrophonic array, which is characterized by the best performance in the absence of sea clutter; in the second case (sea state 4), sea clutter strongly reduces the hydrophone performance, as well as degrading radar performance near the coast (0 ÷ 5 km), whereas the IR sensor continues to ensure a good performance. The C-ESM probability of intercept is limited due to the low percentage of emission time of onboard emitters.

Measurements are also fused to obtain an accuracy that is better than the measurement accuracy of each single sensor. All sensors[3] provide an azimuth estimation with a certain accuracy; the radar also provides a velocity estimation (speed and heading). The speed estimation is useful in recognizing the kind of

3. In this case C-ESM accuracy is the poorest and falls out of the figure.

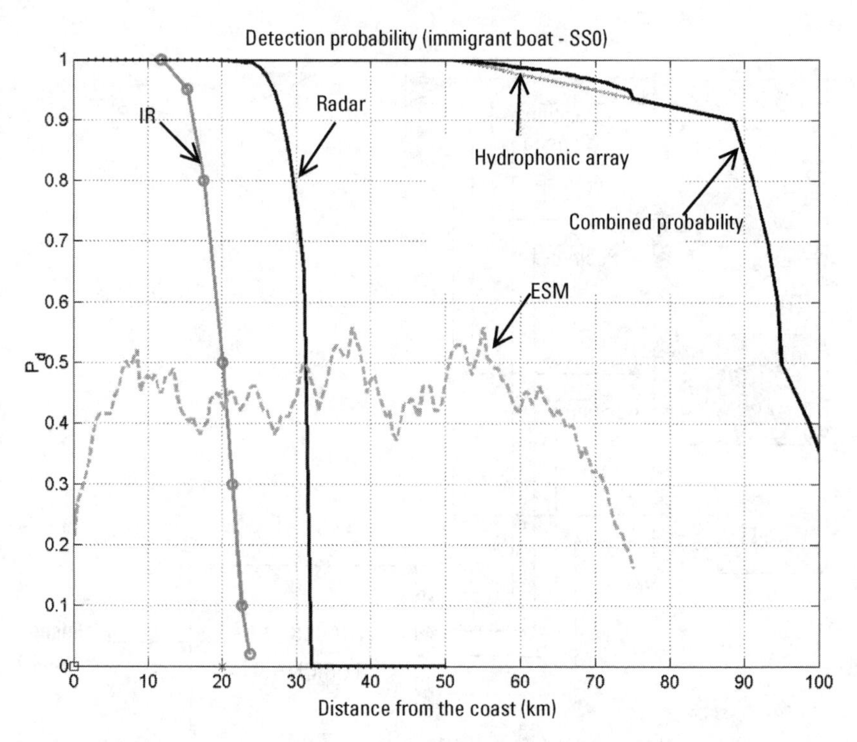

Figure 9.10 Single sensor and combined P_d for hydrophonic array, radar, IR camera, and C-ESM sensor, for an immigrant boat, sea state 0, and 50% emission time of onboard emitters.

target and also to evaluate its threat level. As an example of results Figure 9.12 shows the accuracy of azimuth estimation after fusion for an RHIB in case of sea state 4; it is evident that the radar provides the most accurate measurements.

9.7.2.4 Recognition

The recognition performance of the hydrophonic array, radar, IR, and C-ESM sensors is combined to obtain the overall recognition probability, which is higher than the recognition probability of each single sensor; the basic combination scheme is provided in Figure 9.13. The global recognition probability is:

$$P_{ric}^{comb} = 1 - \prod_{k=1}^{Nsensors} (1 - P_{ric,k}) \tag{9.2}$$

The radar contributes to the recognition process by providing subsequent target measurements that the tracker uses to estimate recursively the target position and speed; the radar also measures target radial speed via the bank of MTD (moving target detector) Doppler filters. In this case target speed is compared with a suitable threshold (λ); if the speed is high enough the target may be recognized, otherwise no information can be derived about its class.

Figure 9.14 depicts the recognition probabilities for the RHIB versus distance from the coast in case of sea state 0. The RHIB is traveling at a speed of 40 knots, which exceeds the threshold speed set at 15 knots, and this fact contributes to the global recognition capability. If the target speed were lower than the threshold, the speed would not give any contribution: it may be an immigrant boat or an RHIB traveling at low speed. Notice that the hydrophonic array recognition probability refers to the ability of discriminating between the RHIB and the immigrant boat by comparing their acoustic spectra.

9.7.2.5 Threat Evaluation

The target threat level is evaluated on the basis of its kinematics. The parameters exploited for the threat evaluation are the distance D between the target and the off-limit zone (in this case it coincides with the shoreline), the speed V, and its heading θ_T, as represented in Figure 9.15. Periodically the target position and speed are updated and the following criteria are applied:

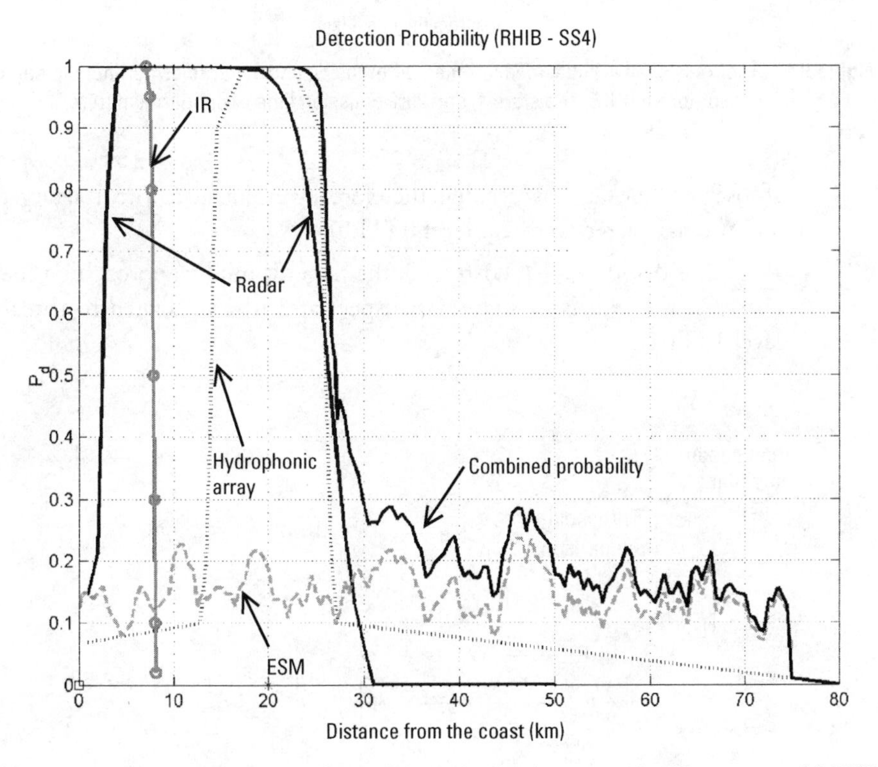

Figure 9.11 Single sensor and combined P_d for hydrophonic array, radar, IR camera, and C-ESM sensor, for an RHIB, sea state 4, and 20% emission time of onboard emitters.

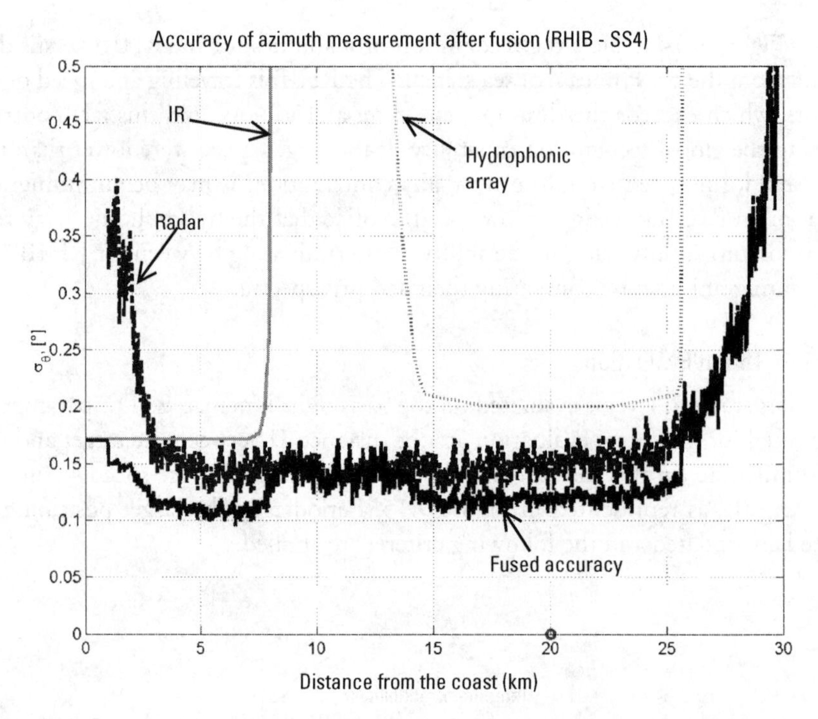

Figure 9.12 Accuracy of an azimuth measurement after fusion and accuracies pertinent to each sensor for an RHIB, sea state 4, and 20% emission time of onboard emitters.

- If $\cos\theta_T < 0$, the target is receding, therefore it is evaluated as a neutral target and it is associated to threat level 0 (TL0).
- If $\cos\theta_T \geq 0$ and $D - V\tau_S \cos\theta_T > 0$, the target is mildly approaching the coast, therefore it is evaluated as suspect and it is associated to threat level 1 (TL1).

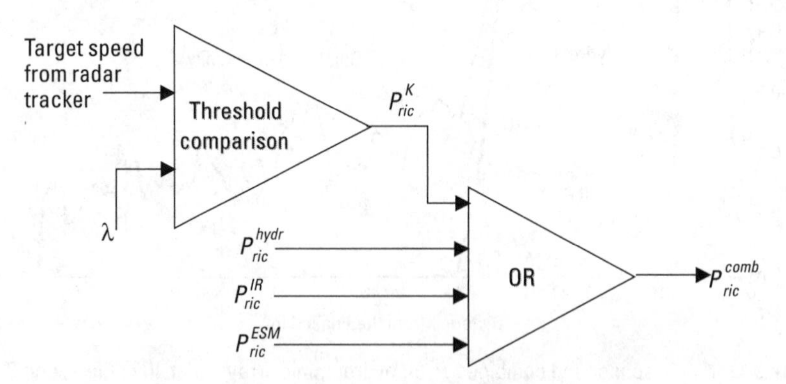

Figure 9.13 Block diagram for the combination of sensor recognition performance.

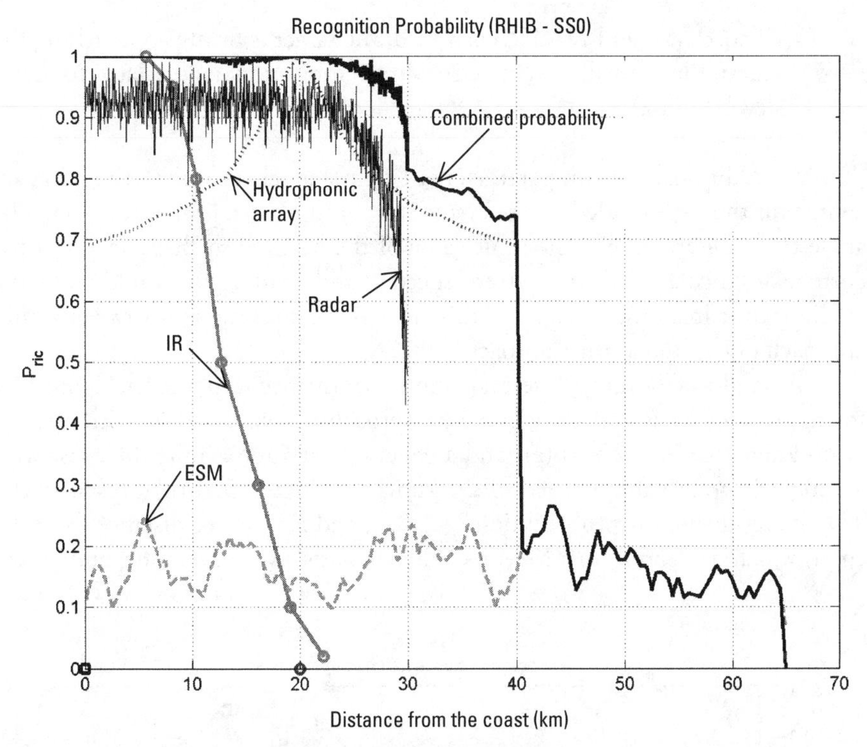

Figure 9.14 Single sensor and combined P_{ric} for hydrophonic array, radar, IR, C-ESM device, and P_{ric} for an RHIB, sea state 0, and 20% emission time of onboard emitters.

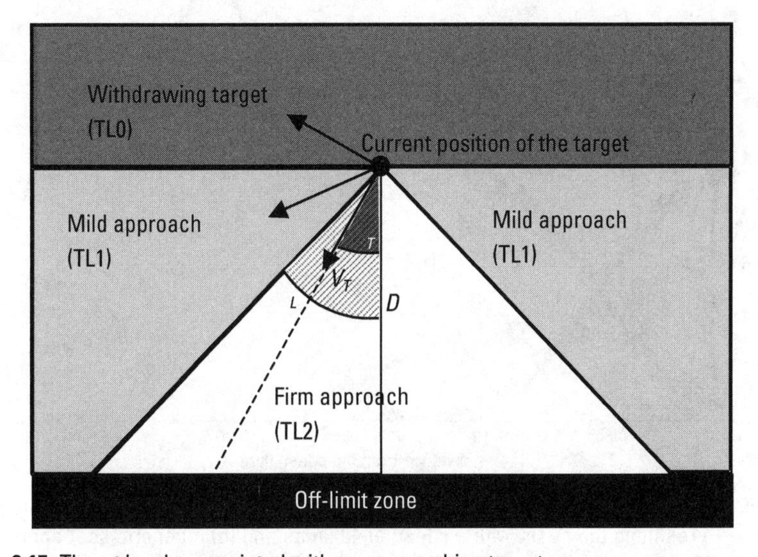

Figure 9.15 Threat levels associated with an approaching target.

- If $\cos\theta_T \geq 0$ and $D - V\tau_S \cos\theta_T \leq 0$, the target is firmly approaching the coast, therefore it is evaluated as threatening and it is associated to threat level 2 (TL2).

An additional relevant parameter is τ_S, a system characteristic time that accounts for the time needed by the system to put in place a proper reaction. The angle θ_L in Figure 9.15 is the limit value of the heading so that a target at the current distance D, with the current speed V and heading θ_L, would reach the off-limit zone in a time equal to τ_S; this angle is the discriminant between a mild approach (TL1) and a firm approach (TL2).

Examples of the threat-level evaluation performance are plotted in Figure 9.16 for the case of an RHIB orthogonally approaching the coastline with a speed of 40 knots (see Figure 9.16(a)) and a generic boat approaching the coast with an angle $\theta_T = 18°$ and a speed of 20 knots (see Figure 9.16(b)), respectively. The figures depict the probabilities P_{TL0}, P_{TL1}, and P_{TL2} versus distance from the coast, which descend from a Monte Carlo simulation with 200 independent tri-

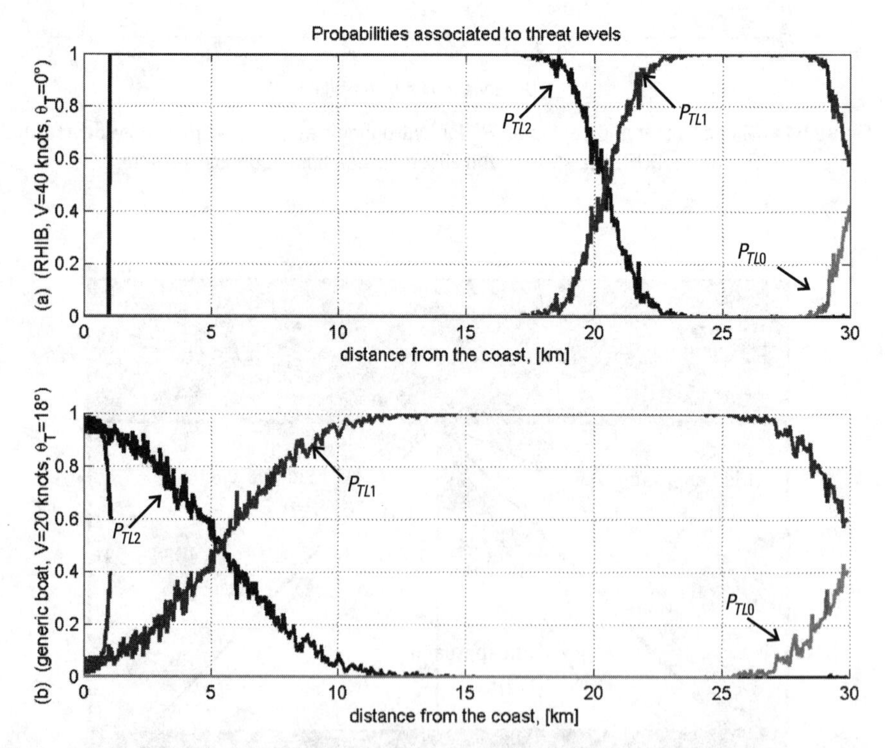

Figure 9.16 Probabilities associated with each threat level for (a) an RHIB orthogonally approaching the coast with a speed of 40 knots and (b) a generic boat approaching the coast with an angle $\theta_T = 18°$ and a speed of 20 knots.

als; the system characteristic time τ_S is fixed equal to 7 minutes. With reference to Figure 9.16(a), at the beginning of its trajectory the RHIB is ranked as TL1 with unitary probability: despite the target firmly approaching, it is far enough and is considered only suspect. When the RHIB approaches the limit of territorial waters, it becomes more threatening and it is ranked as TL2 with increasing probability. The same happens for the generic boat (see Figure 9.16(b)), but it is ranked as TL2 at a shorter distance from the coast (about 6 km) because its approach is mild.

9.7.2.6 Confusion Matrix

A possible synthesis of the previous measures of performance is the confusion matrix, which provides the overall capability of the system to correctly classify and evaluate the vessels; the *i,j*th element of the matrix is the probability that the *i*th target is classified as the *j*th target. The ideal confusion matrix is therefore an identity matrix where all targets are perfectly discriminated with unitary probability. As the values of the off-diagonal entries of the confusion matrix grow, the discrimination performance gets poorer.

Seven vessels are involved in the simulation and are an immigrant boat, a neutral RHIB, a suspect RHIB (TL1), a threatening RHIB (TL2), a neutral generic boat, a suspect generic boat (TL1), and a threatening generic boat (TL2). A vessel cooperating via AIS is identified as neutral, otherwise it is assessed as suspect or threatening on the basis of its kinematics. The 7×7 matrix has been computed at the following ranges from the coast (see Figure 9.17): 88 km, 66 km, 44 km (limit of contiguous waters), 22 km (limit of territorial waters), and 11 km.

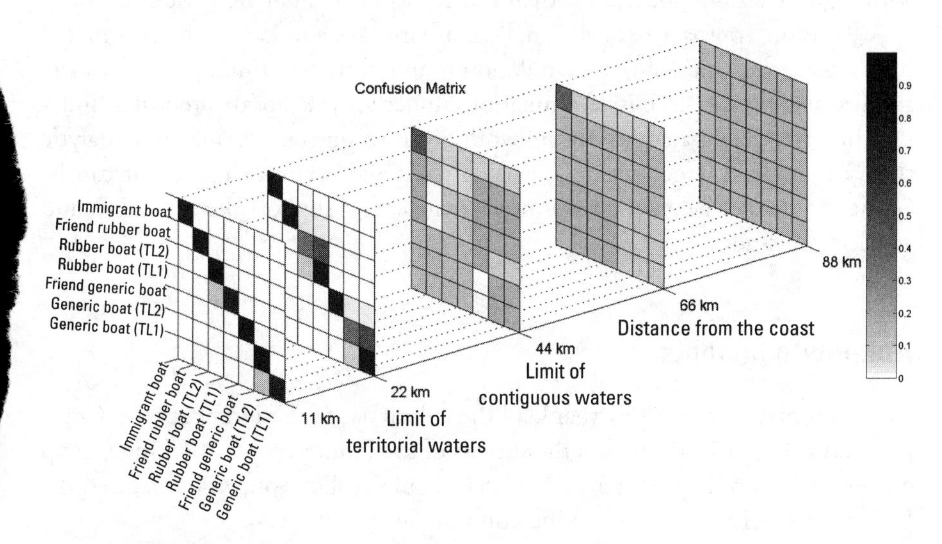

Figure 9.17 Confusion matrix quantification.

At the range of 80 km only target detection is possible, while no kind of recognition can be performed, so recognition probabilities are equally distributed (e.g., all the matrix elements are equal to 1/7). At the ranges of 66 km and 44 km only the immigrant boat can be recognized with a probability of about 50%, while the other kinds of targets cannot be recognized. At the ranges of 22 km and 11 km the matrix is very close to an identity matrix so that it is possible to identify and recognize all vessels.

9.8 Conclusions

Modeling & simulation is a well-known technique in systems analysis, yet new challenges emerge in the context of large systems. Large systems cannot always be studied with a "divide and conquer" approach and large systems modeling requires a significant and time-consuming effort, both during the setup and the maintenance of the simulator; the need emerges to define a suitable trade-off between model accuracy on one side and greater speed and flexibility on the other side.

This chapter provides an overview on this important issue on the basis of the personal experience of the authors that participate to the joint and ongoing effort by Finmeccanica Companies, led by SELEX Sistemi Integrati, to address modeling and simulation in the analysis of large systems. In particular, a multilevel simulator architecture is described where each level provides a different mix between the width of the field of view and the granularity of the models; as the field of view narrows, models get more precise and detailed; synthetic results obtained with high-precision models may then be used as the field of view widens.

Current approaches to the analysis of large systems cannot be considered fully satisfactory: modeling & simulation is an effective technique, yet the massive use of computer-intensive simulations underlies a lack of theoretical foundation in large system analysis; future work needs to pursue a formal and analytic theory to address large systems, devising tools and methodologies that can be applied to predict behavior, assess performance, and support the validation and verification issues.

Acknowledgments

The authors wish to warmly thank all the colleagues of the Finmeccanica Companies that have contributed to the success of the Finmeccanica Working Group on the analysis of large systems. A special thanks to Dr. Sofia Giompapa of the University of Pisa (Italy), for cooperation on Section 9.7.1.

References

[1] Gruppo di Lavoro Finmeccanica sull'Analisi dei sistemi integrati per Homeland Security, SELEX Sistemi Integrati, *Border control: Sintesi dei risultati della cooperazione fra le Aziende del gruppo Finmeccanica,* Rome, Italy, June 2006.

[2] Skinner, C. J., et al. "Defense against Terrorism: The Evolution of Military Surveillance Systems into Effective Counter Terrorism Systems Suitable for Use in Combined Military Civil Environments. Dream or Reality?" *NATO Panel "Systems, Concepts and Integration (SCI) Methods and Technologies for Defence against Terrorism,"* London, UK, October 25–27, 2004.

[3] Alberts, D. S., J. J. Garstka, and F. P. Stein, *Network Centric Warfare: Developing and Leveraging Information Superiority*, CCRP Publ., 2nd ed. (Revised). August 1999, 2nd Print February 2000.

[4] U.S. Office of Homeland Security, *National Strategy for Homeland Security*, Washington, D.C., July 2002, http://www.whitehouse.gov/homeland/book/nat_strat_hls.pdf.

[5] U.S. Environmental Protection Agency, *Strategic Plan for Homeland Security*, Washington, DC, September 2002 (www.epa.gov/epahome/downloads/epa_homeland_security_strategic_plan.pdf).

[6] Moteff, J., C. Copeland, and J. Fischer, *Critical Infrastructure: What Makes an Infrastructure Critical?*, Report for Congress RL31556. The Library of Congress, August 2002, www.fas.org/irp/crs/RL31556.pdf.

[7] U.S. Government, *The National Strategy for the Physical Protection of Critical Infrastructure and Key Assets*, Washington, D.C.: The White House, February 2003.

[8] Bologna, S., and R. Setola, "The Need to Improve Local Self-awareness in CIP/CIIP," *Proclamation of 1st IEEE International Workshop on CIP (IWCIP 2005)*, Darmstadt, Germany, November 3–4, 2005, pp. 84–89.

[9] Rinaldi, S., J. Peerenboom, and T. Kelly, "Identifying, Understanding and Analyzing Critical Infrastructure Interdependencies," *IEEE Control System Magazine*, Vol. 21, No. 6, 2001, pp. 11–25.

[10] Panzieri, S., R. Setola, and G. Ulivi, "An Approach for Modeling Heterogeneous and Interdependent Critical Infrastructure," submitted to *IEEE Transactions on System, Man and Cybernetic, part C.*

[11] Ruberti, A., and S. Monaco, *Teoria dei sistemi. Appunti delle lezioni*, Consorzio Nettuno, Bologna, Italy: Pitagora Editrice, 1998.

[12] Kailath, T., *Linear Systems*, Upper Saddle River, NJ: Prentice-Hall, 1980.

[13] Zadeh, L. A., and E. Polak, *System Theory*, New York: McGraw-Hill, 1969.

[14] Sontag, E. D., *Mathematical Control Theory—Deterministic Finite Dimensional Systems*, New York: Springer, 1998.

[15] Luenberger, D. G., *Introduction to Dynamic Systems: Theory Models and Application*, New York: John Wiley, 1979.

[16] Kalman, R. E., "A New Approach to Linear Filtering and Prediction Problems," *Transactions of the ASME—Journal of Basic Engineering*, Vol. 82, 1960, pp. 35–45.

[17] Yager, R. R., and L. A. Zadeh, *An Introduction to Fuzzy Logic Applications in Intelligent Systems*, Boston: Kluwer Academic, 1992.

[18] Naresh, K., and M. M. Gupta, *Soft Computing and Intelligent Systems: Theory and Applications*, L. A. Zadeh (honorary ed.), San Diego: Academic Press, 2000.

[19] Dumitas, A., and G. Moschytz, "Understanding Fuzzy Logic: An Interview with Lotfi Zadeh," *IEEE Signal Processing Magazine*, Vol. 24, No. 3, May 2007, pp. 102–105.

[20] Gell-Mann, M., "What Is Complexity?" *Complexity*, Vol. 1, No. 1, 1995 pp. 16–19.

[21] Gell-Mann, M., "Simplicity and Complexity in the Description of Nature," *Engineering and Science* LI (3), California Institute of Technology, 1988, pp. 3–9.

[22] Holland, J. H., *Adaptation in Natural and Artificial Systems: An Introductory Analysis with Applications to Biology, Control and Artificial Intelligence*, Cambridge, MA: MIT Press, 1992.

[23] Golino, G., et al. "Surveillance by Means of a Random Sensor Network: A Heterogeneous Sensor Approach," *IEEE 8th International Conference on Information Fusion 2005*, Philadelphia, PA, July 25–28, 2005, pp. 1086–1093.

[24] Capponi, A., et al. "Algorithms for the Selection of the Active Sensors in Distributed Tracking: Comparison between Frisbee and GNS Methods," *IEEE 9th International Conference on Information Fusion 2006*, Florence, Italy, July 10–13, 2006, pp. 1–8.

[25] Wendell, J., *Complex Adaptive Systems. Beyond Intractability*, G. Burgess and H. Burgess (eds.), Conflict Research Consortium, University of Colorado, Boulder, October 2003 http://www.beyondintractability.org/essay/complex_adaptive_systems/.

[26] Bellinger, G., "Modeling & Simulation, an Introduction," September 2007, http://www.systemsthinking.org/modsim/modsim.htm.

[27] Dahmann, J. S., "High Level Architecture for Simulation," *1st International Workshop on Distributed Interactive Simulation and Real-Time Applications (DIS-RT '97)*, Eilat, Israel, January 9–10, 1997, p. 9.

[28] Negroponte, N., *The Architecture Machine: Towards a More Human Environment*, Cambridge, MA: MIT Press, 1970.

[29] Bradshaw, J. M., (ed.), *Software Agents*, Cambridge, MA: MIT Press, 1997.

[30] Epstein, J. M., and R. Axtell, *Growing Artificial Societies: Social Science from the Bottom Up*, Cambridge, MA: MIT Press, 1996.

[31] NATO AAP-31(A), *NATO Glossary of Communication and Information Systems Terms and Definitions*, 2002.

[32] Farina, A., and F. A. Studer *Radar Data Processing: Advance Topics and Applications, Vol. 2*, New York: John Wiley, Research Studies, 1986.

[33] Bell, M. R., and R. A. Grubbs, "JEM Modeling and Measurement for Radar Target Identification," *IEEE Transactions on Aerospace and Electronic Systems*, Vol. 24, No. 1, January 1993, pp. 73–87.

[34] Di Lallo, A., et al. "Bi-Dimensional Analysis of Simulated HERM and JEM Radar Signals for Target Recognition," *1st International Waveform Diversity & Design Conference*, Edinburgh, U.K., November 8–10, 2004.

[35] Martin, J., and B. Mulgrew, "Analysis of the Theoretical Radar Return Signal from Aircraft Propeller Blades," *IEEE International Radar Conference*, RADAR90, Arlington, VA, May 7–10, 1990, pp. 569–572.

[36] Wehner, D. R., *High Resolution Radar*, Norwood, MA: Artech House, 1995.

[37] Lapierre, F., et al., "Naval Target Classification by Fusion of IR and EO Sensors," *SPIE Europe Optics & Photonics in Security & Defence Conf.*, Florence, Italy, September 17–21, 2007.

[38] Benavoli, A., et al., "Reasoning Under Uncertainty: from Bayes to Valuation Based Systems. Application to Target Classification and Threat Assessment," Rome, Italy, SELEX Sistemi Integrati, September 2007.

[39] Shafer, G., *A Mathematical Theory of Evidence*, Princeton, NJ: Princeton University Press, 1976.

[40] Smets, P., "The Combination of Evidence in the Transferable Belief Model," *IEEE Pattern Analysis and Machine Intelligence*, Vol. 12, 1990, pp. 447–458.

[41] Smets, P., *Belief Functions and the Transferable Belief Model*, Technical Report, 2000 (www.iridia.ulb.ac.be/psmets/).

[42] Smarandache, F., and J. Dezert, *Advanced and Applications of DSmT for Information Fusion*, Rehoboth: Am. Res. Press, 2004, www.gallup.unm.edu/smarandache/DSmT.htm.

[43] Roveri, A., et al., *Reti di telecomunicazioni per grandi sistemi*, Rome, Italy, SELEX Sistemi Integrati, March 2006.

[44] Fall, K., and K. Varadhan, *The ns Manual*, www.isi.edu/nsnam/ns/doc/ns_doc.pdf.

[45] Signorini, A. M., A Farina, and G. Zappa, "Applications of Multiscale Estimation Algorithm to SAR Images Fusion," *International Symposium on Radar, IRS98*, Munich, Germany, September 15–17, 1998, pp. 1341–1352.

[46] Gibson, C., et al., *Security Risk Management*, Sydney: Standards Australia, 2006.

[47] Blake, L. V., *Radar Range Performance Analysis*, Dedham, MA: Artech House, 1980.

[48] Barton, D. K., *Modern Radar System Analysis*, Norwood, MA: Artech House, 1998.

[49] Ratches, J. A., R. H. Vollmerhausen, and R. G. Driggers, "Target Acquisition Performance Modeling of Infrared Imaging Systems: Past, Present, and Future," *IEEE Sensors Journal*, Vol. 1, No. 1, June 2001, pp. 31–40.

[50] Vollmerhausen, R. H., and E. Jacobs, *The Targeting Task Performance (TTP) Metric—A New Model for Predicting Target Acquisition Performance*, Technical Report AMSEL-NV-TR-230 Modeling and Simulation Division Night Vision and Electronic Sensors Directorate, U.S. Army CERDEC, Fort Belvoir, VA, 2004.

[51] Crawford, F. J., "Electro-Optical Sensors Overview," *IEEE Aerospace and Electronic System Magazine*, Vol. 13, No. 10, October 1998, pp. 17–24.

[52] Driggers, R. D., *Encyclopedia of Optical Engineering*, Boca Raton, FL: CRC Press, 2003.

[53] Slater, P. N., *Remote Sensing, Optics and Optical Systems*, Reading, MA: Addison-Wesley, 1980.

[54] Kim I. I., B. McArthur, and E. Korevaar, "Comparison of Laser Beam Propagation at 785 nm and 1550 nm in Fog and Haze for Optical Wireless Communications," *Proceedings of SPIE*, Vol. 4214, No. 2, 2000, pp. 26–37.

[55] *IALA Guidelines on AIS as a VTS Tool*, December 2001, http://www.iala-aism.org.

[56] *IALA Guidelines on the Universal Automatic Identification System (AIS)*, Vol. 1, Part II, Technical Issues Edition 1.1, December 2000.

[57] Skolnik, M., *Radar Handbook*, New York: McGraw-Hill, 1990.

About the Authors

Giorgio Franceschetti is a professor at the University Federico II of Napoli, Italy, an adjunct professor at University of California, Los Angeles (UCLA), a distinguished visiting scientist at the Jet Propulsion Laboratory, and a lecturer at Top-Tech Master, Delft University. He has also been a visiting professor in Europe, the United States, and Somalia, and a lecturer in China and India. Dr. Franceschetti is the author of more than 180 peer-reviewed papers and 10 books, and he has been the recipient of several awards, including the gold medal from the President of Italy in 2000 and an elevation to the grade of Officer of Italian Republic in 2001. His most recent recognition has been the prestigious 2007 IEEE GRS-S Distinguished Achievement Award: "For outstanding research in Electromagnetics, Propagation, Remote Sensing and Information Data Processing," which correctly describes his research interests. He is an IEEE Life Fellow.

Marina Grossi is the chief executive officer of SELEX Sistemi Integrati, a Finmeccanica company. Ms. Grossi, a graduate in electronic engineering, was previously the chairman and CEO of Alenia Marconi Systems and the general manager of MBDA Italia, both Finmeccanica companies. She formerly worked in other important defense companies such as SMA, Officine Galileo, and Alenia Difesa. In Finmeccanica, Ms. Grossi was responsible for the Value Creation Project, of which a book, *Value as a Management Model*, was published in 2003. In 2005, Ms. Grossi was also awarded the Marisa Bellisario Award for competitiveness, research, and innovation.

Alessandro Bissacco is currently a software engineer at Google Inc. He received a Laurea summa cum laude from University of Padova, Italy, in 1998, and an M.Sc. and a Ph.D. in computer science from UCLA, in 2002 and 2006 respectively. His research interests include low-level vision, object recognition, face recognition, human motion modeling, synthesis and tracking, human detection and pose estimation, identification and filtering of stochastic dynamical systems, and machine learning.

Alvaro A. Cárdenas received a B.S. with a major in electrical engineering and a minor in mathematics from the Universidad of los Andes, Bogota, Colombia, in 2002, and an M.S. and a Ph.D. in electrical and computer engineering from the University of Maryland, College Park, in 2002 and 2006, respectively. He is currently a postdoctoral scholar at the University of California, Berkeley. His research interests include information security, statistics, and machine learning. He received a two-year graduate school fellowship from the University of Maryland and a two-year distinguished research assistantship from the Institute of Systems Research.

Phoebus Chen is currently an electrical engineering graduate student under Professor Shankar Sastry's guidance at the University of California, Berkeley. He received an M.S. in electrical engineering and B.S. in electrical engineering and computer science at the University of California, Berkeley, in 2005 and 2002, respectively. His research interests are in mesh networking and distributed control systems, particularly control systems running over sensor networks. He is a member of the IEEE, the HKN, and the TBP.

Alfonso Farina is a Fellow of the IEEE and the IEE. He is an International Fellow of Royal Academy of Engineering, United Kingdom. He received a doctor degree in electronic engineering from the University of Rome (I). In 1974 he joined SELEX-SI, where he is the director of the Analysis of Integrated Systems Unit. In his professional life he provided many technical contributions in the area of signal detection and image processing for radar systems designed in his company. Dr. Farina was a part-time professor at the University of Naples. He is also the author of 450 peer-reviewed technical papers, books, and monographs, especially *Radar Data Processing* (translated in Russian and Chinese), *Optimised Radar Processors,* and *Antenna Based Signal Processing Techniques for Radar Systems.* He has received several technology excellence awards including the 2004 first prize award for Innovation Technology of Finmeccanica, and The Fellowship from the Royal Academy of Engineering.

Mario Gerla received an engineering degree from the Politecnico di Milano, Italy, in 1966, and an M.S. and a Ph.D. from UCLA in 1970 and 1973, respectively. He was elected IEEE Fellow in 2002. At UCLA, he was part of the team that developed the early ARPANET protocols under the guidance of Professor Leonard Kleinrock. He joined the Computer Science Department at UCLA in 1976. At UCLA he has designed and implemented several popular protocols for ad hoc wireless networks such as ODMRP and CODECast.

Antonio Graziano received a Degree cum Laude in electronic engineering from the University of Palermo (I) in 1988. He has worked in the R&D Division of

Ericsson-Fatme on the analysis and simulation of communication networks and at SELEX Sistemi Integrati as system analysis engineer, where he has been involved in air defense, surveillance, and command and control systems. Mr. Graziano has been the team leader of several national and international study projects on data fusion and network-centric warfare. He is author of several journal and conference publications. He is currently the head of Systems and C4I Analysis in the Integrated Systems Analysis unit of the Engineering Division.

Uichin Lee is a Ph.D. student in the Department of Computer Science at UCLA. His research interests include wireless networking applications, mobile wireless sensor networks (e.g., vehicular and underwater sensor networks), and user behavior studies [e.g., peer-to-peer (P2P) and Web search].

Paolo Neri graduated as an electronic engineer from Pisa University in 1970. He has worked at Selenia, Elettronica, Fincantieri, Orizzonte, and Selex-Sistemi Integrati, dealing with different systems of growing complexity, from radar to ships and from ships to system of systems. At Selex-Sistemi Integrati Mr. Neri works in the Large Systems Architectures Division. He is skilled in system architectural analysis in the field of command and control systems involving radar, communications, electronic warfare, IR, management control systems, and weapons systems. In the system analysis field Mr. Neri has much experience in operational system performance evaluation and the reliability, availability, and maintainability of ILS system design.

Songhwai Oh is an assistant professor of electrical engineering and computer science in the School of Engineering at the University of California, Merced. His research interests include wireless sensor networks, robotics, networked control systems, estimation and learning, and computer vision. He received a B.S. in 1995, an M.S. in 2003, and a Ph.D. in 2006 in electrical engineering and computer sciences (EECS) at the University of California at Berkeley. In 2007, he was a postdoctoral researcher in EECS at the University of California at Berkeley. Before his Ph.D. studies, Dr. Oh worked as a senior software engineer at Synopsys, Inc., and a microprocessor design engineer at Intel Corporation.

Luciana Ortenzi received a doctor degree in telecommunications engineering in 2001 and a Ph.D. in remote sensing from the University of Rome (I) in 2007. In 2004 she joined AMS, now SELEX-Sistemi Integrati. Her present areas of investigation are adaptive signal processing for detection, estimation, and tracking with application to passive and active phased array radars. She received the 2003 AMS MD Award and the 2006 SELEX-Sistemi Integrati CEO Award for Innovation Technology. Dr. Ortenzi authored and coauthored several peer reviewed journal and conference papers.

Rafail Ostrovsky received a Ph.D. from MIT in 1992 in computer science with a minor from the Sloan School of Management. He is a professor of computer science and Mathematics at UCLA. He is also a director of the Multidisciplinary Center of Information and Computation Security at UCLA. Dr. Ostrovsky came to UCLA from Telcordia Technologies, an SAIC subsidiary. He authored more than 100 published papers in refereed journals and conferences and 8 patents. He is a member of the editorial boards of *Algorithmica* and the *Journal of Cryptology* and the editorial and advisory board of the *International Journal of Information and Computer Security*. He is a recipient of multiple academic awards and honors.

Tanya Roosta is in the last year of her Ph.D. studies in electrical engineering and computer science at the University of California at Berkeley, after having received a B.S. in EECS with honors at the University of California at Berkeley in 2000 and an M.S. in 2004. She also holds an M.A. from the University of California at Berkeley in statistics. She received the 3-year National Science Foundation fellowship for her graduate studies. Her research interests include sensor network security, fault detection, reputation systems, privacy issues associated with the application of sensors at home and health care, and sensor networks used in critical infrastructure. Her additional research interests include: robust statistical methods, outlier detection models, statistical modeling, and the application of game theory to sensor network design.

Shankar Sastry received a B.Tech. from the Indian Institute of Technology, Bombay, in 1977, and an M.S. in EECS, an M.A. in mathematics, and a Ph.D. in EECS from the University of California at Berkeley, in 1979, 1980, and 1981, respectively. Dr. Sastry is currently the dean of the College of Engineering. He was formerly the director of CITRIS (Center for Information Technology Research in the Interest of Society) and the Banatao Institute. He served as the chair of the EECS Department, as the director of the Information Technology Office at DARPA, and as the director of the Electronics Research Laboratory at Berkeley, an organized research unit on the Berkeley campus conducting research in computer sciences and all aspects of electrical engineering. He is the NEC Distinguished Professor of Electrical Engineering and Computer Sciences and holds faculty appointments in the Departments of Bioengineering, EECS, and Mechanical Engineering.

William E. Skeith III received a Ph.D. in mathematics in 2007 from UCLA, specializing in algebra and cryptography. He has five publications in international conferences and journals. He is a winner of the UC Chancellor Presidential Dissertation Year Fellowship and the recipient of a National Science Foundation VIGRE Fellowship. In his current employment, Dr. Skeith develops high-performance software for cryptography and privacy preserving data-mining at an early-stage company of which he is a cofounder.

Stefano Soatto is a professor of computer science and electrical engineering at UCLA where he directs the Vision Laboratory. He received the Marr Prize and the Siemens Prize for his work on 3D reconstruction, the Okawa Foundation grant, and the NSF Career Award. Dr. Soatto is a member of the editorial board of the *International Journal of Computer Vision.* Prior to UCLA, he held positions at Washington University, Harvard University, and the University of Udine, Italy. He received a Ph.D. from the California Institute of Technology and a D. Ing. from the University of Padova, Italy.

Emidio Spinogatti received a university degree in electronic engineering from the University of Ancona in 2000. In 2002 he joined AMS J.V. in the Land Systems Division. He worked as system engineer on Italian Army Command and Control Information System and on Peace Support Operations Tools Italy-France joint effort project. Mr. Spinogatti has been a member of the MIP (Multilateral Interoperability Programme) System Engineering and Architecture Working Group (SEAWG). Since 2005 he has been an analyst in the Systems and C4I Analysis group of Selex Sistemi Integrati S.p.A, with his present focus on the performance evaluation of large systems for homeland protection.

Mani Srivastava is a professor and the vice chair of electrical engineering and a professor of computer science at UCLA. He leads the systems area research at UCLA's Center for Embedded Networked Sensing (CENS). Dr. Srivastava received a Ph.D. from the University of California at Berkeley in 1992 and has worked at Bell Labs Research. His research interests are in power-aware computing and communications, wireless and mobile systems, embedded computing, pervasive sensing, and their applications in urban, social, and personal contexts. He is the editor-in-chief of for the *IEEE Transactions on Mobile Computing* and an associate editor of the *ACM Transactions on Sensor Networks.*

Gelareh Taban is a Ph.D. student in electrical and computer engineering at the University of Maryland, College Park. She received an M.S. from the University of Maryland and a B.S. in computer engineering from the University of Wollongong, Australia. Before continuing her graduate studies, she was an engineer at Nortel Research and Development Laboratory in Wollongong. Ms. Gelareh's research interests include security and resiliency in sensor and ad hoc networks, threat modeling, applied cryptography, and digital rights management.

Hetal Thakkar is currently a Ph.D. candidate in the UCLA Computer Science Department. Since 2003, he has been performing research with Professor Zaniolo on data stream management systems and their languages. He has also worked as a research intern at Google and IBM Almaden Research Center, investigating click streams, data stream languages, RFID data, and Web 2.0 applications. His current research interests include data stream mining systems and data stream languages. He received an M.S. in computer science in 2005 and expects to complete his Ph.D. in 2008.

Carlo Zaniolo is a professor of computer science at UCLA, which he joined in 1991 as the N. Friedmann Chair in Knowledge Science. Before that, Dr. Zaniolo was a researcher at AT&T Bell Laboratories and the associate director of the Advanced Computer Technology Program of MCC, a U.S. research consortium in Austin, Texas. He received a degree in electrical engineering from Padua University, Italy, in 1969 and a Ph.D. in computer science from UCLA in 1976. A prolific author and program chair or cochair of major IS conferences, his recent research focuses on data stream management systems, data mining, and archival IS.

Index

The Artech House Intelligence and Information Operations Series

Concepts, Models, and Tools for Information Fusion, Éloi Bossé, Jean Roy, and Steve Wark

Electronic Intelligence: The Analysis of Radar Signals, Second Edition, Richard G. Wiley

Electronic Warfare for the Digitized Battlefield, Michael R. Frater and Michael Ryan

Electronic Warfare in the Information Age, D. Curtis Schleher

Electronic Warfare Target Location Methods, Richard A. Poisel

EW 101: A First Course in Electronic Warfare, David Adamy

Homeland Security Technology Challenges: From Sensing and Encrypting to Mining and Modeling, Giorgio Franceschetti and Marina Grossi, editors

Information Warfare and Organizational Decision-Making, Alexander Kott, editor

Information Warfare Principles and Operations, Edward Waltz

Introduction to Communication Electronic Warfare Systems, Richard A. Poisel

Knowledge Management in the Intelligence Enterprise, Edward Waltz

Mathematical Techniques in Multisensor Data Fusion, Second Edition, David L. Hall and Sonya A. H. McMullen

Modern Communications Jamming Principles and Techniques, Richard A. Poisel

Principles of Data Fusion Automation, Richard T. Antony

Strategem: Deception and Surprise in War, Barton Whaley

Statistical Multisource-Multitarget Information Fusion Ronald P. S. Mahler

Tactical Communications for the Digitized Battlefield, Michael Ryan and Michael R. Frater

Target Acquisition in Communication Electronic Warfare Systems, Richard A. Poisel

For further information on these and other Artech House titles, including previously considered out-of-print books now available through our In-Print-Forever® (IPF®) program, contact:

Artech House
685 Canton Street
Norwood, MA 02062
Phone: 781-769-9750
Fax: 781-769-6334
e-mail: artech@artechhouse.com

Artech House
46 Gillingham Street
London SW1V 1AH UK
Phone: +44 (0)20-7596-8750
Fax: +44 (0)20-7630-0166
e-mail: artech-uk@artechhouse.com

Find us on the World Wide Web at: www.artechhouse.com